AMBIENT INTELLIGENCE:
IMPACT ON EMBEDDED SYSTEM DESIGN

T0138047

Ambient Intelligence: Impact on Embedded Sytem Design

Edited by

Twan Basten
Eindhoven University of Technology,
Eindhoven, The Netherlands

Marc Geilen
Eindhoven University of Technology,
Eindhoven, The Netherlands

and

Harmke de Groot
Philips Research Laboratories Eindhoven,
Eindhoven, The Netherlands

KLUWER ACADEMIC PUBLISHERS
BOSTON / DORDRECHT / LONDON

A C.I.P. Catalogue record for this book is available from the Library of Congress.

ISBN 978-1-4419-5407-7   e-ISBN 978-0-306-48706-4

Published by Kluwer Academic Publishers,
P.O. Box 17, 3300 AA Dordrecht, The Netherlands.

Sold and distributed in North, Central and South America
by Kluwer Academic Publishers,
101 Philip Drive, Norwell, MA 02061, U.S.A.

In all other countries, sold and distributed
by Kluwer Academic Publishers,
P.O. Box 322, 3300 AH Dordrecht, The Netherlands.

Printed on acid-free paper

Contents

Part II Developments

Foreword

Hugo de Man
Professor Katholieke Universiteit Leuven
Senior Research Fellow IMEC

The steady evolution of hardware, software and communications technology is rapidly transforming the PC- and dot.com world into the world of Ambient Intelligence (AmI). This next wave of information technology is fundamentally different in that it makes distributed wired and wireless computing and communication disappear to the background and puts users to the foreground. AmI adapts to people instead of the other way around. It will augment our consciousness, monitor our health and security, guide us through traffic etc. In short, its ultimate goal is to improve the quality of our life by a quiet, reliable and secure interaction with our social and material environment.

What makes AmI engineering so fascinating is that its design starts from studying person to world interactions that need to be implemented as an intelligent and autonomous interplay of virtually all necessary networked electronic intelligence on the globe. This is a new and exciting dimension for most electrical and software engineers and may attract more creative talent to engineering than pure technology does.

Development of the leading technology for AmI will only succeed if the engineering research community is prepared to join forces in order to make Mark Weiser's dream of 1991 come true. This will not be business as usual by just doubling transistor count or clock speed in a microprocessor or increasing the bandwidth of communication.

Indeed, this book shows clearly that progress in this next wave of information technology can only result from merging previously unconnected domains and techno-cultures to create the ultra-low power, low cost, always connected devices that autonomously and seamlessly adapt to social human behavior.

Hence, building ambient intelligent systems requires a team effort of sociologists, system architects, computer scientists, network and communication specialists, electrical and mechanical engineers as well as device physicists with an open mind for interdisciplinary dialogue. Progress will result from theme driven demonstrator experiments that identify the relevant challenging problems to be solved by the best, deeply specialized but open-minded, academics. This may require novel ways of organizing industrial and academic research.

So, more than ever before, we must bring system design oriented people together, join in acts of creativity and find ways how to organize system design methodology research crossing the discipline boundaries to solve the societal challenges stated above. The best way of developing the basic AmI technology is to bring the best engineers together in open, long term, team based AmI prototype projects with a clear goal to create innovation by merging their expertise.

In that respect this book is truly an eye-opener as it is the first one that relates the dream scenarios of Ambient Intelligence quantitatively to the technical challenges and requirements of the huge distributed and interoperable embedded systems needed to implement AmI systems in the real world. It provides a connection between AmI architects and the specialized engineers in the fields of computer architecture, software architecting, chip and sensor design. In addition, it shows partial practical experimental answers to the challenges and what can be learned from them for further research.

The editors succeeded marvelously to connect the previously unconnected. As such the reader of this book should carefully avoid just reading what belongs to his/her domain of expertise but rather read the book from A-Z because only then he/she will grasp the awareness that the added value comes from putting it all together into the context of the global AmI system rather than optimizing one single part of it.

Therefore this book is strongly recommended to a wide spectrum of engineers interested to embark in this rapidly emerging and fascinating technology. For engineering managers and educators it provides not only a great vision on the technical challenges ahead but it also shows the need to reflect on new organizational structures for theme and team based research necessary to make rapid progress in a domain that promises to be of great economical and societal impact.

Leuven, August 2003

Omnia Fieri Possent

Everything may happen

Seneca, Ad Lucilium Epistulae Morales, Epistle 70

Twan Basten, Marc Geilen

Eindhoven University of Technology, Eindhoven, The Netherlands

{ a.a.basten, m.c.w.geilen } @tue.nl

Harmke de Groot

Philips Research Laboratories Eindhoven, Eindhoven, The Netherlands

harmke.de.groot@philips.com

Tom is in a good mood this morning. It's Friday and tonight he's going to the movies with Amy. After a shower, he briefly reviews the movies that are showing tonight. Info and clips are displayed by his new bathroom mirror. He requests a preview of "You've got mail," which is a rerun of some old movie about an e-mail romance. Personally, he would like to see some more action, but Amy will definitely love this one.

After breakfast he hops into a PeopleMover to go to the office. It's quiet this morning. He requests some music. The audio system of the PeopleMover starts playing one of his favorite songs, using his preferences from his SmartId. He decides to book "You've got mail." Payment goes through his SmartId, and tickets are downloaded onto it. He also makes dinner reservations. [...]

At ten past five, Tom hurries out of his office to a PeopleMover that will take him to the city center. Silly staff meetings. They always run late. When he arrives, Amy is not yet at the restaurant so he goes to the ReadingDesk to catch up on the latest news. Strange, the Desk doesn't show him his preferred news items. Damn! He forgot his SmartId at the office. And it's too late to pick it up. But that means he's Disconnected! No-one will be able to contact him. And he'll have to ask Amy to pay for dinner, and to buy new tickets for the movie! He'll refund her, of course, but it's definitely not the best way to start the evening. [...]

It's a bit past eleven when Tom arrives at home. Because he doesn't have his SmartId, he is glad he left open a window at his balcony this morning. He manages to enter his home through this window. Before he forgets, he quickly refunds Amy through the HomeSystem. He decides to go to the office to pick up his SmartId. He doesn't want to wait till Monday before getting it back. There are no PeopleMovers this late in the evening, so he takes his car. He (manually!) searches some music. He relaxes; he had a wonderful evening after all. Suddenly, a police car appears, flashing a stop sign ... .

T. Basten et al. (eds.), Ambient Intelligence: Impact on Embedded System Design, 1-8.

The setting in the story of Tom and Amy is clearly not today's world. However, in the not so far future, the world may look a bit like the world of Tom and Amy. In fact, the story fits perfectly in the emerging Ambient-Intelligence (AmI) vision, as outlined in [1]. AmI is essentially a vision on tomorrow's world and the role of technology in that world. It builds on ubiquitous computing, as originally envisioned by Mark Weiser [2], but in comparison it is more human-centric. It integrates concepts from natural user interaction and autonomous and intelligent systems. The AmI vision is gaining traction both in industry and in academia, and it is often illustrated via scenarios as the one above. These scenarios typically sketch a world that is great to live in, suggesting that technology has a positive effect on quality of life. However, the story of Tom is not exactly typical in all respects. In typical scenarios, the main characters do not forget crucial items, and they don't enter houses through windows. And why is it that the police car appears at the end? Well, ... this is what actually happened.

It's 8:03am and HomeSecurity detects that Tom leaves the house. It switches to security level orange. [...]

At 11:23pm, HomeSecurity detects movement at the balcony. Someone enters the house. It doesn't appear to be Tom; HomeSecurity doesn't detect Tom's SmartId. The intruder immediately activates the HomeSystem. Since one minute has passed and HomeSecurity has not yet been switched to green, it goes to red. It immediately contacts Tom.

The intruder tries to transfer money from Tom's bank account, but HomeSecurity smoothly intercepts the transfer, of course without alerting the intruder. Since Tom doesn't respond within 60 seconds, HomeSecurity alarms the police. The intruder removes Tom's car keys from the SecurityBox, leaves the house, and starts the car. HomeSecurity reacts seamlessly by sending video images of the intruder, a picture of the car, and its license plate to the police. Of course, HomeSecurity tracks the car and keeps the police up-to-date. When it receives the expected SecurityCode from the police, it knows that everything is ok. It switches back to orange, waiting for the return of Tom.

The story of Tom and Amy illustrates the AmI vision and its potential, but it also shows the precarious balance between success and failure. With AmI, our daily lives may become very exciting and fun, or, without proper care, they may just as easily become a nightmare. In our opinion, the AmI vision has a lot of potential. It is up to us to realize it, and to make sure that nightmare scenarios are avoided. But as is often the case with predicting the future, in the end, only time can tell what AmI will bring us.

So what is this volume all about? AmI scenarios usually focus on the user and application perspective of technology. Users play a central role in the AmI vision. Nevertheless, in this volume, we turn our attention to the underlying technology, in particular, the underlying embedded hardware, software, and communication infrastructure. In other words, the volume explores some of the essential enabling and support technologies needed for a successful realization of the AmI vision. Before going into more details about the contents, we first provide some background information about this volume.

The basis for this volume was unknowingly already laid in 1999, when Jean Gelissen at Philips Research initiated a European project called Ozone. This project, which is ongoing at the moment of writing, aims at the development of platform and middleware technology for AmI. Ozone is thus an early initiative that explores the impact of AmI on embedded hardware and software. A second crucial step was the decision of the executive committee of the Design, Automation and Test in Europe (DATE) conference of 2003 to organize a special day on Ambient Intelligence. Diederik Verkest, the chair of DATE 2003, invited Twan Basten and Menno Lindwer from the Ozone project to organize this special day. A final step was the invitation by Mark de Jongh of Kluwer Academic Publishers to publish a book about the DATE special day on AmI. Menno unfortunately had to end his work as an editor in an early phase due to other obligations. Nevertheless, he certainly had an impact on the contents of this volume.

Our first decision was to not base the volume entirely on the DATE 2003 special day. The AmI vision is still too recent and the relation between AmI and embedded hard- and software still too fresh to draw sufficient relevant and high-quality material from a specialized conference like DATE. Therefore, we combined selective invitations for contributions to the volume with an open call for papers. To allow authors to explore the relation between AmI and embedded systems in the widest possible sense, we decided not to pose any restrictions but to simply give prospective authors a brief description of the AmI vision and its relation to embedded systems. To maintain some coherency, we asked all prospective authors to submit a two-page abstract first. Based on these abstracts, we made a selection and invited authors to write full papers. To guarantee the quality of the final contributions, all full papers were reviewed by two to four reviewers, and revised according to the reviews. The result is this volume with fifteen contributions. Together, they provide an interesting view on the impact of the AmI vision on embedded system design. Of course, this view cannot be expected to be complete in all aspects, nor will it be final in any sense. On the contrary, it's only a starting point, but a challenging and inspiring one.

After this brief history, it's time to have a closer look at the contributions. As mentioned, the fifteen papers in this volume have a common basis in the form of a brief description of the AmI vision and its impact on embedded systems:

Ambient intelligence (AmI) envisions the complete integration of technology into our environment so that people can freely and interactively utilize it. It is the combination of casually accessible networked digital devices performing all kinds of functions in entertainment, education, professional applications, etc. Technology shall be made invisible, embedded in our natural surrounding, present whenever we need it, enabled by simple and effortless interactions, and attuned to all our senses. The systems shall be intelligent in the sense that they are aware of and autonomously react to their contexts, in particular the users within those contexts.

Ambient Intelligence will have a major impact on embedded systems, software and silicon design. It introduces many new media applications and new user interface concepts. It requires the design of very powerful compute machinery for demanding media and user interface operations, as well as a powerful, omnipresent communication infrastructure. However, it will simultaneously require extremely low-cost and low-power designs for the compute and communication machinery that seamlessly and fully surrounds the users.

The volume is divided into two parts, Challenges and Developments. The first part, consisting of six papers, contains visionary papers and papers identifying challenges to be found in realizing the AmI vision. The second part, consisting of nine papers, contains concrete developments and first results towards the realization of the AmI vision.

The first paper in the Challenges part, *Embedded System Design Issues in Ambient Intelligence*, is based on the keynote speech of Emile Aarts at DATE 2003. Emile Aarts and Raf Roovers, both of Philips Research, explore the impact of the AmI vision on embedded systems in general. Their paper is a fitting start of this volume. They partition AmI devices into microWatt, milli-Watt, and Watt devices, discuss the mapping of AmI functionality onto these devices, and emphasize the major design challenges.

In *Ambient Intelligence: A Computational Platform Perspective*, Luca Benini, University of Bologna, and Massimo Poncino, University of Verona, explore the architectural requirements for future generations of computational platforms. They start from a similar partitioning as Aarts and Roovers. Their conclusion in one word: Parallelism!

The third paper in the Challenges part investigates the embedded-system-software point of view. Anu Purhonen and Esa Tuulari argue in *Ambient Intelligence and the Development of Embedded System Software* that customizability, adaptability, and flexibility of embedded software will soon become the key issues.

Peter van der Stok of Philips Research continues in *Preparation of Heterogeneous Networks for Ambient Intelligence* with the networking and communication perspective. Despite (or should we say 'due to'?) the many standards that are available, means to cope with heterogeneity and interoperability turn out to be the real challenges.

The Challenges part concludes with two papers on security and privacy issues. This is not a coincidence. Security and privacy are hot topics and potentially the biggest challenge in the realization of AmI. If security and privacy are not guaranteed, AmI will fail without a doubt. The unfortunate story of Tom is just one example of what may go wrong when security systems are not carefully designed or when they are not functioning properly. In *The Butt of the Iceberg: Hidden Security Problems of Ubiquitous Systems*, Frank Stajano and Jon Crowcroft of the University of Cambridge argue that the mere quantity of devices in an AmI world is already sufficient to open up a whole new range of challenging problems in security and privacy. Srivaths Ravi and Anand Raghunathan from NEC, Jean-Jacques Quisquater from Université Catholique de Louvain, and Sunil Hattangadi from Texas Instruments focus in *Emerging Challenges in Designing Secure Mobile Appliances* on security of mobile devices. Mobile devices are obviously key components in realizing the AmI vision. They identify so many challenges that one is easily left with the impression that secure mobile devices are an oxymoron.

The Developments part continues with the remaining nine contributions. Fiora Pirri, Ivo Mentuccia, and Sandro Storri of "La Sapienza" University of Rome start this part with a description of an intelligent domestic robot positioned in an AmI home environment in their paper *The Domestic Robot - A Friendly Cognitive System Takes Care of your Home*. The paper is an interesting mixture of concrete achievements and results, and views on future developments in cognitive robotics.

The next two contributions discuss a topic that is a very good example of the impact of AmI on embedded system design, namely Quality of Service (QoS). To a user, the quality of a requested service, for example image quality of a requested video, is one of the most important concerns. An AmI environment should provide the best possible quality at all times. The highly dynamic environment with often limited and continuously changing processing and communication resources requires therefore adaptability and flexibility in the services. In particular, it must be possible to make quality-resource trade-offs at any point in time. In *QoS-based Resource Management for Ambient Intelligence*, Clara M. Otero Pérez, Liesbeth Steffens, Peter van der Stok, Sjir van Loo, Alejandro Alonso, José F. Ruíz, Reinder J. Bril, and Marisol García Valls of Philips Research, Universidad Politécnica de Madrid, Eindhoven University of Technology, and Universidad Carlos III de Madrid present a unified approach to QoS in an AmI setting. In *Terminal QoS: Advanced Resource Management for Cost-Effective Multimedia Appliances in Dynamic Contexts*, Jan Bormans, Nam Pham Ngoc, Geert Deconinck, and Gauthier Lafruit of IMEC and Katholieke Universiteit Leuven focus on QoS-instigated resource management for multimedia services running in conjunction with 3D rendering applications. It is shown that software/hardware reconfigurability provides an important degree of freedom in the quality-adaptation process.

The fourth paper in the Developments part, *Scalability and Error Protection - Means for Error-Resilient, Adaptable Image Transmission in Heterogeneous Environments*, by Adrian Chirila-Rus, Gauthier Lafruit and Bart Masschelein of IMEC, also relates to quality. It presents a practical wavelet-based coding solution for the automatic, quality-based adaptability of image transmission over ubiquitous networks in the dynamic, heterogeneous AmI environment, with sometimes unreliable network connections.

Metaprogramming Techniques for Designing Embedded Components for Ambient Intelligence, by Vytautas Štuikys and Robertas Damaševičius of Kaunas University of Technology, is the fifth contribution in the Developments part. It describes how software-engineering techniques are adapted to the development of software and hardware components for AmI, taking into account the specific needs resulting from the AmI context. Three case studies demonstrate the validity of the presented approach. Zbigniew Chamski, Marc Duranton, Albert Cohen, Christine Eisenbeis, Paul Feautrier, and Daniela Genius of Philips Research, INRIA, Ecole Normale Supérieure de Lyon, and Université Paris 6 consider in *Application-Domain-Driven System Design for Pervasive Video Processing* also system-design techniques but they focus their attention on video processing, one of the core application domains in AmI. Video applications typically need many processing and communication resources, often leading to dedicated solutions. The paper describes a complete development chain for video applications, including methods and prototype tools.

The last three contributions in this volume bring us to the realm of high density networks, that is, networks with large numbers of often (very) small processing nodes. Such networks are essential in realizing an omnipresent AmI infrastructure that 'seamlessly and fully surrounds the users'. In *Collaborative Algorithms for Communication in Wireless Sensor Networks*, Tim Nieberg, Stefan Dulman, Paul Havinga, Lodewijk van Hoesel, and Jian Wu of the University of Twente present a new approach to communication in sensor networks. Since energy is the essential resource in such networks, the approach focuses on energy efficiency. Rex Min and Anantha Chandrakasan of MIT discuss energy-efficient wireless communication in more detail in *Energy-Efficient Communication for High Density Networks*. They emphasize the importance of accurate energy models, and they argue the need for performance-energy trade-offs and for application-specific solutions. They do so by presenting five myths of energy-efficient wireless communication. The final paper, *Application Re-mapping for Fault-Tolerance in Ambient Intelligent Systems*, by Phillip Stanley-Marbell, Nicholas H. Zamora, Diana Marculescu and Radu Marculescu of Carnegie Mellon University, investigates techniques to achieve fault-tolerance in high-density networks. Their techniques use the abundance of processing elements in such networks to cope with the run-time failures that will be unavoidable in networks of very many, very small, low-cost, low-energy elements.

What conclusions can be drawn from this volume? From the perspective of computational devices, the basic infrastructure enabling AmI can be partitioned into three classes of devices, based on energy consumption. The classes roughly correspond to mains-powered devices, wireless handheld devices, and wireless sensor-type devices. The characteristics of devices in these classes are fully determined by the available energy sources. Energy will become the most critical resource in an AmI environment. Wireless devices have an inherently limited power supply, and even the energy consumption of mains-powered devices is limited, mainly by the cooling capacity that can be built in.

The AmI infrastructure is characterized by the presence of large numbers of various kinds of devices, of all three classes, all connected and communicating with each other, and all fully integrated in our natural surrounding. Distribution, heterogeneity, and dynamism are key characteristics of the infrastructure. Distribution is obvious in the basic networked infrastructure, but future embedded systems will also contain many processor cores, memories, and other components on a single chip. Heterogeneity is present in the large variety of devices, in communication standards, media coding formats, processor cores, screen sizes, and so on. Typical sources of dynamism are the mobility of devices and users, applications that compete for shared processing and communication resources, and changing conditions of wireless networks.

Applications will run on top of the sketched infrastructure with all its characteristics. Traditionally, there is a tight coupling between applications and devices. It is clear that this one-to-one link will disappear. Application designers and system engineers will have to re-invent their design flows. To date, for example, designers often spend a lot of time optimizing applications for specific processors. But is that still meaningful if an application hops from one device to another from time to time?

The solutions presented in this volume to cope with particular AmI characteristics have one thing in common. They all focus on flexibility and adaptability. Prominent examples are QoS techniques, error-resilient and scalable image coding, and adaptive planning strategies used in intelligent robots. Also fault-tolerance and component-based design techniques fit the trend, and many other examples can be found throughout the book. At a first glance, the approach towards flexibility and adaptability may seem straightforward, but it turns out that a reliable and efficient realization is difficult. Another important observation in this volume is that security and privacy are crucial. Without secure infrastructure and privacy-respecting applications, AmI is deemed to fail. In conclusion, it is clear that many basic ingredients of the infrastructure needed to enable AmI are present today; the major challenge is to integrate all components into a smoothly operating whole.

Throughout history, technological advances have brought about many revolutionary changes in our daily lives. The last few decades, we have seen the rise of the personal computer, the breakthrough of Internet, and the introduction of mobile telephony. Other revolutions will follow, and the realization of the Ambient-Intelligence vision could certainly lead to one. Everything may happen. If sufficiently many people believe in it, we can make sure that Tom's evening will get a positive ending.

Eindhoven, August 2003

Acknowledgements

Many people have contributed to this volume in one way or the other, directly or indirectly, and sometimes even unknowingly. We are thankful to all of them. We are particularly grateful to the reviewers that helped us to guarantee the quality of the final result, and of course to the authors for sharing with us their views and results. We explicitly want to mention the following persons:

Emile Aarts, Anastase Adonis, Jayaprakash Balachandran, Lakshmanan Balasubramanian, Giorgio Da Bormida, Giel van Doren, Joan Figueras, Jean Gelissen, Manuel Gericota, Kees Goossens, Sumit Gupta, Hendrik Wietze de Haan, Dimitri van Heesch, Wim van Houtum, Dirk-Jan Jongeneel, Mark de Jongh, Willem Jonker, Jeffrey Kang, Bart Kienhuis, Thomas Kunz, Menno Lindwer, Jie Liu, Sue Menzies, Yasser Morgan, MohammadReza Mousavi, Laurentiu Papalau, Clara Otero Pérez, Piet van der Putten, Isabelle Reymen, Martijn Rutten, Sandeep Shukla, Amjad Soomro, Liesbeth Steffens, Sander Stuijk, Rineke Verbrugge, Diederik Verkest, Erik de Vink, Jeroen Voeten, Clemens Wüst, and Cindy Zitter.

We want to thank the Ozone project team, the DATE 2003 executive committee, and Kluwer Academic Publishers for providing us with the opportunity to publish this volume. As a final remark, we would like to acknowledge the financial support of the European Commission. The editors are supported through the IST-2000-30026 project Ozone.

References

[1] E. Aarts, R. Harwig, M. Schuurmans, "Ambient Intelligence." In P.J. Denning, editor, *The Invisible Future: The Seamless Integration of Technology in Everyday Life*, pp. 235–250. McGraw–Hill, New York, NY, USA, 2002.

[2] M. Weiser, "The Computer for the 21st Century." *Scientific American*, 265(3): 94–104. Sep. 1991.

Part I
Challenges

The best way to predict the future is to invent it.

Alan Kay, Xerox Palo Alto Research Center, 1971

Embedded System Design Issues in Ambient Intelligence

Emile Aarts and Raf Roovers

Philips Research Laboratories Eindhoven, Eindhoven, The Netherlands
{ emile.aarts, raf.roovers } @philips.com

Abstract   The vision of ambient intelligence opens a world of unprecedented experiences: the interaction of people with electronic devices is changed as context awareness, natural interfaces and ubiquitous availability of information are realized. We analyze the consequences of the ambient intelligence vision for embedded systems by mapping the involved technologies on a power-information graph. Based on differences in power consumption, three types of devices are introduced: the autonomous or microWatt-node, the personal or milliWatt-node and the static or Watt-node. Ambient intelligent functions are realized by a network of these devices with the computing, communication and interface electronics realized in Silicon IC technologies. The tri-partition of the ambient intelligence system hierarchy into microWatt nodes, milliWatt nodes, and Watt nodes introduces four major issues for embedded systems design.

Keywords   integrated circuits, ambient intelligence, energy, power, system architecture, wireless communication, sensors

1.   Introduction

Ambient intelligence refers to electronic environments that are sensitive and responsive to the presence of people [2, 9]. Ambient intelligent systems consist of many distributed devices that interact with users in a natural way. The concept builds on the early ideas of ubiquitous computing introduced by the late Marc Weiser [17] and anticipates a digital world in which electronic devices are the embedded parts of fine-grained distributed networks. An extensive overview of recent developments and challenges in this extremely interesting field of research is given by the research agenda compiled by the national Research Council [13]. MIT's Oxygen project [8] and IBM's effort on pervasive computing [14] are similar approaches addressing the issue of integration of embedded networked devices into peoples' backgrounds.

Ambient intelligence aims at taking the integration even one step further by realizing environments that are sensitive and responsive to the presence of people. The focus is on the user and his experience from a consumer electron-

T. Basten et al. (eds.), Ambient Intelligence: Impact on Embedded System Design, 11-29.
© 2003 *Kluwer Academic Publishers. Printed in the Netherlands.*

ics perspective, which introduces several new basic problems related to natural user interaction and context aware architectures supporting human centered information, communication, service, and entertainment. For a treatment of these novel distinguishing factors we refer to the book The New Everyday [3]. The technology of ambient intelligence builds on Moore's law and the resulting consequence that electronics have become so small and powerful that it can be integrated into every physical object. The introduction of ambient intelligence will have a tremendous impact on IC design, and planar technology development, which can lead to a paradigm shift in these worlds [6, 11]. The impact on systems design might be even larger because of the high degree of hybridization and complexity of ambient intelligent systems. Below, we further elaborate on these issues, addressing the impact of ambient intelligence on the design and embedding of integrated devices.

2. Ambient Intelligent Environments

In an ambient intelligent environment, people are surrounded with networks of embedded intelligent devices that provide ubiquitous information, communication, services, and entertainment. Furthermore, these devices adapt themselves to users, and even anticipate their needs. Ambient intelligent environments present themselves quite differently compared to contemporary handheld or stationary electronic boxes and devices. Electronics will be integrated into clothing, furniture, cars, houses, offices, and public places, introducing the problem of developing new user interface concepts that allow natural interaction with these environments. A promising approach is the one in which users interact with their digital environments in the same way as they interact with each other. Reeves and Nass were the first to formulate this novel interaction equivalence, and they called it the Media Equation [16]. Ambient intelligence covers a whole world of underlying technologies used to process information: software, storage, displays, sensors, communication, and computing. To identify the different devices that are needed to realize ambient intelligent environments we first introduce a scenario that facilitates the elicitation of a number of ambient intelligent functions from which device requirements can be determined.

3. The New Everyday [Scenario]

Returning home late after a days work, Tom approaches the front door of his home where he is recognized by the 3D animated dog Bello that appears tail wagging on the door, which is also a flat display. After authentication the door alerts that it is open by changing its color. Tom enters the hall where he can see in a single glance on the ambient home flow management system that his daughter Kathy is out with friends, that there are a few new messages, and that the domestic robot Dusty is cleaning the living. Dusty approaches him when he enters the living, and through a single phrase he sends Dusty off to the kitchen to continue its work there. He walks to the interactive table and starts handling his messages through a combination of touch and speech control. One of the massages indicates that the shopping list composed by the intelligent larder needs confirmation before it is sent to the e-supermarket. It also lists a range of menus that can be cooked with the food that is currently in the larder. Another message tells him that the home information system has located the information he requested about affordable holiday cottages with sea views in Spain. A glow tag in the photo frame on the buffet indicates that Kathy, who in the mean time has been notified by her private mobile communicator that dad is home, wants to contact her dad when he is available. Tom asks the home communicator to contact Kathy, and after a few seconds her image appears on a display in the window screen. After a chat with Kathy Tom starts browsing in his personalized video systems using a mobile touch pad integrated into the armrest of his chair. After some time Tom decides to stop video browsing and calls for a nice view. The home information system then switches the window screens to display the life sunset at the beach of the Cafe del Mar and starts playing easy listening music.

4.   Basic Functions of Ambient Intelligence

The above scenario contains three basic ambient intelligent functions. Firstly, the environment is context aware, which means that there are sensors integrated into the environment which communicate events that can be combined to determine meaningful states and actions, like person identification, position detection, and query interpretation. Secondly, audio, video, and data can be streamed wirelessly to any access device present in the environment, thus enabling ubiquitous wireless access to information, communication, services, and entertainment. Thirdly, users in an ambient intelligent environment interact with their surrounding through natural modalities such as speech, gesture, and tactile movements, thus enabling hands free interaction with their environment. These basic ambient intelligence functions not only apply to home environments; they also apply to other environments such as mobile spaces, i.e., car, bus, plane, and train, public spaces, i.e., office, shop, and hospital, and private spaces, i.e., clothing. They support a variety of human activities including work, security, healthcare, entertainment, and personal communications.

5.   Basic Devices for Ambient Intelligence

All devices that process information need energy to do so. Depending on the availability of energy resources in a device, the amount of information processing for a given technology is constrained. The availability of energy in a device is therefore the discriminating factor for the distribution of ambient intelligence functionality in a network of devices. We define three generic classes of devices:

1. Autonomous devices that empower themselves autonomously over a full lifetime. They extract the required energy from the environment by scavenging light or electro-magnetic energy, mechanical energy from vibrations or from temperature differences by means of thermo-electric generator. Examples are all kinds of tags and sensors. These autonomously-empowered devices are called microWatt nodes.

2. Portable devices that use rechargeable batteries with typical autonomous operational times of a few hours, and standby time of several days. Examples are personal digital assistants, mobile phones, wireless monitors, portable, storage containers, and intelligent remote controls. These battery-powered devices are called milliWatt nodes.

3. Static devices that have quasi-unlimited energy resource, e.g., mains powered or combustion engine. Examples are large flat displays, recording devices, (home) servers, and large storage and computing devices. These net-empowered devices are called Watt nodes.

Examples of these device classes are already found today, but are not yet ambient intelligent because they are not part of an integrated intelligent network. The energy availability as well as the power dissipation of a device is not constant over time: different operating modes are defined depending on the device activity and energy resources. Large variations in peak to average power dissipation are encountered. It is clear that for all types of devices the energy management is a key function as it determines the potential for information processing.

6. Information Processing and Power Dissipation

A combination of technologies is required to realise ambient intelligence: in an attempt to classify these technologies we differentiate between communication, computing, storage, sensor and actuator technologies. All these technologies that have two parameters in common: the amount of information that is processed and the associated power dissipation. In order to illustrate the impact of these parameters, some examples of the technologies involved in ambient intelligence are combined in a single graph shown in Figure 1, where the x-axis represents power dissipation, and y-axis represents the information processing capacity, given by

- bytes of memory for storage,

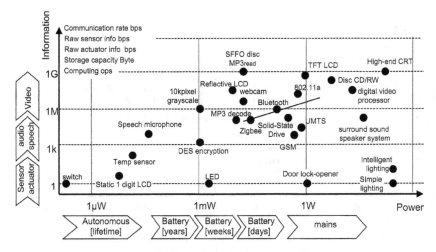

Figure 1.   Power and information processing.

- displayed raw bit rate for displays,

- raw bit rate of captured information for sensors,

- communicated bit rate for communication, and

- operations per second for computing.

The upper left area of the figure is rather empty, because the combination of high information processing capacity and low power is difficult to achieve. For some of the technologies, like communication and computing, the power scales about linearly with information processing capacity.

The power needed for information processing technologies has a major impact on the appearance of ambient intelligence in devices. It will largely influence the functions that are allocated in different types of devices and how the intelligence stack in the network is organized: the distribution of computation, storage, and intelligence over the devices in the system.

Based on this graph, the technologies that can be used in different node types are identified. Silicon IC technology can be used to implement computing, communication and interface electronics in all three devices types. The performance of IC technology spans orders of magnitudes, ranging from a few logic gates to GOPS for computing, from kb/s to Gb/s for communication and in interface electronics from pA, low bandwidth sensor signal to 100V high bandwidth signals in display drivers. This wide range of information processing capability of IC technology opens up opportunities for implementing ambient intelligence. An analysis of the computational-platform aspects of ambient

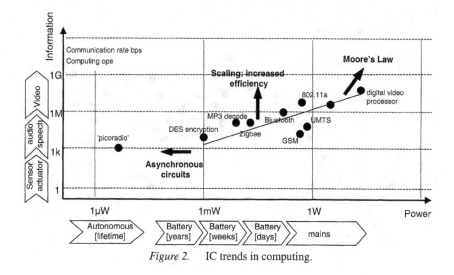

Figure 2.   IC trends in computing.

intelligence is given in [4]. Below we elaborate on the resulting IC design trends for computing, communication and interfacing.

IC Trends in Computing

The IC technology roadmaps for computing are well described by the ITRS roadmaps [12]. Some of the trends are indicated in the graph shown in Figure 2. The continued device scaling of CMOS technology results in an further increasing availability of computational power As described by Moore's Law. This increased computational power is exploited in all kind of parallel architectures. The increased power efficiency of new technologies is particularly interesting for the battery-powered and autonomous devices but is countered by the increased leakage currents in these technology generations. For ultra low power asynchronous circuits are very attractive [5].

IC Trends in Communication

In communication circuits the quest for increased data rates hasn't stopped. However this is not the only trend that can be seen in Figure 3. New IC technologies provide options to increase the communication data rates: new spectrum at higher radio frequencies combined with new techniques like multiple antenna systems, space-time coding or ultra wideband (UWB) make Gb/s wireless communication feasible for consumer applications. However the energy efficiency of communication circuits scales different as for computing: the energy efficiency of communication (energy/bit) not only depends on the digital signal processing efficiency but also on the RF and mixed-signal processing. The power required for RF and mixed-signal circuits relates to noise

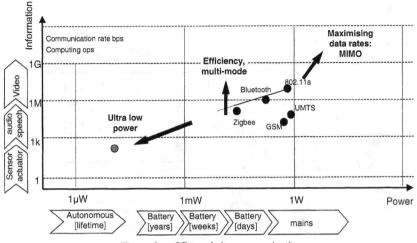

Figure 3.   IC trends in communication.

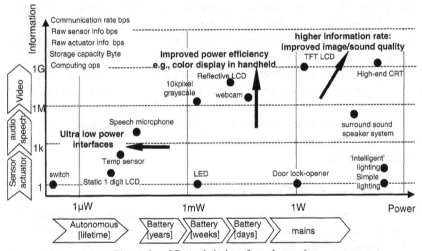

Figure 4.   IC trends in interface electronics.

and distortion parameters of devices and does not directly scale down with IC technologies. Nevertheless in milliWatt node devices the trends is towards increased efficiency and multi-mode, multi-standard capabilities. A relative new area is arising in the microWatt area, aiming for low data rate, low power communication.

Trends in Interfacing

The interface electronics follows closely the developments in sensor and actuator technology. The trends in interface electronics are indicated in Figure 4.

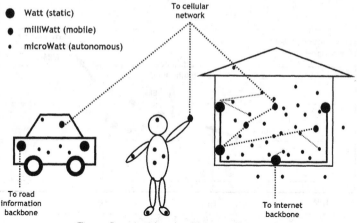

Figure 5.   Ambient intelligent environments.

Higher quality of audio and video input and output devices require more band-width and dynamic range from the interface electronics: amplifiers, filters, analog-to-digital and digital-to-analog converters. The power efficiency of these circuits becomes more important as low power high quality displays be-come available in battery-powered devices. Ultra low power sensor interfaces are needed for autonomous devices. For interface electronics, dedicated de-vices are often required to handle high voltages and/or currents. These devices can be monolithically integrated with computing and communication electron-ics in dedicated IC technologies or as separate die combined in a multi-die package or module.

7.   Architectures for Ambient Intelligence

Ambient intelligent environments are created by the interaction of many de-vices in a network. All these devices have a communication capability to re-alize the interaction and require digital computing to embed the intelligence stack. An ambient intelligence environment is shown in Figure 5 and is based on many networked intelligent devices of the three types.

For each of these device types an area in the information power graph is allocated as shown in Figure 6.

The first device type is the autonomous device: the available power is in the microWatt node. Therefore the communication is limited to kb/s and com-puting to few kops/s. The available power limits dramatically the sensor and actuator functionionality of the microWatt node.

In the milliWatt node, more functionality can be added as communication extends to Mbit/s. More sensor and actuator technologies fit in this power range resulting in large functionality.

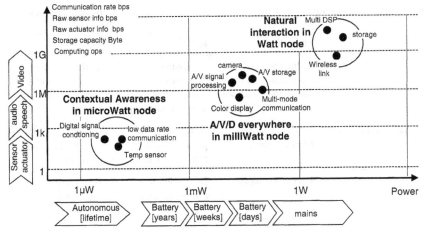

Figure 6. Clustering of information processing technologies in devices.

The Watt node can cover all technologies indicated in the graph and hence process Gbit/s of information. The penalty is the power dissipation that limits this device to static implementation.

Any ambient intelligent functionality has to be realised by optimally combining information processing technologies distributed over different devices types. Starting from the communication and computing limitations for each node, the network of devices is extended with sensor, actuator and storage technology to make ambient intelligence alive.

The main question is: what technologies need to be added to these networked nodes to create an ambient intelligent home? How can awareness, information everywhere, and natural interfaces be implemented? This exercise will point towards the IC design challenges for ambient intelligence. For the indicated themes a possible distribution of functionality is shown in Table 1. Apart from technical feasibility other criteria are taken into account: the cost of devices, privacy and interoperability issues, size and design aspects, etc. Below we elaborate on the IC design challenges for three separate cases. For each of the nodes a combination of communication, computing and sensor/actuator is made and the challenges for ambient intelligence are shown.

Context Awareness in the MicroWatt Node

The functional block diagram for a microWatt node is illustrated in Figure 7. The awareness functionality is realised by combining a sensor with an mixed signal interface, digital signal conditioning and control functions, a wireless transceiver and an energy supply. Rabaey et al., have formulated various IC-design related research activities related to this field of research [15]. The design challenges for the microWatt node are at multiple levels. Although

Table 1.   Ambient intelligence functionality distribution over micro-, milli- and Watt node.

	microWatt	*milliWatt*	*Watt*
Awareness	Gathering information from environment: sensor	Accessing awareness information: actuator, memory	Building awareness intelligence with information: memory
Information everywhere	Data (e.g. smart label): storage	A/V/data access, personal: sensor, actuator, memory	A/V/data content: sensor, actuator, memory
Natural interfaces	Interaction input: sensor	Interaction I/O: sensor, actuator	Interaction intelligence: memory

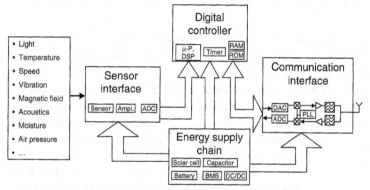

Figure 7.   MicroWatt node for contextual awareness.

the complexity of a single device is rather low, the combination of various technologies with ultra low power dissipation is a challenging task. The design of these nodes requires

- Accurate modeling of the various technologies e.g. power dissipation has to be characterized for all technologies, with very high accuracy over wide operating conditions, including lifetime degradation.

- Multi-technology device simulation: power optimization requires combined simulation of multi technologies probably on a very low abstraction level to achieve the highest power accuracy e.g. a combination of MEMS for sensor, solar cell, digital and RF technology.

- Network simulation: the combination of a large number of microwatt nodes in a network requires network modeling and simulation to study

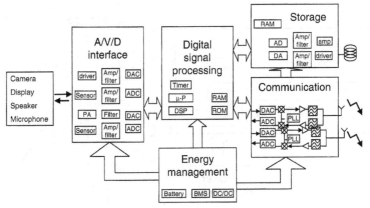

Figure 8.　MilliWatt node for information everywhere.

the effects of redundancy, defective nodes, protocol aspects, etc. of multi devices in network.

Information Everywhere in the MilliWatt Node

A multitude of wireless connections guarantees access to different types of information with different data rate and quality of service. Interface circuits are processing audio and video signals and a local storage device that adds functionality. Energy and battery management play an essential role in these devices. The design needs to be flexible as it has to cope with multi-standards and multi- modes, changing network environments and upgrading of the functionality during lifetime. The IC design challenges in this milliWatt node are:

- The flexibility/power trade-off: accurate power estimation of digital circuit and architecture implementations are needed to make optimal design choices.

- Integration of multi-mode/standards communication interfaces require low power mixed-signal and RF design tools.

- On system level a trade-off has to be made between the increased amount of compression of the A/V/data stream with related power dissipation and the reduced data rate of the communication link.

Natural Interfaces in the Watt Node

Figure 9 shows the Watt node for natural interfaces. The natural interfaces requires the extraction and of information from various information streams (audio and video signals) and setting up an intelligent dialogue with the user. The information input streams are received from milliWatt nodes. The extrac-

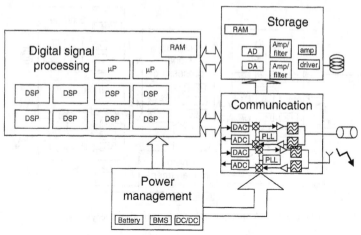

Figure 9.   Watt node for natural interfaces.

tion of information from these data streams requires large amounts of memory e.g. content databases, language information, etc. This memory is provided by local storage combined with internet-like distributed storage. However the major IC content of these nodes will be the computational functionality needed for the intelligence stack: huge amounts of digital signal processing have to be organized to realize the natural interface functionality. This node will benefit the most from Moore's Law. Below we mention some of the IC design challenges for the Watt node; see also [7].

- IC design complexity management: integrating billions of transistors into a working IC. All design tools will have to deal with these giga-scale designs.

- Energy-flexibility trade-off: the designer has to make choices within the range from fully hard-wired to full programmable solutions, and needs tools that support to explore the resulting design space.

- Ensuring the system modularity: systems should be open to support addition and development of new system functionality.

- Architectural issues of the intelligence stack: the development of intelligent software components requires the introduction of adequate software interfaces that support distributed processing.

- Testability: modular test approaches are needed that support system test and evaluation based on test specifications of individual components.

These case studies illustrate the variety of IC design challenges that are found in ambient intelligent devices and the IC design tools needed to realize them.

8. Design Issues

The tri-partition of the ambient intelligence system hierarchy into microWatt nodes, milliWatt nodes, and Watt nodes as defined in the previous sections introduces four major challenges for embedded systems design, which can be formulated as follows.

1. Design for variation

2. Design for distribution

3. Design for intelligence

4. Design for experience

Below, we briefly elaborate on each of these challenges.

Design for Variation

Embedded systems consisting of microWatt nodes, milliWatt nodes, and Watt nodes must support large variations in system parameter specifications in different dimensions. The following three dimensions are of particular interest.

- Communication requirements will range from 1 bit/s up to several Gbit/s resulting from the low bit rate information exchange between the micoWatt nodes as well as from the ultra large bit rates that are required to transport multiplexed real-time video streams between Watt nodes.

- Computational requirements will range from 1 kops up to several Tops resulting from the execution of communication protocols at the microWatt nodes as well as the execution of extremely demanding algorithms for interfacing or video rendering such as 3-dimensional TV.

- Power requirements will range from 1 microWatt up to several Watts resulting from the power demands needed to process information in the microWatt nodes as well as the power demands resulting from the ultra high speed Watt nodes.

The various nodes in the tri-partite systems architecture can only operate within the limits set by the values of the parameters determined by the operational intervals of the node. As a consequence of the large variation in performance quantifiers of the various nodes, designers need to have deep understanding of the physical limitations of the various nodes. The nodes will combine hybrid elements, e.g., a combination of analogue and digital design, possibly augmented with physical sensors and electromechanical actuators. As the characteristic dimensions of silicon devices approach sizes of tens of nanometers, they become within the ranges of the typical sizes of micro electronic mechanical devices (MEMS), leading to a potential combination of elements from the classical silicon technology with elements from the newly developed nano-technology. For the design of embedded system devices that opens new opportunities, but it also combines the complexity of ultra-large scale integration with the complexity of ultra-small device design [7].

Also the design time and effort for the various nodes may vary substantially. The microWatt nodes should be extremely cheap. In some cases, as with tagging devices, the nodes will become disposable, and consequently their prices should not exceed a few eurocents. Their quantities however may become extremely large, for instance in the case where each piece of package or garment would carry a few tags, the number of devices would easily range up to a few billion. In case production volumes are low, design efforts need to be low cost. In case production volumes are high, design efforts may be high cost. The design complexity of milliWatt or Watt nodes, however, will continue to grow exponentially as it has done over the past thirty years. Consequently, we can only resort to the use of modular systems design techniques supported by sophisticated tool suites in order to cope with the growing complexity of the individual nodes. A true leap forward in the design of complex embedded systems can be made by resorting to methods that support distributed systems design in an effective and efficient way.

Design for Distribution

A tri-partite ambient intelligence systems architecture should support execution of distributed applications; i.e., applications consisting of different modular tasks that can be executed by different nodes. An example of such an application is the use of distributed document technology supported by HTML, SMIL, and timesheets, By using a combination of these technologies it becomes possible produce multimedia documents only once without adhering to fixed formats in terms of layout and timing. The document then can be displayed later using any suitable device or set of devices applying the specific device formats upon rendering. One way to meet the design challenge resulting from the need to support distributed applications is to use networks on silicon consisting of four different types of units, i.e., input-output units, storage units, routing units, and processing units. The processing units are the computationally demanding parts of the network. They typically have a size of a few million transistors and they have their own operating system. They in turn consist of four elements, i.e., a memory element, a processing element, a configuration element, and a communication element. The memory element acts as an embedded local storage device; the processing element accounts for the arithmetic computing; the communication element deals with the communication between the processing unit and other units in the network; and the configuration element executes an embedded real-time operating system that can reconfigure properties of the processing element, such as its internal clock. For a more extensive description of such processor networks we refer to Goossens et al. [10].

A next step is to design tri-partite systems architectures consisting of nodes that follow the networks on silicon design approach. For the microWatt nodes this means that they predominantly will act as input units. For the milliWatt and Watt nodes, one may think of devices consisting of networks on silicon that all apply similar systems architectures. Consequently, adding new devices would then simply imply an enlargement of the network without a major change of its constituting architecture. This facilitates interconnectivity and interoperability, and thus supporting the execution of distributed applications. Design for distribution calls for novel middleware constructs that enable multiple networked and distributed applications and services to co-exist and co-operate. The three main issues in this respect relate to interoperability, heterogeneity, and dynamics. Interoperability refers to the ability to exchange devices and application code in networked systems. The issue includes the development of communication protocols that support plug and play such as HAVi and UPnP. Heterogeneity refers to the ability to run software applications on devices with markedly different performance characteristics, i.e., static versus mobile or general purpose versus dedicated. This calls for middleware that is scalable to meet different footprints and operating systems. It also calls for bridging and gateway concepts that support data exchange between networks applying different communication protocols. Dynamics refers to the ability of networked systems to adapt to changes of the environment, such as position, context, configuration, data, and others. To be able to cope with the issues mentioned above, ambient intelligent middleware should support the following functionalities: device abstraction, resource management, stream management, content and asset management, personalization, collaboration, multimodal interaction, context awareness, and security. All these functionalities impose specific, possibly conflicting requirements, and the definition of appropriate APIs and their integration within one software layer imposes major research challenges on the design of an ambient intelligent middleware layer.

Design for Intelligence

Ambient intelligent environments are networked systems of embedded devices. They are however not standalone environments, but they will be part of larger networks that eventually will connect the entire world. Clearly the Internet is already an example of such a network supporting ubiquitous communication all over the world. The Oxygen architecture proposed by researchers from MIT is another example of a ubiquitous communication infrastructure, which is still under development [8]. Ambient intelligent environments can be seen as clusters of computing, storage, and input-output devices that are connected through ubiquitous broadband communication networks allowing data rates up to a few Gigabytes. The ubiquitous network contains servers that act as routers and internal storage devices. This implies that one can trade-off the location of the software that provides the system intelligence. Roughly speaking there are two extremes, which are referred to as *ambient intelligence inside* versus *ambient intelligence outside.* In the inside case the system intelligence resides as embedded software in the terminals. They are big footprint devices that can efficiently process large software stacks that implement sophisticated computational intelligence algorithms. The network that connects the devices is rather straightforward from a functional point view just allowing data communication, possibly at high bit rates. In the opposed view of ambient intelligence outside the terminals may be small footprint devices just allowing for the data communication that is required to generate output or to take new input. The system intelligence is residing at the powerful servers of the ubiquitous communication network where it can be accesses by the terminals.

As an example we consider a surround lighting system for home theatres. Such a system consists of a central flat screen with a backlighting element, four lighting satellites that can be positioned at the corners of a room, and a sublighting element that can be positioned under the coach. Furthermore, there is a lighting-surround player that can play a DVD in such a way that along with the movie that is displayed on the screen and the music that is produced by the surround sound system, light effects are generated by the surround lighting system that enhance the viewing experience. In the ambient intelligence inside case the lighting script is generated on-line by the surround lighting systems. This requires the use of extensive video signal processing through sophisticated content analysis algorithms that generate the lighting script on-line and real time. Rough estimates indicate that this will require between ten and fifty Gops of computational effort calling for powerful digital signal processors. In the outside case the lighting script is computed of-line by a server in the ubiquitous communication network and delivered to the surround lighting system upon request. The script is then synchronized with the movie and its sound, and executed by the surround lighting system. Clearly the trade-off between ambient intelligence inside and out side relates to complex design decisions that need to be evaluated at system level. The resulting design problem is referred to as "inside-outside co-design" and imposes several new challenges for systems design. The trade-off is not only determined by system related performance issues; it also requires the involvement of business models that can quantify the costs of the various ambient intelligence inside versus outside models, because the use of such systems in daily life will determine its design.

Design for Experience

Ambient intelligent environment are networked embedded systems that are sensitive to the presence of people and that can respond accordingly. Consequently, electronics will disappear into peoples' background moving social interaction and functionality to the foreground resulting in experiences that enhance peoples' lives. This not only requires insight into the design of sophisticated distributed systems as sketched above; it also requires deep understanding of the translation of user needs into functional requirements. The design of ambient intelligent systems is markedly different from the design of classical one-box systems. Ambient intelligent environment introduce many new options for services and applications. The fact that no boxes will be present anymore introduces the need to come up with novel interaction concepts that allow the user to communicate with their electronic environment in a natural way. Only little is known about the system requirements such novel environments should meet, and there is only one way to find out and that is through building prototypes that allow people to interact directly with the novel concepts and that facilitate user studies in a scientific way. This is to a large extent an underestimated field of research, which undoubtedly will become of great significance in the near future. Requirements engineering for embedded systems design can no longer be seen as a task that can be accomplished through the development of scenario's and the translation of use cases into system requirements. System functionalities that generate true user experiences can only be determined in a reliable way from feasible prototypes providing proofs of concept. These are called experience prototypes, and in order to support their effective and efficient development prototyping environments are needed that that facilitate both feasibility and usability studies. More specifically this means that laboratories are needed which contain infrastructures that support fast prototyping of novel interaction concepts and that resemble natural environments of use. Moreover, these experience prototyping centers should also be equipped with an observation infrastructure that can capture and analyze the behavior of people that interact with the experience prototypes.

As an example we mention Philips' HomeLab, which is a combined feasibility and usability laboratory that is aimed at investigating through extensive empirical research, which embedded interaction technologies are really perceived by people as life enhancing experiences [1]. HomeLab is an advanced research center in which engineers build prototypes of embedded distributed systems that implement novel interaction concepts developed by industrial designers, and which are subsequently evaluated by cognitive psychologist through extensive user tests. HomeLab consists of a feasibility nucleus surrounded by a usibility shell. The feasibility nucleus is identical to an ordinary house with a hall, a living, and a kitchen on the ground floor, and a parents' bedroom, a children's bedroom, a den, and a bathroom on the first floor. The usability shell consists of a sophisticated observation system with cameras, microphones, display monitors, and a storage systems that van be used to record observed behavior that can be analyses to validate certain working hypothesis supporting the use of novel interaction paradigms.

Through the use of HomeLab we have developed the strong belief that killer applications do not exist by nature; they can only be developed through the use of artificial means. HomeLab has proved very instrumental in this respect and as a consequence it has become a major instrument in the development of insight into user needs and their translation to functional requirements for distributed embedded systems. So, experience prototyping facilities such as HomeLab are the instruments through which functional requirements of ambient intelligent systems are obtained.

9.   Conclusions

Ambient Intelligence is presented as a vision for the future of consumer electronics. It is shown that within the ambient intelligence concept the energy resources are a main criterion in the design of devices resulting in three generic types: microWatt, milliWatt and Watt nodes. For these devices, silicon IC is the basic technology for computing, communication and mixed-signal interfaces. Ambient intelligent functionality is mapped onto a network of devices of the three types. This is applied to context awareness, information everywhere and natural interfaces and results in three case studies indicating the various challenges for IC design. The tri-partition of the ambient intelligence system hierarchy into microWatt nodes, milliWatt nodes, and Watt nodes introduces four major challenges for embedded systems design: design for variation, design for distribution, design for intelligence and design for experience.

Acknowledgements

The authors thank Twan Basten, Henk Jan Bergveld, Kees van Berkel, Neil Bird, Marc Johnson, Menno Lindwer, Marcel Pelgrom, and Tony Sayers for their contributions to this paper.

References

[1] E. Aarts and B. Eggen. *Ambient Intelligence Research in HomeLab.* Neroc Publishers, Eindhoven, The Netherlands, 2002.

[2] E. Aarts, R. Harwig, and M. Schuurmans. "Ambient intelligence." In J.Denning, editor, *The Invisible Future*, pages 235–250, McGraw–Hill, New York, 2001.

[3] E. Aarts and S. Marzano, editors. *The New Everyday: Visions of Ambient Intelligence*. 010 Publishing, Rotterdam, The Netherlands, 2003.

[4] L. Benini and M. Poncino. "Ambient Intelligence: A Computational Platform Perspective," this volume, 2003.

[5] C.H. van Berkel, M.B. Josephs, and S.M. Nowick. "Scanning the technology: Applications of asynchronous circuits." *Proceedings of the IEEE*, 87(2):223–233, 1999.

[6] F. Boekhorst. "Ambient intelligence, the next paradigm for consumer electronics: How will it affect silicon?" *Proceedings of the International Solid State Circuit Conference*, pages 20–27, 2002.

[7] H. De Man. "Nanoscale system design challenges: business as usual." *Proceedings of the ESSCIRC*, pages 3–10, 2002.

[8] M. Dertouzos. "The future of computing." *Scientific American*, 281(2):52–55, 1999.

[9] K. Ducatel, M. Bogdanowicz, F. Scapolo, J. Leijten, and J-C. Burgelman. *Scenarios for ambient intelligence in 2010*. IST Advisory Group final report, 2002.

[10] K. Goossens, J. Dielissen, J. van Meerbergen, P. Poplavko, A. Radulescu, E. Rijpkema, E. Waterlander, and P. Wielage. "Guaranteeing the quality of services in Networks on a Chip." In A. Jantsch and H. Tenhunen, editors, *Networks on Chip*, pages 61–82. Kluwer Academic Publishers, 2003.

[11] R. Harwig and E. Aarts. "Ambient intelligence: Invisible electronics emerging." *Proceeding of the 2002 International Interconnect Technology Conference*, pages 3–5, 2002.

[12] ITRS roadmap. http://public.itrs.net/.

[13] National Research Council. *Embedded Everywhere*. National Academy Press, Washington DC, 2001.

[14] A. Mukherjee and D. Sata. "Pervasive computing, a paradigm for the 21st century." *IEEE Computer*, 36(3):25–31, 2003.

[15] J. Rabaey, J. Ammer, T. Karalan, S. Li, B. Otis, M. Sheets, and T. Tuan. "Picoradio for wireless sensor networks: the next challenge in ultra low power design." *Proceedings of the International Solid State Circuit Conference*, pages 200–201, 2002.

[16] B. Reeves and C. Nass. *The Media Equation*. Cambridge University Press, Cambridge, Massachusetts, 1996.

[17] M. Weiser. "The computer for the twenty-first century." *Scientific American*, 165(3):94–104, 1991, reprinted in IEEE Pervasive Computing, march 2003, 19-25.

Ambient Intelligence: A Computational Platform Perspective

Luca Benini

Università di Bologna, Bologna, Italy

lbenini@deis.unibo.it

Massimo Poncino

Università di Verona, Verona, Italy

massimo.poncino@univr.it

Abstract   Computational platforms are a key enabling technology for materializing the Ambient Intelligence vision. Ambient intelligence devices will require a widely ranging computational power under widely ranging system-level constraints on cost, reliability, power consumption. We coarsely group computational architectures in three broad classes, namely: fixed-base network (the *workhorses*), wireless base network (the *hummingbirds*), and wireless sensor network (the *butterflies*). Speed and power requirements for devices in these three classes span six orders of magnitude. In this paper, we analyze commonalities and differences between these three classes of computational architectures, and moving from the analysis of representative state-of-the-art devices, we survey design trends directions of research.

Keywords   platform-based design, architectures, microprocessors, power consumption, wireless LANs, systems-on-chip, networks-on-chip, wireless sensors, MEMS

1.   Introduction

Ambient Intelligence devices are expected to provide scalable processing power at every level of the network infrastructure. As outlined in [2], the range of performance and power requirements spanned by these devices is truly impressive: data rates range between the tens of Kb/s and the hundred of Gb/s; power consumption ranges between the tens of microWatts and the hundred of Watts. Developing devices spanning ranges of six orders of magnitude in performance and power is a daunting task, but it can be done, and it has been done. In fact, natural evolution has produced life forms that span an even wider range in size and weight, based on a common "technology", i.e, carbon biochemistry.

The technology of ambient intelligence is silicon CMOS; even though it is not as flexible and highly "integrated" as carbon biochemistry, its flexibility is

T. Basten et al. (eds.), Ambient Intelligence: Impact on Embedded System Design, 31-50.

still remarkable. The minimum device size is close to $0.1\mu m$, and it is possible to integrate hundred of millions of these minuscule devices on a $6mm^2$ die. Clock frequency of a complex CMOS circuit can easily exceed the 1THz mark, or be lower than 1kHz, according to the needs. We claim that six orders-of-magnitude ranges in performance and power are indeed achievable in CMOS technology. This paper focuses on the main challenges that must be addressed to meet this objective from an architectural viewpoint.

Even assuming no slowdown in technology scaling, designing devices that meet cost, reliability and energy-efficiency requirements is going to be extremely challenging. In fact, as technology scales down, it becomes increasingly unwieldy [4]: (i) logic gates get faster, but wires do not, and performance is increasingly interconnect-dominated; (ii) power density (in active state) and leakage power (in idle state) increase significantly even if supply voltage is down-scaled aggressively; (iii) signal-to-noise ratio and circuit reliability decrease; (iv) design complexity scales up.

Besides taming technology, architects are faced with additional challenges. Flexibility, both at deployment time (i.e., programmability) and in-field (i.e., re-configurability) is required to support scalability. Here, scalability refers to the capability of providing adequate performance under widely varying operating conditions, such as those required by AmI devices. Therefore, most AmI devices will typically be programmable and reconfigurable. Furthermore, truly scalable computation must be cheap, reliable and energy-efficient.

To address the scalability challenge in view of technology and design complexity limitations, hardware architectures for AmI will assume widely varying characteristics [1]. Pursuing a biological analogy, we grouped AmI architectures in three classes, which roughly match the node type classification introduced in [2]:

- *workhorses*, top-of-the line processors powering the high-speed fixed network backbone, exceeding 10Gb/s performance and 100W power consumption;

- *hummingbirds*, complex systems-on-chip (SoCs) for high-bandwidth wireless networking and multimedia, with 10Mb/s performance and 100mW power consumption;

- *butterflies*, highly integrated wireless micro-sensor nodes for pervasive sensor networks, with 10Kb/s performance and $100\mu W$ power consumption.

For each class of devices, in the next sections we first survey design requirements and constraints, then we analyze in detail an existing representative device from an architectural viewpoint, and finally we survey directions of future evolutions. We have chosen to initially put the emphasis on architectures (and specifically, on existing architectures), and to analyze platform-related issues in perspective, when discussing the future challenges that ambient intelligence implies.

2. Workhorses: Fixed Base Network Devices

High performance microprocessors are the workhorses of the fixed base network infrastructure. These chips sell in extremely large volumes and with high margins, hence research and development investments are massive. Developing a new processors takes several years (three to five) and very large design teams (several hundred of engineers). These unique characteristics contribute to making high performance microprocessor the drivers for advances in technology and circuits, which then percolate to less aggressive designs.

Performance is by far the most important design metric for this class of architectures. Several standardized execution time performance indicators on benchmark suites (e.g. SpecINT and SpecFP) are now routinely used. A fairly weak performance indicator which still has high marketing value is clock speed. State-of-the-art general purpose processors in 2003 are indeed clocked at frequencies exceeding by a significant amount the 2GHz mark. Such fast clock speeds put microprocessors well ahead of any other large-scale digital VLSI design, especially for what concerns the techniques developed to overcome the formidable clock distribution challenges. To meet cycle time constraints, micro-architecture has evolved toward deeply pipelined solutions, where only a few (8-12) stages of logic are available on the critical path of a stage.

Fast clocks and high complexity imply high power consumption. Power consumption is indeed viewed as one of the most significant limitation to current evolutionary trends in microprocessor design [18]. Lowering power supply does not solve the problem, because design complexity and clock speed scale faster, and their compound effect swamps the benefits given by voltage scaling. Average power consumption, routinely exceeding 100 Watts, is just one facet of the problem. Power density and power delivery are even more serious short-term challenges. In fact, a 150-Watt microprocessor with a voltage supply set at 1.5V, needs a whooping 100A of current, to be delivered on a silicon surface smaller than a post stamp.

Even though these design challenges are indeed formidable, high-performance processors have successfully scaled for the last twenty years. In the following sub-section we analyze a leading microprocessor, which represents a pivotal example of the "workhorse" family of ambient intelligence devices.

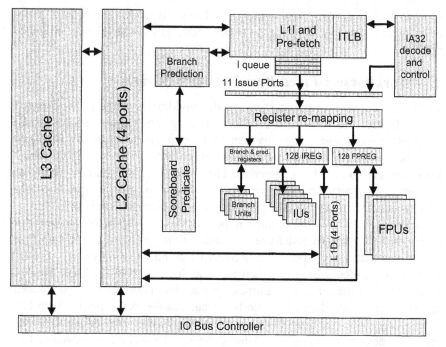

Figure 1.   Itanium II micro-architecture.

Case Study

Itanium II is the second-generation IA-64 microprocessor developed by Intel and Hewlett-Packard that was introduced in 2002 [25]. This chip targets the high-end technical and commercial workstation and server market. However, it also nicely fits into an ambient intelligence scenario, where workhorses may include traditional consumer electronics products with high processing requirements (such as high-end TV sets, set-top boxes, or game consoles), that need to be connected to a fixed supply network and have non-negligible costs.

It is based on a 0.18μm, 6-metal CMOS process; the die size is 19.5×21.6mm (421mm^2) and it is clocked at 1GHz. Itanium II represents a state-of-the-art top of the line general purpose microprocessor; its organization is depicted in Figure 1.

From the micro-architectural viewpoint, the emphasis of the entire design is to provide extensive support for instruction-level parallelism discovery and exploitation. With a remarkable deviation from the trend set by 32-bit Intel architectures, the IA-64 family adopted an explicitly parallel instruction set architecture (code-named EPIC), where compilers are given the task of explicitly scheduling two instruction bundles (of 3 instructions each) for execution. In other words, the architecture can dispatch up to six instructions per cycle, but it requires compiler support to achieve the top dispatch rate. However, backward compatibility (at the price of significantly reduced performance) with IA-32 code is guaranteed.

Another striking architectural feature is the inclusion of a three-level cache hierarchy on the CPU chip, for a total cache size of 3.3 MB. As a result, almost 3/4 of the total chip area is dedicated to caches. The L3 cache is 3-MB, with a 12-cycle access latency, and dominates the chip floor-plan. The L2 cache is much smaller (256-KB), but it features 6-cycle latency, and it is multi-ported (4 read and 2 write ports). The 4-ported L1 cache is extremely fast (1/2 cycle latency) and small (16-KB data and 16-KB instruction). Relying on large on-chip caches and non-blocking low-level caches is a common design choice in high-performance processors, and Itanium II pushes this concept to an unprecedented level: it has at least twice as much on-chip caches than the closest competing general-purpose processor.

The execution pipeline of Itanium is highly parallel and decoupled from the instruction fetch stages via an eight-bundle queue. Execution issue is in-order, completion is out-of order, and the pipeline is fully interlocked. Overall, the pipeline depth is 8. The most significant features of the execution pipeline are extensive support for speculation past branches, based on aggressive branch prediction. Prediction support, coupled with an extremely fast I-cache, allows zero pipeline overhead (no bubbles) and 95.2% prediction accuracy on standard benchmarks. Execution parallelism is supported via a large number of functional units of three different types: 6 integer units, 6 multimedia units and 2 floating point units. Such an exasperate parallelism puts significant pressure on register files, which are highly multi-ported (integer RF is 20 ported, and floating point register file is 14 ported).

Another key point in Itanium's architecture is the emphasis on a fast off-chip interface. External memory and I/O are accessed through a 128-bit front-side bus (FSB), clocked at 400-MHz. Thus, the bus achieves a 6.4 GB/s bandwidth (a two-fold improvement with respect to the first-generation Itanium), which is required to support scalability and multi-processor configurations. Providing adequate bandwidth to the execution units is a cross-cutting theme in the design of the overall Itanium architecture. In fact, I/O bandwidth optimization is probably the most obvious improvement from Itanium I to Itanium II.

Itanium II is obviously a full custom design and implements a number of innovative and technology and circuit techniques to achieve higher speed. Even though the details of circuit-level design are outside the scope of our overview (the interested reader is referred to [25] and references therein), three important points are worth mentioning. First, dynamic logic is used extensively throughout the chip; second, a great deal of attention is dedicated to clock distribution (which uses a balanced H-tree architecture); third, custom self-timed and clock domain interface circuits are common in various parts of the processor. All these design efforts provide evidence of "synchronization bottleneck": clocking is the toughest challenge for high-performance microprocessors, and many advanced solutions are finding their way to product implementation.

Power dissipation of Itanium is high: the chip consumes 130 Watts. The analysis of power breakdown is very illuminating. Roughly half of the power is dissipated in "switching logic", which includes caches. The chip contains 221M transistors, and only 25M is dedicated to computational logic. Hence, the vast majority of transistors are in memories. Even though, memory power density is notoriously smaller than logic power density, such a disproportion between memory transistors and logic transistors leads us to conjecture that at least half of the "switching logic" power is consumed in memories and memory buses. Furthermore, 33% of the total power is consumed by clock generation and distribution, and another 8% is consumed by global repeaters and I/O drivers. Leakage is still marginal (only 2%). We can therefore conclude that approximatively 80% or more of the power in Itanium is *not* consumed in computation, but in bringing data through space (with communication circuits and buses) and time (with memories) to the computational units!

Challenges Ahead

Overall, Itanium is a highly-complex, but evolutionary architecture, which appears from outside as a single execution unit. The limited scalability of this approach, as pointed out by various authors [18, 29, 3], is apparent even from our brief analysis. Most of the performance enhancement features adopted in Itanium are already in the region of diminishing returns: there is a growing disproportion between the number of transistors (and power consumption) dedicated to computation and those dedicated to storage and communication.

Furthermore, even though huge on-chip caches and very high performance I/O circuits can ease the latency and throughput bottlenecks, there is limited parallelism left to be exploited in single instruction stream. Simultaneous multi-threading (SMT) is emerging as the new architectural leit-motif [9], which can push the discovery of new sources of parallelism one step further. SMT enables a single processor to run instructions from multiple parallel streams in a time-multiplexed fashion. Unfortunately, from a microarchitecture viewpoint SMT processors are slightly more complex than standard microprocessors, and therefore they suffer from the same scalability limitations.

For this reason, high-performance devices for the fixed network infrastructure are rapidly evolving toward explicitly parallel architectures. Single-chip multiprocessors [13] are a natural evolution of current single-processor solutions in an effort to support truly scalable data processing. With respect to high-complexity single stream execution engines, like Itanium, single-chip multiprocessors have the advantage of easier scaling. In fact, communication bottlenecks are not hidden in highly shared storage resources (e.g. register files, execution queues), but they are explicit in the on-chip interprocessor communication network.

Ultimately, the ever-increasing ratio between wire delay (which remains constant, even in the most optimistic assumptions) and device switching speed (which scales with technology) has become a fundamental bottleneck in designing high-performance integrated systems [4]. The only viable answer to this challenge is to localize computation and emphasize parallelism. For this reason, even general-purpose single-chip processors for high-end servers are becoming increasingly distributed and highly parallel. Explicit parallelism is sought to avoid architectural bottlenecks and to ensure long-term scalability across several technology generations.

Most newly-designed high-performance processors are highly parallel architectures, with multiple program counters [29]. Moreover, hardware designers routinely resort to distributed implementation styles: multiple clock domains, multiple execution units, distributed register files are representative instances of this trend [29]. One step further, next-generation architectures are focusing on parallelism not only at the micro-architectural level, but also across the entire memory hierarchy. Multi-processing becomes multi-streaming: streaming supercomputers are under development [16] where computation and storage are seen as successive stages of an information processing pipeline (a stream), and communication is made explicit as much as possible.

3. Hummmingbirds: Wireless Base Network Devices

The wireless base network occupies the middle range of computational power and energy budget. The characterizing feature of this class of devices is their wireless nature, that causes energy to become a design constraint. Besides energy, however, there are three other issues to be considered when evaluating possible candidates for this part of the communication infrastructure.

The first issue, and probably the most important is related to the available *bandwidth*. Current wireless technologies cover a wide bandwidth range: from low-bandwidth services like voice and messaging services of current 2G cell phones (in the order of tens of kbps) to high-bandwidth, point-to-point optical connections (in the order of Gbps). In an effort towards an ubiquitous distribution of wireless devices, high bandwidth is clearly a desirable feature. The ubiquity constraints imposed by ambient intelligence, however, limit the possible choices, and higher bandwidth is not always the best choice. For example, high-bandwidth optical connections which require line-of-sight operations are clearly not suitable for outdoor connectivity.

In addition, since bandwidth is related to the frequency range, it also affects coverage. For example, lower frequencies are less sensitive to the presence of physical obstacles (e.g., walls), while high frequencies require clear line-of-sight. For each particular application, throughput decreases as distance from the transmitter or access point increases.

Given these constraints, wireless LAN (WLAN) technology appears to be a good compromise between bandwidth requirements and environmental constraints. The IEEE 802.11 group of standards [15] and the ETSI HiperLAN/2 [12] standard are currently filling the gap between short-distance point-to-point technologies like BlueTooth [8] and 3G mobile phone systems like UMTS. The most recent and powerful versions of these two standards operate in the license-exempt 5GHz band, and are able to provide maximum physical layer bandwidths of 54 Mbps.

Due to its simpler protocol specification, the 802.11 standard is emerging as the dominating solution. This standard comes into several variants (802.11 a/b/g) with different features, including various operating frequencies and bandwidths. However, the HiperLAN/2 standard, that consists of a single specification, is still receiving some attention as an European standard. Furthermore, HiperLAN/2 has the advantage of offering QoS features as well as higher security in the base standard. The 802.11 standard has recently introduced the 802.11e and 802.11i drafts, in an effort towards filling this gap. For a comprehensive comparison between the two standards the reader is referred to [14].

The second issue is related to the *security* of the communication. This is a major issue for wireless due to the nature of the transmission mechanism (electromagnetic signals traveling through the air). In WLANs, authentication and encryption provide data security. Clearly, encryption protects data only between the wireless adapter and the fixed base. Application level encryption mechanisms, (such as SSH) are responsible for further protection of data on the wired LAN.

The third issue is concerned with *cost* of the device. The current cost of a typical WLAN (or similar) card is definitely too high to allow a true ubiquity. For future applications, the card size may also become an issue. This constraint translates into the adoption of more efficient architectures that allow to put most functionalities on a single chip. In particular, at least the MAC and the baseband processing should be implemented in a single-chip. In some cases, even the transceiver section can be implemented in CMOS technology and integrated onto the same chip. Fewer components translate into reduced bill-of-materials required for a multi-mode implementation of the device, and cost is reduced as a consequence.

The previous analysis identifies the essential requirements of a wireless base station suitable for a pervasive computing paradigm:

- Low-energy consumption, in the order of the hundreds of mW;

- High-bandwidth, in the range offered by WLAN technologies (in the order of tens of Mbps);

- Security features to protect the transmitted data.

- Low-cost (in the order of few dollars), small size (in the order of cubic centimeters).

While the first and last constraints push towards custom-designed devices, the second and the third ones force to resort to programmable architectures. The wide variety of WLAN standards, in particular, makes programmability mandatory. Besides programmability, some degree of reconfigurability is also desirable, particularly in the case of small, incremental upgrades of the system.

In the following subsection we discuss a commercial product as an example of a device that includes, to some extent, all the above features.

Case Study

The TNETW1130 by Texas Instruments is one of the most interesting commercial solutions. While there are many similar chips on the market, this one represents probably the best synthesis of the four constraints defined above.

The TNETW1130 is a single-chip MAC and baseband processor that is compatible with *all* versions and additions of the 802.11 standards. It supports the three base standard 802.11a, 802.11b and 802.11g draft standards. To achieve compatibility with these three standards, the device features seamless operation in the 2.4 GHz and 5.2 GHz bands. This is achieved by the proprietary Auto-Band^{TM} technology, that allows automatic switch between different transmission modes and frequency bands.

Besides the three basic standard, the TNETW1130 also handles the three draft extensions of 802.11: the 802.11e draft standard (for QoS support of for real-time applications with fixed bandwidth and timing requirements), the 802.11h draft (that provides support for global certification), and the 802.11i draft (for multi-mode security and privacy features).

Figure 2 shows a block diagram of the TNETW1130, where some details contained in the data-sheet have been abstracted [31]. The schemes shows two main blocks: the one on the left, implementing the MAC layer interface, and the smaller one on the right implementing the baseband (BB) interface. Both are integrated onto the same die.

The MAC portion of the design consists of five main blocks: The central processing engine consists of an ARM9 core, that executes less time-critical tasks of the MAC operations. The choice of a relatively powerful core allows to maintain high throughput rates.

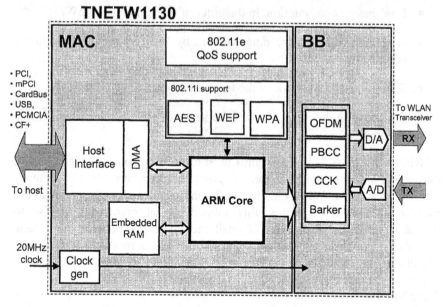

Figure 2.   Block diagram of the TNETW1130 chip.

The core communicates with a 128K embedded SRAM; the presence of on-chip memory differentiates this product from many similar devices that require external memory chips. Besides providing a higher processor-memory bandwidth, this solution is consistent with a solution with minimum number of components.

The third block concerns the interface towards the host. The interface is DMA-based: this allows data in the host memory to be stored in noncontiguous memory buffers, thus reducing the number of copy operations that the host must perform. Various interface types are provided, including 32-bit Cardbus, PCI, MiniPCI, USB1.1, and CompactFlash+, that allows to choose various form factors to fit the application.

The other two blocks contains the hardware support for the 802.11 extensions. The first one implements the security features of the the 802.11i draft: as the latter recommends, such feature are realized by hardware accelerators. More specifically, it offers supports for Advanced Encryption Standard (AES) in both the mandatory (AES-CCM, Cipher block Chaining Message authentication code) and optional (AES-OCB, Offset Code Book) modes, legacy support for WEP (Wired Equivalency Protocol, the basic security mechanism of 802.11), and additional authentication mechanism such as Wi-Fi Protected Access (WPA).

The second blocks is relative to the 802.11e extension, relative to QoS features such as *enhanced distributed coordination* function (EDCF) and *hybrid coordination function (HCF)*. This allows a WLAN to dedicate bandwidth for emerging real-time applications like voice-over-WLAN, broadcast video, video conferencing and others, or time-sensitive applications in general.

Not explicitly visible in the figure are the enhanced low-power features of the design, consisting of special power-management techniques that effectively exploit 802.11 standby modes for reducing power consumption.

The baseband section is conceptually described as a set of blocks corresponding to the various coding schemes used by the different 802.11 frequency bands: OFDM (Orthogonal Frequency Division Multiplexing) for 802.11a, and CCK (Complementary Code Keying), PBCC (Packet Binary Convolutional Code), and Barker for 802.11 b/g. Further functionalities of this section include forward error correction coding, signal detection, automatic gain control, frequency offset estimation, receive/transmit filtering, frame encryption/decryption, and error recovery operations according to the definitions of the standards. Finally, the chip includes a section for interfacing with the WLAN transceiver, basically some A/D channels (for receiving) and D/A channels (for transmitting).

Challenges Ahead

The chip described in the previous section represents a first solution towards the computational demand required by ambient intelligence. In a near future, however, ubiquity of computation will require further capabilities.

In particular, we envision two main, related trends. The first is of technological nature: the power, cost and size constraints push towards single-chip integration of wireless hardware. For instance, a complete WLAN chip-set consists today of no less than four chips, even using advanced solutions such as the TNETW1130: one for the power amplifier, one for the RF Interface, one for the RF Modem, and one for the MAC and Baseband processing. Some WLAN chip manufacturers are proposing direct conversion technology to combine the RF/IF and Modulator/Demodulator chips into one, thus reducing the chip count to three. Yet, a true single-chip solution has still to come.

The second trend is of architectural nature. The quest for higher integration, coupled with the quickly evolving wireless standards (as for the 802.11) and the introduction of new ones, push for new architectural paradigms. Furthermore, the integration of higher-level services in the chip such as network, transport and even application level (e.g. an MPEG2 decoding) are even more challenging requirements.

Energy efficiency and cost constraints are much tighter than for high-performance servers, and the computational requirements can be matched only by resorting to heterogeneous architectures (in contrast with homogeneous general-purpose processors), which provide computational power tailored to a specific class of applications [17, 37]. Designing such heterogeneous, application-specific multi-processor system-on-chip (MPSoC) is a challenging task because their complexity is comparable to that of most aggressive general-purpose platforms, yet their time-to-market and profitability windows are much shorter and focused.

Next-generation wireless base network devices are likely to include several application-specific processors (e.g., MAC accelerators, digital MODEMS, cryptoprocessors), as well as several general-purpose processing cores (e.g., FPGA fabrics, DSP cores, VLIW multimedia processors, FP coprocessors, RISC controllers, etc.), a significant amount of on-chip storage (both volatile and non-volatile), various peripheral units (e.g., external DRAM controllers, interfaces to on-board buses, RF front-ends, BIST units, etc.). The connective fabric for this multitude of devices will be a *Network-on-Chip* (NoC), most likely a packet-based network with QoS guarantees [7].

Last, but not least, all this computational power translates into large memory bandwidth requirements: on-chip memories will thus be ubiquitous, especially if many application-level multimedia functions will be supported. Therefore, energy-efficient memory techniques will be essential [21], and the I/O paths to memories will need to be very carefully optimized, through high performance I/O interfaces [20], typically using low swing and signal enhancement techniques [22].

Finally, these platforms will heavily rely on technology support for aggressive power management. Multiple and variable clock-frequencies, supply voltages, transistor thresholds will be supported on the same chip [19].

4.   Butterflies: Sensor Network Devices

The wireless sensor network is the most critically power- and cost-constrained component of the communication infrastructure. Individual nodes will be extremely simple objects, but a sensor network will contain a large number of them, tightly connected to the wireless base network infrastructure. Wireless sensors leverage recent advances in miniaturization and the integration of micro-electro-mechanical systems (MEMS) to build small sensors, optical communication components, and power supplies.

From the operational point of view, wireless sensor requirements are quite atypical within the wireless domain, in particular when compared to usual wireless technologies: sensor networks have very high density (several nodes per square meter), low-range communication (100 meters range is typically long enough), and especially very low data rates (few computations per second), and consequently very low bandwidth (fraction of Kbps) [24].

Sensors, however, share with other wireless devices the problem of limited power supply, that makes power definitely the most stringent constraint for the

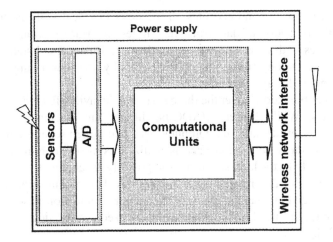

Figure 3.   Conceptual architecture of a typical wireless sensor.

sensor network. Such constraint is further enforced by even more stringent *size* and *weight* constraints. In fact, the ambient intelligence paradigm envisions micro-sensors no larger than a few cubic millimeters, and no heavier than a few grams.

These size and weight ranges automatically rule out the possibility of adopting conventional power sources: any type of battery would be too large and heavy. Furthermore, the sensors will not necessarily be located in fixed positions, and their tiny size may even make impossible to physically handle them. Therefore, sensors may rely on some alternative energy source that allows them to be autonomous, such as solar cells. The latter, however, provide modest energy densities, when compared to conventional batteries. For instance, a typical AAA re-chargeable lithium battery roughly provides 10 times more power than a $1cm^2$ centimeter solar cell (in direct sun) and 1000 times more than the same cell in shade [28]. Since current solar cell technology provides between $1mW/mm^2$ (in full sunlight) and $1\mu W/mm^2$ (bright indoor illumination), we can assume a rough upper bound of power consumption of $100\mu W$, for autonomous sensor operations in all conditions.

Fortunately, sensor nodes do not need to be awake all the time; their activity factor is typically less than 1%. This implies intrinsically low average power, yet meeting such tight power constraint may be difficult.

To understand whether this power budget is really a constraint, we must define what functionalities a typical sensor should provide. Figure 3 shows the main components of a typical wireless sensor.

The architecture obviously include the actual sensor section with its relative interface. The type and number of hosted sensors will depend on the target application. The sensors provide inputs to the computational units, that process data from the sensors and executes all the tasks that deal with the transmission and the reception of data from the wireless interface section. The latter communicates with other sensors or to other wireless of fixed stations using the proper technology (radio vs. optical). Finally, the power supply section provides energy for all other units.

The power breakdown over the three sections is heavily affected by their implementation. While the sensor/ADC has limited implementation flexibility, the wireless interface and the computational units offer several alternatives. Intuitively, if the wireless interface is realized using RF communication, it would definitely be the largest contributor, possibly exceeding the whole power budget alone. For this reason, optical, laser-based communication should be preferred; to give a rough idea, a RF-based communication has a transmission efficiency around $1 \mu J/bit$, whereas laser-based communication is around 1 nJ/bit [33].

Concerning the computational units, the options for their implementation range from a full-custom implementation to the use of a general-purpose processor. While power reduction pushes for architectures that are as much application-specific as possible, programmability might be desirable for execution of communication-related tasks or for the implementation of power management techniques. In fact, state-of-the-art low-power processors may be much more energy-efficient than the wireless interface hardware. For instance the CoolRisc core [27]) requires about 20 pJ/instruction, thus allowing the execution of 50 instructions for each (optical) transmission.

Another advantage of programmability is flexibility. Flexibility can also be achieved by using re-configurable hardware instead of a programmable core. In both cases, *remote* programmability or re-configurability plays an important role. Given their small size and large numbers, it is preferable to program these devices avoiding direct connections. This also avoids the costs of recollecting and reprogramming devices after deployment. This aspect, however, could sensibly complicate the adoption of flexible architectures and push towards custom implementations.

This analysis point out what are the essential requirements of a wireless sensor network in an ambient intelligence context

- Low-energy consumption, in the order of the tens of μW;

- Low-cost (in the order of few cents), small size (in the order of cubic millimeters).

Requirements related to sensor re-configurability and/or adaptive features should be considered as secondary issues. The very limited cost would make disposing old sensors and replacing them with new ones cheaper than reprogramming them.

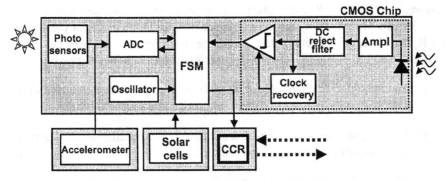

Figure 4.   Architecture of the Golem Dust Mote.

Case Study

As an example of wireless sensor that closely matching the previous constraints, we discuss the latest wireless sensor developed at the University of California at Berkeley in the *"Smart Dust"* project [30, 33]. Among the many similar research projects on wireless sensors (μamps at MIT [23], PicoRadio at Berkeley [26], and WINS Project at UCLA [35]), its target seems to be better centered around ambient intelligence needs.

The choice of an academic, prototype-quality design as a case study is not incidental: current commercial products (such as the MICA wireless sensors by Crossbow [10]) are meant for different targets: they are quite far from the requirements, for both application and constraints. For instance, the cost of a single sensor exceeds 100$.

According to the project mission, the term "smart dust" should denote a cloud of tiny elements of active silicon, called *motes*, each one of few cubic millimeters size. Each mote senses, communicates with the wireless base station and/or with the other motes and is self-powered; In addition, it converts sunlight into electricity, locally elaborates information, and localizes itself. Some quantitative features of smart dust are: free-space communication at optical frequencies, bit rates in the order of 1 kbps or less, power consumption below 1 mW, and directional transmission within a narrow beam.

Concerning the specific implementation, we focus on the the latest mote prototype developed by the Smart Dust group [34]: it is a $16mm^3$ autonomous, solar-powered sensor node with bi-directional optical communication, that represents the evolution of previous efforts [6, 32]. Figure 4 shows a block diagram of the sensor.

The device digitizes integrated sensor signals and transmits and receives data optically. The demonstrated system consists of four dies: a $0.25\mu m$ CMOS ASIC, a trench-isolation SOI solar cell array, a capacitive accelerometer fabricated in SOI process, and a micro-machined four-quadrant corner-cube retroreflector (CCR) [38], for optical communication. The author claim that a new MEMS process is being developed that will integrate the solar cells, the CCR, and the accelerometer onto a single die, yielding a $6.6mm^3$ device.

The CMOS die includes most of the functionalities, and in particular:

- An ambient light photo-sensor consisting of a photo-diode in series with a resistor;

- An ADC based on an effective successive-approximation algorithm that makes it suitable for low-power applications;

- An optical receiver used for downlink communication, consisting of a photo-detector, and the relative analog signal processing circuitry;

- A finite state machine (FSM) that controls the entire system, by multiplexing sensors, directing the ADC to take samples, and sending data to the CCR transmitter.

The optical communication mechanism used by the mote is very interesting. The uplink section is realized through a passive reflector (the CCR) that requires an incoming optical laser signal from an *interrogator*. Light entering the CCR is normally reflected back parallel to the incoming beam, but it can be deflected by electrically actuating the CCR so that less light is reflected. The former case represents the communication of a "1", the latter case a "0". The CCR is able to transmit data at 4 Kbps with a range over 100m. The downlink section uses the same laser beam used to interrogate the CCR, yet in a different frequency band.

Concerning power consumption, the two largest contributors are the optical receiver that consumes $26\mu W$ at 375Kbps, and the digital section of the CMOS chip, that consumes $22\mu W$. Including the remaining blocks, the overall power consumption of the system is around $75\mu W$.

Challenges Ahead

Technology scaling and advances in MEMS will allow such devices to become more and more integrated, and their size will soon fall below the cubic millimeter mark. It will be thus possible to deploy such tiny sensors as real "dust", or hidden in fabrics or paint. Such tiny sizes will require alternative power sources, since their minuscule surfaces will not be guarantee enough exposure to light (for solar cell supplies); therefore, energy scavenging techniques will be required to harvest energy from vibrations or environmental noise [5]. Micro-machining technology will allow to build micro-actuators to attach legs or wings to such devices, so that they can be transformed into autonomous micro-robots [36].

These visionary issues, however, do not address the real challenge from the computational point of view, that is, what degree of computational complexity these devices will be likely to have. For instance, the computational power of the mote of Figure 4 is limited to the FSM; in addition, there is absolutely no degree of re-configurability. The question is then whether miniaturization will be achieved only by sacrificing flexibility.

Current research efforts seem to push towards an increase of the "intelligence" in the wireless sensors, by either adopting very energy-efficient processor core (such as the CoolRisc [27]), or by increase the degree of re-configurability (for instance, re-configurable hardware as in the PicoRadio sensors [28]). The Smart Dust architects are also claiming the future adoption of an ultra-low energy core that consumes less than 20 pJ/instruction, roughly one orders of magnitude less than current low-power microprocessors (e.g., the StrongArm consumes approximately 1 nJ/instruction).

Our vision is partly in contrast with this trend. We believe that future wireless sensor for ambient intelligence *will not be general-purpose*; ubiquity constraints will definitely sacrifice power efficiency and cost for flexibility. The deployment of a huge number of sensors in the environment will reduce the importance of issues such as location awareness; similarly, their almost invisible size will make very difficult to re-program them remotely. Therefore, the availability of a very large number of sensors will be more important than their individual computational power. High availability translates into low cost: sensor should be inexpensive, or even disposable. In other terms, flexibility will migrate from the device to the design flow adopted for their production.

The "chip-in-a-day" design paradigm proposed by the Berkeley Wireless Research Center [11] is an example of such design flow flexibility. Besides the appealing short turnaround times suggested by its name, this design paradigm offers an alternative solution to re-configurable and programmable platforms. It can be thought of as a sort of platform-based design, where a highly parallel architectural template and a set pre-characterized hardware components drastically simplify the way to layout.

Our vision of future wireless sensors does not rule out the current efforts towards sensors with high computational power. The target for this type of sensors will be closer to what can be done today with commercial sensors, such as interactive buildings or environmental control. The node's intelligence, for instance, could be used to provide compatibility with different communication protocol standards.

5. Conclusions

Ambient Intelligence devices are expected to provide scalable processing power at every level of the networked infrastructure. Such scalability must be matched by scalability in the energy consumption, bandwidth, and size of the devices.

In this paper, we have explored this scalability range from the device perspective; we identified three main classes of devices, covering six orders of magnitude of computational power, energy consumption, bandwidth and size, and described their main features as well as the challenges posed by future technological trends. We have analyzed the architectural features of existing devices that are good candidates for becoming Ambient Intelligence devices.

From the analysis of these devices, a clear common trend emerges, across all the levels of the Ambient Intelligence hierarchy, towards higher parallelism, as an effort to catch up with scalability challenges. For the higher levels of the hierarchy (the fixed base network and the wireless base network devices) higher parallelism will translate into explicitly parallel architectures (in the form of on-chip multiprocessors, or networks-on-chip). For the wireless sensor networks, conversely, parallelism will translate into a larger number of simpler sensors, since to have a large number of sensors will be more important than their individual computational power.

References

[1] E. Aarts, R. Roovers, "IC Design Challenges for Ambient Intelligence," *IEEE Design Automation and Test in Europe Conference*, pp. 2-7, March 2003.

[2] E. Aarts, R. Roovers, "Embedded System Design Issues in Ambient Intelligence", this volume, 2003.

[3] V. Agarwal, M. Hrishikesh, S. Keckler, D. Burger, "Clock Rate Versus IPC: the End of the Road for Conventional Micro-Architectures," *ACM International Symposium on Computer Architecture*, pp. 248-259, 2000.

[4] A. Allan, D. Edenfeld, W. Joyner, A. Kahng, M. Rodgers, Y. Zorian, "2001 Technology Roadmap for Semiconductors," *IEEE Computer*, Vol. 35, No. 1, pp. 42-53, January 2002.

[5] R. Amirtharajah, A. Chandrakasan, "Self-Powered Signal Processing Using Vibration-Based Power Generation," *IEEE Journal of Solid-state Circuits*, May 1998, pp. 687-695,

[6] B. Atwood, B. Warneke, K.S. Pister, "Preliminary circuits for Smart Dust," *SSMS D'00: IEEE Southwest Symposium on Mixed-Signal Design, 2000.*, San Diego, CA, February 2000, pp. 87–92.

[7] L. Benini, G. De Micheli, "Networks on Chip: A New SOC Paradigm," *IEEE Computer*, Vol. 35, No. 1, January 2002, pp. 70–78.

[8] Bluetooth V1.1 Specification, http://www.bluetooth.org.

[9] J. Borkenhagen, R. Eickenmeyen, R. Kalla, S. Kunkel, "A Multi-Threaded PowerPC Processor for Commercial Servers," *IBM Journal of Research and Development*, Vol. 44, No. 6, Nov. 2002.

[10] Crossbow, MICA and MICA2 Motes & Sensors, http://www.xbow.com/Products/Wireless_Sensor_Networks.htm.

[11] W.R. Davis, N. Zhang, K. Camera, F. Chen, D. Markovic, N. Chan, B. Nikolic, R.W. Brodersen, "A Design Environment for High Throughput, Low Power, Dedicated Sig-

nal Processing Systems," *CICC'01: IEEE Custom Integrated Circuits Conference*, San Diego, CA, May 6–9, 2001, pp. 545–548.

[12] ETSI, *Broadband Radio Access Network (BRAN); HyperLAN Type II; Physical Layer*, TR 101-475, 2000.

[13] P. Gelsinger, "Microprocessors for the New Millennium: Challenges, Opportunities and New Frontiers," *IEEE International Solid-State Circuits Conference*, pp. 22-25, Jan. 2001.

[14] A. Hettich, M. Schrother, "IEEE 802.11 or ETSI BRAN HIPERLAN/2: Who Will Win the Race for a High Speed Wireless LAN Standard?", *European Wireless Conference*, Munich, Germany, Oct. 1999, pp. 169–174.

[15] IEEE, *Standards for Information Technology – Part 11: Wireless LAN Medium Access Control (MAC) and Physical Layer (PHY) Specifications*, 1999.

[16] B. Khailany et al., "Imagine: Media Processing with Streams," *IEEE Micro*, Vol. 21, No. 2, pp. 35-46, March 2001.

[17] T. Koyama, K. Inoue, H. Hanaki, M. Yasue, E. Iwata, "A 250-MHz Single-chip Multi-processor for Audio and Video Signal Processing," *IEEE Journal of Solid-State Circuits*, Vol. 36, No. 11, November 2001, pp. 1768–1774.

[18] R. Krishnamurthy, A. Alvandpour, V. De, S. Borkar, "High-Performance and Low-Power Challenges for Sub-70nm Microprocessor Circuits," *IEEE Custom Integrated Circuits Conference*, pp. 125-128, May 2002.

[19] D. Lackey, et al., "Managing Power and Performance for Systems-on-chip Designs using Voltage Islands," *ICCAD'02: ACM/IEEE International Conference on CAD,* November 2002, pp. 195–202.

[20] M.-J.E. Lee, W.J. Dally, P. Chiang, "Low-Power Area-Efficient High-Speed I/O Circuit Techniques,' *IEEE Journal of Solid-State Circuits*, Vol. 35, No. 11, November 2000, pp. 1591–1599.

[21] A. Macii, L. Benini, M. Poncino, "Memory Design Techniques for Low Energy Embedded Systems," *Kluwer Academic Publishers*, 2002.

[22] K.W. Mai, T. Mori, B.S. Amrutur, R. Ho, B. Wilburn, M.A. Horowitz, I. Fukushi, T. Izawa, S.S. Mitarai, "Low-Power SRAM Design Using Half-Swing Pulse-Mode Techniques," *IEEE Journal of Solid-State Circuits*, Vol. 33, No. 11, November 1998, pp. 1659–1671.

[23] μAmps Project Home Page, `http://www-mtl.mit.edu/research/icsystems/uamps`.

[24] R. Min, M. Bhardwaj, S.-H. Cho, N. Ickes, E. Shih, A. Sinha, A. Wang, A. Chandrakasan, "Energy-Centric Enabling Technologies for Wireless Sensor Networks," *IEEE Wireless Communications*, Vol. 9, No. 4, August 2002, pp. 28-39.

[25] S. Naffzinger et al., "The Implementation of the Itanium 2 Microprocessor," *IEEE Journal of Solid-State Circuits*, Vol. 37, No. 11, pp. 1448-1460, Nov. 2002.

[26] PicoRadio Project Home Page, `http://bwrc.eecs.berkeley.edu/Research/Pico_Radio`.

[27] C. Piguet et al., "Low-Power Design of 8-bit Embedded CoolRisc Microcontroller Core," *IEEE Journal of Solid-State Circuits*, Vol. 32, No. 7, July 1997, pp. 1067-1078.

[28] J. Rabaey, M.J. Ammer, J.L. da Silva, D. Patel, S. Roundy, "PicoRadio Supports Ad-Hoc Ultra-Low Power Wireless Networking" *IEEE Computer*, Vol. 33, No. 7, pp. 42–48, July 2000.

[29] R. Ronen et al., "Coming Challenges in Microarchitecture and Architecture," *Proceedings of the IEEE*, Vol. 89, No. 3, pp. 325-430, March 2001.

[30] "Smart Dust" Project Home Page, http://robotics.eecs.berkeley.edu/~pister/SmartDust.

[31] Texas Instruments, "WLAN Solutions: TNETW1130 Converged Single-Chip MAC and Baseband Processor for IEEE 802.11 a/b/g," http://www.ti.com/wlan

[32] B. Warneke, B. Atwood, K.S.J. Pister, "Smart Dust Mote Forerunners," *MEMS'01, IEEE International Conference on Micro Electro-Mechanical Systems*, Interlaken, Switzerland, January 2001, pp. 357–360.

[33] B. Warneke, M. Last, B. Liebowitz, K.S.J. Pister, "Smart Dust: Communicating with a Cubic-Millimeter Computer," *IEEE Computer*, Vol. 34, No. 7, pp. 44–51, January 2001.

[34] B. Warneke, M. Scott, B. Leibowitz, L. Zhou, C. Bellew, J.M. Kahn, B.E. Boser, K.S.J. Pister, "An Autonomous 16 mm^3 Solar-Powered Node for Distributed Wireless Sensor Networks," *IEEE International Conference on Sensors*, Orlando, FL, June 2002, pp. 1510–1515.

[35] WINS (Wireless Integrated Sensor Network) Project Home Page, http://www.janet.ucla.edu/WINS/.

[36] R. Yeh, S. Hollar, K.S.J. Pister, "Design of Low-Power Silicon Articulated Microrobots," *Journal of Micromechatronics*, Vol. 1, No. 3, pp. 191–203, 2001.

[37] H. Zhang, et al., "A 1-V Heterogeneous Reconfigurable DSP IC for Wireless Baseband Digital Signal Processing," *IEEE Journal of Solid-State Circuits*, Vol. 35, No. 11, November 2000, pp. 1697–1704.

[38] L. Zhou, K.S.J Pister, J.M. Kahn, "Assembled Corner-Cube Retroflector Quadruplet," *MEMS'02: IEEE International Conference on Micro Electro-Mechanical Systems*, Las Vegas, NV, January 2002, pp. 556–559.

Ambient Intelligence and the Development of Embedded System Software

Anu Purhonen and Esa Tuulari

{ anu.purhonen, esa.tuulari } @vtt.fi

VTT Electronics, Embedded Software, Oulu, Finland

Abstract   Traditionally, many embedded software products have been developed without support from system software. When system software has been used it has consisted of simple device drivers and an operating system. With an increasing demand for wired and wireless communication, embedded software has started to use middleware to hide the implementation details of low-level communication. The vision of Ambient Intelligence is that applications will be more and more distributed and will run on platforms offering dynamically varying resources. Moreover, this vision claims that applications should be adaptive to changes in the applications environment and adjust according to different users' preferences. In this paper, we discuss the requirements that Ambient Intelligence will set to the system software. We also present some of the solutions that have been proposed to address the increasing demand for system software.

Keywords   system software, software architecture, embedded software

1.   Introduction

Currently one could arguably say that Ambient Intelligence is most developed in cars, where different systems such as engine control, navigation, and entertainment have been successfully integrated to benefit the user. This success has been possible because each car manufacturer has complete control over all subsystems that form the operational application [36]. Unfortunately, this is not the case with the home and office domain where there are many independent players and integration of subsystems forms a serious bottleneck in the development of Ambient Intelligent applications.

The starting point for designing and implementing embedded software for Ambient Intelligence does not differ from traditional embedded software design. Our experience with the SoapBox [33] platform has shown that, indeed, in order to build embedded software for Ambient Intelligence we need everything that is usually needed in designing embedded software, like designers who are familiar with low-level programming, assembler-language, programming for minimal resources (memory, performance) etc. But although the

T. Basten et al. (eds.), Ambient Intelligence: Impact on Embedded System Design, 51-67.
© 2003 *Kluwer Academic Publishers. Printed in the Netherlands.*

SoapBox-platform has been successfully used in implementing several Ambient Intelligence applications there are things that could be improved. For example, if we want to change from one application to another we have to program the processor inside the SoapBox with an in-circuit programmer, which will take up-to 10 minutes. This is clearly in contradiction with the vision of Ambient Intelligence where applications should change and adapt instantly to changing contexts without bothering the user. On the other hand, the applications often consist of several co-operating parts running on diverse computing platforms. According to the Ambient Intelligence vision the communication should be seamless as all objects are interoperable. In our current system this is only a dream, as we have to manually verify that all parts of the application running on two or three computers are updated at the same time.

From a general point-of-view there are two aspects that cause the main challenge in building applications that obey the Ambient Intelligence vision. Firstly, the applications are usually distributed and run on several objects at the same time. For example, a simple tag, which transmits its ID periodically, does not have any value without the corresponding receiving part. But together they can be an important part of a larger infrastructure that monitors, for example, the status of a stock. Secondly, the run-time environment of the application is extremely dynamic. This is caused partly by the mobility of the objects that host the application and partly by the application's sensitivity to the environment. These issues are extensively discussed e.g. in [8, 18, 27].

The purpose of system software is to support application development. Within distributed systems, familiar topics are remote communication, fault tolerance, high availability, remote information access, and security. Whereas within mobile computing are inherited mobile networking, mobile information access, support for adaptive applications, system-level energy saving techniques, and location sensitivity. Those functions are the common services that can be used by several applications and thus need to be implemented only once. Sometimes the system software is called run-time environment meaning that together with the hardware platform it provides the environment in which applications are run. From a software development point of view, system software is often called software platform. The same software platform can be reused in different products with different applications.

The normal engineering solution for dealing with complexity is to use layers. This is also the case with software where more and more layers have been invented between the hardware and the application. Adding layers has also meant that the boundary between system software and application software is continuously changing and varies from case to case. For example, until recently the software that created mobile phone connection was the application in the handsets, but at the moment it is merely a system software service that the applications can use. This is a considerable change from previously, when most embedded systems did not have any layers and did not even distinguish the operating system from the application software.

Embedded software for Ambient Intelligence is going to be distributed and mobile consisting of several software layers that belong either to the system software or to the application software. To cope with this complex situation the role of software architecture is growing more and more important as it defines the fundamental organization of the software. Specifically, it defines the system components, their relationships to each other and the environment in which the system will operate [14]. In order to be able to define the system services one has to understand the purpose of the system and how it will be used. Software architecture is a way to get this big picture and to make sure that the various requirements of the stakeholders are taken care of [1, 24]. Although the functionality is the reason why a system exists, the basis of architectural design decisions are the non-functional requirements [1]. The non-functional requirements of distributed applications are related to the application's quality-of-service (QoS). These requirements include, among other things, reliability, availability, security, real-time guarantees, and synchronization of data streams in multimedia delivery [32, 35].

In this paper, we discuss Ambient Intelligent systems from the point of view of system software development. The paper is organized as follows: Section 2 describes the responsibilities of traditional system software. The new challenges introduced by Ambient Intelligent requirements are discussed in Section 3. Section 4 discusses the implications for software development and solutions that have been proposed for some of the specific problems so far. Section 5 concludes with some summarizing statements.

In this paper, we use the terms Ambient Intelligence, Ubiquitous Computing and Pervasive Computing interchangeably. In general, these terms are very closely related and share a same vision of an environment with smart spaces and smart objects.

2. Traditional System Software

System software has been traditionally divided into layers of middleware, operating system (OS) and network protocol stacks. In recent years embedded systems have started to support various kinds of communication with the outside world. For example, in addition to supporting communication in one or more cellular networks, mobile phones can support short-range wireless communication such as Bluetooth [4] or Infrared [15]. Therefore, the system software often includes several network protocol stacks.

The purpose of operating systems is to make easier and more efficient use of underlying hardware by providing a set of services that are generally needed by all or most of the application developers. An operating system can include the following kinds of services: program execution, I/O operations, file-system manipulation, communications, error detection, resource allocation, and accounting and protection [31]. There are different kinds of operating systems starting from the tiny real-time kernels to general-purpose full operating systems depending on the needs of the application developers and capabilities of the platform. Sometimes different types of OSs are combined e.g. a real-time kernel takes care of the operation needing real-time response and a general-purpose OS takes care of user-interfaces and so on.

Middleware is defined as a reusable system software that bridges the gap between the end-to-end functional requirements and the lower-level OSs and network protocol stacks [28]. Middleware hides distribution and composition of the underlying infrastructure from the application developers. In addition, it enables the applications and the infrastructure to evolve independently. Typical middleware services are name service, resource directories, discovery protocols, resource-access protocols, public-key certification, key management, and event notification [32]. While the operating system hides the platform of one device from the applications, the middleware layer was developed to hide many co-operating devices to provide the services a user needs. Schmidt has divided the middleware for real-time and embedded systems into further layers starting from the lowest layer [30]:

- *Host infrastructure middleware* encapsulates and enhances native operating system communication and concurrency mechanisms to create portable and reusable network programming components. Typically the implementation of this layer is a virtual machine.

- *Distribution middleware* defines higher-level distributed programming models whose reusable APIs and mechanisms automate and extend the capabilities of the underlying host infrastructure middleware. This is the layer that hides the distribution from application developers. The object request brokers (ORBs) implement this layer.

- *Common middleware services* augment distribution middleware by defining higher-level domain independent components to facilitate the work of the application developers in managing global resources. Common middleware services focus on allocating, scheduling, and coordinating various end-to-end resources throughout distributed systems.

- *Domain-specific middleware services* are tailored services to the requirements of particular system domains.

3.   New Requirements

When discussing the various constraints for system software we partly apply the conceptual model of pervasive computing [7]. It has been inspired by the

OSI-model, but some new layers and especially the model of the user, which is important for managing the concepts of pervasive computing as well as Ambient Intelligence, are added. Ciarletta and Dima define five layers, but for our purposes we utilize only the following three [7]:

- The purpose of the *physical layer* is to guarantee that its entities are physically compatible with one other, i.e. electrical, mechanical or physical characteristics. The device has to have the means to communicate with the user so that the user and the environment are capable of receiving and sending signals.

- The *resource layer* defines what logical resources are available i.e. entities for processing, storing and I/O. The user entities on the resource layer relate to the personal characteristics of the user such as language and other skills and temperament.

- The *abstract layer* contains the application software and the user mental models; thus the aim is to maintain consistency between the user's reasoning and expectations and the logic and state of the application.

The system software is situated in the resource and physical layers. The physical layer components control the hardware components and the resource layer components hide the details of the platform from the application developers. The system software has thus four types of interfaces, as shown in Figure 1. In addition to the two interfaces within the same device, the system software takes care of the responsibilities of the resource and physical layers in communicating with the other devices in the network and with the user.

According to Ciarletta and Dima's model, the operating system is situated in the resource layer and the physical layer is hardware. We use a modified version of this model, in which the physical layer also includes the software functionality that is needed to interface with the hardware. In addition, because in embedded systems the hardware/software partition often varies depending on the requirements of the system the physical layer has functionality that can be implemented either by software or hardware. Consequently, we do not make assumptions as to what level the OS is situated on because operating systems are different and moreover, the implementation of the system may not necessarily need a traditional OS but use other methods in implementing the same functionality.

The approach in the following sections is to study which features are required to the system software because of changes in the relationships with the actors with which the system software has to co-operate. The reasons behind the changes may come from the same source but we are trying to look at them only from the viewpoint of how they affect the system software.

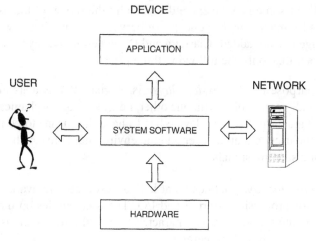

Figure 1.   System software interfaces.

3.1   Interaction with the Applications

The special needs of Ambient Intelligent systems with respect to connections to the applications come from the increased variation in what kinds of needs applications have. Also, these needs can change during runtime. One of the main design decisions is where to put the boundary in the responsibilities between system software and applications. Because the decision is usually made by the application designers, system software should be able to adapt to whatever the applications want. Traditionally, complete transparency of the underlying technology has been the ultimate goal for middleware solutions. For ambient intelligent systems such an approach is not adequate because the applications have to be aware of the surrounding environment [37].

The interaction with the applications results in one or more of the following responsibilities of the system software:

- Is able to run one or more applications and satisfy their quality of service (QoS) requirements.

- Is able to run a distributed application of which other parts are running in other parts of the network.

- Depending on the needs of the application, supports centralized or distributed control of devices.

- Hides the details of the platform from the application so that the application can be driven regardless of the capabilities of the platform and the environment.

- Gives a possibility to the application to select and tune the services to it needs.

- Is able to reconfigure the system to run new applications.

- Is able to download new applications during runtime.

3.2 Interaction with the Hardware Platform

Want *et al.* [36] claim that in order to make hardware disappear, we also have to make software disappear. Today's software is too monolithic and written with too many assumptions about hardware and software resources. In Ambient Intelligence, this situation will change as devices, objects and people will be constantly moving around, creating an ever-changing set of resources. Want offers three complementary requirements for solving the problem. First, applications should be decoupled into small pieces of code and distributed across ubiquitous hardware. Secondly, there should be data interchange formats in which the data is self-describing and the message finds its way from sender to the receiver autonomously and securely. Thirdly, there should be mechanisms for devices to advertise their capabilities so that applications can become aware of available resources.

Developments in electronics have been one of the driving forces towards Ambient Intelligence, for example improvements in integration density of systems, bandwidth of communication systems, storage capacity and cost per bit of I/O processing [11]. Some believe that within five years low-cost systems on a chip (SoC) will be available that include a pico-cellular wireless transceiver and a sufficiently rich runtime environment to be capable of running sophisticated virtual machines [23]. Despite all the advances there will still be platforms where the traditional concerns of embedded software development such as resource usage and power consumption are still valid. For example, new kinds of ultra-low-power platforms are being developed [17]. At the other end of the platforms are PDAs that can run similar applications as current PCs and more. Whereas those kinds of platforms have more resources to share they also have more complex applications to support. In these kinds of systems it is difficult to validate all the possible conditions beforehand, so some sort of runtime resource management is needed to provide users with an agreeable level of performance. Furthermore, when the applications are distributed their performance is not only determined by the performance of one device but also determined by a sum/combination of the individual performances. For example, the increased bandwidth in one device is not important if other devices communicating with it cannot communicate with the same increased level. The same is true for power consumption. If one part of the system can extend its operating time it does not come for any use if other parts have run out of power.

One important aspect in Ambient Intelligence vision is that the devices will be more aware of their environment. There are two possibilities to do this without requiring the user to give the necessary information: either the device gets the information by communicating with other devices or by using its own sensors for detecting, sensing and measuring the environment. Dealing with sensors is very much time critical: samples should be taken in predetermined intervals, all fluctuations and jitter in the timing will cause problems in the signal processing. Either the intended measurement is not possible at all, or there will be erroneous readings because of noise.

The challenge with hardware platforms is to provide the means to control the physical entities in the device. Thus the system software should be:

- Able to control different types of sensors.

- Scalable to different types of platforms because hardware platforms may have different capabilities.

- Able to manage or prevent overload conditions when the requirements of the applications vary during runtime.

- Able to take advantage of special purpose hardware when it is available.

3.3   Interaction with Other Devices

System software takes care of the physical transmission of data between other devices in the network with the hardware. In addition, it hides the details of the distributed computing of the other layers such as reserving communication resources and other responsibilities of secure connections. The problem with connecting to other devices is that the user can move and therefore new connections can be created and old connections can be lost. In addition, one should consider how secure the connections are. System software developers should not make any assumptions of the capabilities of the other devices or of the network.

The challenges induced by connections with other devices are:

- Connecting to different types of networks.

- Maintaining several connections at the same time.

- Maintaining security over the connections.

- Maintaining connections when the user (and device) goes out of the range of the network that is being used in an active connection.

- Handling loss of connections or faults in the other devices.

- Discovery of available services and devices when going to a new environment.

3.4   Interaction with the User

The physical implementation of a device determines what user interface modalities are available. For instance, if the requirements include speech recognition, then there has to be a microphone in the physical layer. In Ambient Intelligence, many objects are operable by users. According to the vision of Ambient Intelligence, computing elements (processor, memory, etc) will be integrated into all kinds of objects like chairs, tables and doors. These user interfaces will not be visible but the use of the embedded computer/software is implicit and takes place when the object/artifact is used as part of normal activities, e.g. at home.

On a higher abstraction level, the recognition of users becomes important, as access rights are often based on the identity of the user. User-recognition should preferably be based on biometric methods so that the need for the user to remember passwords, pin-codes etc. is minimized. Besides access rights, the user's identity is also important from the point-of-view of adaptation, as the user interface should adapt to each user's personal preferences.

The most simple case for solving the preference problem is personal devices, used only by a single person. Most mobile phones fall into this category. The device can stay adjusted to the user all the time and the user can control these adjustments as he or she wishes. The second category is devices that are in public use but are only used by one person at a time. Fitness equipment in fitness centers belong to this category. Here the main problem is to remember each user's preferences. There are two main approaches: a) each equipment recognizes users and stores the preferences and specific settings to that particular equipment and b) each user carries a memorycard (of any type) which stores the preferences and settings and transmits these to each piece of equipment as the user begins to use them. The third and most difficult category is formed by devices that are in public use and are used by several people at the same time. The train departure screens at railway stations, for example, belong to this category and without the use of some augmented reality technology it seems impossible to adapt the interface to each user simultaneously.

The challenge in connecting with users is that the system software should allow personalized services to the user so that it does not make unnecessary assumption of the capabilities of the user.

3.5   Discussion

The division of responsibilities is the main concern in all four interfaces of system software. On the one hand, system software should provide some services to applications automatically, but on the other hand the application should be able to adjust the platform according to its needs. Sometimes the user wants to be in control and say how s/he wants to communicate with the device. However, sometimes the user expects that some adaptations will happen automatically. Similarly, in communication with other devices and the hardware, what is assumed and who is in control are important questions.

4.   System Software Development

Real world experiences [22] indicate that there are a couple of subjects that need to be carefully considered while designing applications that realize the Ambient Intelligence vision. Firstly, the traditional development model where components are developed first and then integrated together does not work. Integration should be kept in mind from the very beginning both in implementing easy-to-use APIs to components and in using some standardized middleware solution. Secondly, failure should be treated as a normal part of software execution and not as an exception. Failures will occur more often than in traditional systems as applications are distributed and dynamic. Sensors will fail, messages will be lost during communication and some elements will run out of power. Thirdly, the traditional metrics for system's performance, like bytes-per-message, latencies and throughput do not reveal the true performance of the system. Likewise, more user-oriented metrics like ease of installation and time it takes to restart after loss of power are more difficult to measure. The true value of the system from an Ambient Intelligence point-of-view should be measured by its impact on users, but there is not yet any direct measure for such a parameter.

Because these issues affect the whole system and not just the system software, they should be solved when designing the software's architecture. In traditional embedded system development the experts of different areas, such as application software or digital signal processing, have developed the systems quite independently. This has been possible because the interfaces between these areas have been quite static. The system software is also one domain that cannot be developed in isolation. Increased complexity and variation emphasize that the systems should be thought of as a whole. Software architecture is one way to manage this. As long as the purpose of the system and its software is clarified, the selection of techniques for the system software becomes much easier and systematic.

Software architecture is also a starting point when deciding what is actually the supporting functionality that is needed in the system i.e. what belongs to the system software and what are the applications. In order to be able to manage these different types of problems, software architectures are usually divided into views. For example, the division between applications and system software components can be made using the so-called development view, which is typically formed using layers [21]. On the other hand, in order to be able to make, for example, performance trade-offs, a run-time view of the system is needed. Most of the new services induced by Ambient Intelligence will be implemented in the middleware layers. QADA is a software architecture development method designed especially for developing middleware services [24].

Although Ambient Intelligence systems are in many ways challenging, one thing that especially differentiates them from other systems is the need for different forms of adaptation. Adaptivity is part of the normal operation of the software. The architecture has to change or reconfigure itself based on user's actions. In Ambient Intelligence this is sometimes referred to as latent functionality. It simply means that the user can activate some features, which were set to inactive during shipment. However, sometimes more freedom is needed and there will be a need for the user to actually program new functionality into the system or application.

The need for the application to adapt can arise from different sources: there could be some changes in the platform where the application is running that needs adaptation, such as when the battery runs low, or the amount of free memory gets too low. The user of the application might change or the user 'carries' the application to a new location. There can also be changes in the infrastructure as some network resources become available or others cease to exist. The environment could also change as nodes that should or could be communicated with come and go. The infrastructure can also change the services, like printers, it offers.

The purpose of system software in adapting is to provide the services, which allow the above-mentioned adaptivity to happen. In the following sections, we cover some of the techniques that have been used to solve different aspects of adaptation.

4.1 Adapting to Changing Application and User Needs

The dynamic nature of pervasive computing means that applications/systems have to adapt to changing operating conditions. Traditionally there are two basic strategies for controlling changes: distributed control and centralized control. The third alternative that is extensively researched at the moment is the middleware solution, which lies between these two traditional strategies.

Reflective middleware is one solution to the problem of what should and what should not be hidden from the application developers. This has been considered the opposite of the traditional black-box approach, for it, unlike the black-box approach, is a principled way to achieve openness, configurability and reconfigurability [2]. The reflective middleware is implemented as a collection of components that can be configured and reconfigured by the application [19]. Thus the application developer can decide how much control over the underlying platform s/he wants to have.

Another solution promoted by IBM is autonomic computing [13]. This is based on the assumption that individual devices, objects and artifacts have to be autonomic if we want to make their networking possible. In a way this is very much a distributed control solution. IBM's Autonomic Computing mission statement states that the only way to deal with the ever increasing complexity of computing systems is to make them self-regulative. While not directly dealing with Ambient Intelligence, IBM addresses several important issues, like self-healing and reconfiguration.

In addition to control issues, adaptation brings another problem, namely adding and removing applications and services. Dynamic software architectures have been especially studied in the areas of applications, where systems cannot afford to be down for a long period of time, so updates and modifications should be performed during runtime without affecting the normal operation. One approach to managing dynamic adaptation is the use of software architecture models to monitor an application and guide dynamic changes to it [5]. In order to achieve this, the runtime system needs to include monitoring mechanisms and low-level operations to adapt the system. The basic operations for adaptive systems are adding and removing components.

4.2 Adapting to the Amount of Hardware Resources

In ambient intelligent systems similar services should be provided with devices with different capabilities. System software should also be able to fit into devices with critical resource constraints. The solution for control issues is also one solution to customization: reflective middleware [19]. This approach permits the configuration of very small middleware engines that interoperate with conventional middleware. These kinds of reflective ORBs can have memory footprints as small as 6 KB [26]. Adaptation in reflective middleware is supported with dynamic customization of object behavior and fine-grain resource management through meta-interfaces, for example DynamicTAO [20] and Open ORB [3].

TinyOS is a simple, component-based operating system, which is developed for systems with limited resources [6]. TinyOS has tiny networking stacks, abstractions of physical devices and highly tuned implementations of common functions. Another example is Gaia, which is a metaoperating system that extends the reach of traditional operating systems to support the development and execution of portable applications for programmable ubiquitous computing environments [25]. The Gaia kernel consists of a component management core (CMC) and an interrelated set of basic services used by all Gaia applications. The CMC can, for example, dynamically load and unload Gaia components and applications. The current implementation of Gaia uses Corba but other systems and middleware can also be used.

A portable runtime environment that is capable of executing across a wide variety of 8, 16, and 32-bit processors and that offers mobile code and high-level language support has been proposed in [23]. It would work under a variety of operating systems and it should also run on bare hardware. They have also defined a software construct, Flit, which is similar to an applet but tailored to the needs of lower resource solutions to pervasive computing. For example, portable device drivers for embedded devices could be implemented as flits loaded in ROM.

CentaurusComm is the transport protocol in Centaurus [16]. It has two levels of protocol modules and an application program interface that is responsible for accepting the objects from the application layer for transmission and notifying when messages are received. CentaurusComm has been designed to run on a wide range of low power systems. Therefore, it does not depend on any advanced operating system features such as signaling and multithreading, unlike TCP, which requires substantial support from the OS for signaling. Centaurus minimizes resource consumption on a user device by avoiding the need to have the services installed on each device that wishes to use them.

Although flexible and resource saving approaches are needed, this is not enough. Especially in the more complex systems with many applications and a lot of interaction with users, it is difficult to estimate resource consumption beforehand. Therefore a method for monitoring resource consumption during runtime is needed. For example, Spectra is a remote execution system designed for pervasive systems that monitors resources such as battery energy and file cache state [11]. Furthermore, it can dynamically balance power use and performance to decide where to locate functionality.

4.3   Adaptation to Changing Environment

Adapting to changing conditions means that one has first to be able to realize that conditions have changed. Loss of connection is easy to notice, but sensors may be needed to discover what other changes may have occurred. There are several levels on how to use the information obtained by sensors. In the most simple case, raw sensor readings are usable as such, for example when the value exceeds a predefined limit. In a bit more complicated case, some forms of signal processing are needed to extract useful information from the raw sensor readings. In the most complicated case, one has to apply artificial intelligence and machine learning methods to be able to make the sensor readings useful and meaningful for the application. The amount of needed resources increases accordingly.

Context-aware computing [29] belongs to the third category. Before applications can adapt to changing environments they have to be aware of the context in which they run. Context is "any information that can be used to characterize the situation of an entity. An entity is a person, place, or object that is considered relevant to the interaction between a user and an application, including the user and the application themselves" [10].

There are two different approaches to context-awareness. Sensing and processing can occur either in infrastructure [9, 29] or in the mobile device [12]. When using common sensors provided by the infrastructure the mobile devices can be simpler and less expensive and also consume less power. On the other hand, when devices obtain context autonomously they are independent of any specific smart environments.

System software should also provide for the developers of the context-aware applications the common functionality that is used by several applications. Context Toolkit [9] is an approach of this kind. It is a research testbed and a framework for developing context-aware applications. Context Toolkit provides architectural building blocks, which can also be used in developing common context-aware services that multiple applications can use.

In addition to context-awareness, there are also more traditional problems with a changing environment; for example, how is communication handled. Centaurus [16] is an example of an infrastructure that can be used to connect sensors and other mobile clients with different modes of communication (e.g. Bluetooth [4], Infrared [15]) in office environment. When considering wireless communications and sharing information it is natural that the system software should provide services for ensuring security. For example, Centaurus2 extends the work of Centaurus by considering the security aspects of communication specifically [34].

5.   Conclusions

In this paper, we discussed the challenges and available solutions for system software development in ambient intelligent systems. There are already several approaches available for the various design problems of future system software. However, because they are not yet applied to real, commercial systems, it is difficult to see how useful they will be. One of the main sources of new features is an increased need for customization and adaptation. Combining and creating these different features is a challenge to software architecture development. Software architecture is the starting point in defining a software platform for a system and making trade-offs between different requirements. Layering has been used for separating concerns and it seems to also be an effective approach in this case. However, because of the extensive needs of adaptation there is also a need for new methods.

Acknowledgements

This work has been partially conducted in two projects, Moose and Ambience, under ITEA cluster projects of the EUREKA network, and financially supported by Tekes (the National Technology Agency of Finland). Ms Purhonen would also like to acknowledge the financial support of the Nokia Foundation.

References

[1] L. Bass, P. Clements, and R. Kazman. *Software Architecture in Practice*, Addison-Wesley, Reading, Massachusetts, USA, 1998.

[2] G. Blair, F. Costa, R. Campbell and F. Kon, "Reflective Middleware", *IEEE Distributed System Online*, vol. 2, no. 6, Sep. 2001, http://dsonline.computer.org.

[3] G. Blair, G. Coulson, A. Andersen, L. Blair, M. Clarke, F. Costa, H. Duran-Limon, T. Fitzpatrick, L. Johnston, R. Moreira, N. Parlavantzas, and K. Saikoski, "The Design and Implementation of Open ORB 2", *IEEE Distributed System Online*, vol. 2, no. 6, Sep. 2001, http://dsonline.computer.org.

[4] Bluetooth, http://www.bluetooth.com.

[5] S.-W. Cheng, D. Garlan, B. Schmerl, J.P. Sousa, B. Spitznagel, P. Steenkiste, and N. Hu. "Software Architecture-Based Adaptation for Pervasive Systems", in *Trends in Network and Pervasive Computing - ARCS 2002, International Conference on Architecture of Computing Systems, Karlsruhe, Germany, Proceedings*, H. Schmeck, T. Ungerer and L.C. Wolf, Eds., Apr. 2002, vol. 2299 of Lecture Notes in Computer Science, pp. 67-82, Springer, Berlin, Germany, 2002.

[6] D.E. Culler, J. Hill, P. Buonadonna, R. Szewczyk, and A. Woo, "A Network-Centric Approach to Embedded Software for Tiny Devices", in *EMSOFT 2001, Proceedings*, 2001, pp. 114-130.

[7] L. Ciarletta and A. Dima. "A Conceptual Model for Pervasive Computing", in *the 2000 International Workshop on Parallel Processing, Toronto, Canada, Proceedings*, Aug. 2000, pp. 9-16, IEEE Computer Society, 2000.

[8] N. Davies and H.-W. Gellersen, "Beyond Prototypes: Challenges in Deploying Ubiquitous Systems", *IEEE Pervasive Computing*, vol. 1, no. 1, pp. 26-35, 2002.

[9] A.K. Dey, *Providing Architectural Support for Building Context-Aware Applications*, Ph.D. thesis, College of Computing, Georgia Institute of Technology, 2000.

[10] A.K. Dey and G.D. Abowd, "Towards a Better Understanding of Context and Context-Awareness", in *CHI2000 Workshop on the What, Who, Where, When, and How of Context-Awareness, The Hague, Netherlands, Proceedings*, Apr. 2000.

[11] J. Flinn, D. Narayanan, and M. Satyanarayanan, "Self-Tuned Remote Execution for Pervasive Computing", in *the 8th Workshop on Hot Topics in Operating Systems (HotOS-VIII), Oberbayen, Germany, Proceedings*, May 2001.

[12] H.W. Gellersen, A. Schmidt, and M. Beigl, "Multi-Sensor Context-Awareness in Mobile Devices and Smart Artifacts", vol. 7 of *Mobile Networks and Applications*, Kluwer Academic Publishers, pp. 341-351, 2002.

[13] IBM. http://www.research.ibm.com/autonomic/manifesto/.

[14] IEEE, IEEE Recommended Practice for Architectural Description of Software-Intensive Systems, *IEEE Std 1471-2000*.

[15] Infrared, http://www.irda.org.

[16] L. Kagal, V. Korolev, S. Avancha, A. Joshi, T. Finin, and Y. Yesha, "Centaurus: An Infrastructure for Service Management in Ubiquitous Computing Environments", vol. 8 of *Wireless Networks*, Kluwer Academic Publishers, pp. 619-635, 2002.

[17] J.M. Kahn, R.H. Katz, and K.S.J. Pister, "Next Century Challenges: Mobile Networking for 'Smart Dust'", in *the fifth annual ACM/IEEE International Conference on Mobile Computing and Networking, Seattle, Washington, USA, Proceedings*, 1999, pp. 271-278, ACM Press, 1999.

[18] T. Kindberg and A. Fox, "System Software for Ubiquitous Computing", *IEEE Pervasive Computing*, vol. 1, no. 1, pp. 70-81, 2002.

[19] F. Kon, F. Costa, G. Blair, and R.H. Campbell, "The Case for Reflective Middleware", *Communications of the ACM*, vol. 45, no. 6, pp. 33-38, 2002.

[20] F. Kon, M. Roman, P. Liu, J. Mao, T. Yamane, L. Magalhaes, and R. Campbell., "Monitoring, security, and dynamic configuration with the dynamicTAO ORB", in *the IFIP/ACM International Conference on Distributed Systems Platforms and Open Distributed Processing (Middleware2000), Palisades, NY, Proceedings*, Apr. 2000, pp. 121-143, Springer, Berlin, Germany, 2000.

[21] P. Kruchten, "The 4+1 View Model of Architecture", *IEEE Software*, vol. 12, no. 6., pp. 42-50, 1995.

[22] A. LaMarca, W. Brunette, D. Koizumi, M. Lease, S.B. Sigurdsson, K. Sikorski, D. Fox, and G. Borriello, "PlantCare: An Investigation in Practical Ubiquitous Systems", in *UbiComp 2002, Proceedings*, G. Borriello and L.E. Holmquist, Eds., vol. 2498 of Lecture Notes in Computer Science, pp. 316-332, Springer, Berlin, Germany, 2002.

[23] W. Majurski, A. Dima, and M. Laamanen, "Flits: Pervasive Computing for Processor and Memory Constrained Systems", in *the 2000 International Workshop on Parallel Processing, Toronto, Canada, Proceedings*, Aug. 2000, pp. 31-38, IEEE Computer Society, 2000.

[24] M. Matinlassi and E. Niemelä, "Quality-Driven Architecture Design Method", in *ICSSEA '2002 - International Conference of Software & Systems Engineering and their Applications, Paris, France, Proceedings*, Dec. 2002.

[25] M. Román, C. Hess, R. Cerqueira, A. Ranganathan, R.H. Campbell, and K. Nahrstedt, "A Middleware Infrastructure for Active Spaces", *IEEE Pervasive Computing*, vol. 1, no. 4, pp. 74-83, 2002.

[26] M. Román, D. Mickunas, F. Kon, and R. Campbell, "LegORB and Ubiquitous CORBA", in *the IFIP/ACM (Middleware2000) Workshop on Reflective Middleware, Palisades, NY, Proceedings*, Apr. 2000, pp. 1-2.

[27] M. Satyanarayanan, "Pervasive Computing: Vision and Challenges", *IEEE Personal Communications*, vol. 8, no. 4, pp. 10-17, 2001.

[28] R. Schantz and D. Schmidt, "Middleware for Distributed Systems: Evolving the Common Structure for Network-centric Applications", *Encyclopedia of Software Engineering*, Wiley & Sons, 2001.

[29] B.N. Schilit, N.I. Adams, and R. Want, "Context-Aware Computing Applications", in *IEEE Workshop on Mobile Computing Systems and Applications, Santa Cruz, CA, Proceedings*, Dec. 1994, pp. 85-90.

[30] D.C. Schmidt, "Middleware for Real-time and Embedded Systems", *Communications of the ACM*, vol. 45, no. 6., pp. 43-48, 2002.

[31] A. Silbershatz and P. Galvin, *Operating System Concepts*, 5th ed., Addison-Wesley, Boston, 1998.

[32] A. Tripathi, "Challenges Designing Next-Generation Middleware Systems", *Communications of the ACM*, vol. 45, no. 6, pp. 39-42, 2002.

[33] E. Tuulari and Ari Ylisaukko-oja, "SoapBox: A Platform for Ubiquitous Computing Research and Applications", in *Pervasive Computing, First International Conference, Pervasive 2002, Zürich, Switzerland, August 2002, Proceedings*, F. Mattern, and M. Naghishineh, Eds., Aug. 2002, vol. 2414 of Lecture Notes in Computer Science, pp. 125-138, Springer, Berlin, Germany, 2002.

[34] J. Undercoffer, F. Perich, A. Cedilnik, L. Kagal, and A. Joshi, "A Secure Infrastructure for Service Discovery and Access in Pervasive Computing", vol. 8 of *Mobile Networks and Applications*, Kluwer Academic Publishers, pp. 113-125, 2003.

[35] N. Venkatasubramanian, "Safe 'Composability' of Middleware Services", *Communications of the ACM*, vol. 45, no. 6, pp. 49-52, 2002.

[36] R. Want, T. Pering, G. Borriello and K.I. Farkas, "Disappearing Hardware", *IEEE Pervasive Computing*, vol. 1, no. 1, pp. 36-47, 2002.

[37] S.S. Yau, F. Karim, Y. Wang, B. Wang, and S.K.S. Gupta, "Reconfigurable Context-Sensitive Middleware for Pervasive Computing", *IEEE Pervasive Computing*, vol. 1, no. 3, pp. 33-40, 2002.

Preparation of Heterogeneous Networks for Ambient Intelligence

Peter van der Stok

Philips Research Laboratories Eindhoven, Eindhoven, The Netherlands

peter.van.der.stok@philips.com

Abstract   Ambient Intelligence assumes the existence of ubiquitous networked computing. Networking is supported in the home by a so-called home network. Currently, the home network emerges from the purchase of a second PC in the home. Extending the home network confronts the prospective buyer with a multitude of standards. The more important representatives of these standards are discussed and an integration path is described.

Keywords   networking, home networks, middleware standards, High Level Architecture (HLA)

1. Introduction

Ambient Intelligence relies for a large part on communication between functional entities running on devices that are interconnected via some network. The DICE aspects of such a network are:

- *Domotica*, allowing the control of the house (e.g. setting of the heating).

- *Information*, allowing the searching of documents (e.g. web browsing).

- *Communication*, allowing the interchange of information between individuals (e.g. teleconferencing).

- *Entertainment*, allowing gaming and picture viewing (e.g. video on demand)

Assuming that pictures, and sound are transported digitally, data transport communication media are essential. Middleware provides the functionalities needed to support seamless access to services and multimedia content, anytime, anywhere. The home is the area of choice to deploy the Ambient Intelligence concept. An infrastructure of connected devices (such as microphones and cameras) is needed to correlate images and sounds within the home to deduce user wishes. Ubiquitous home networking is one of the key enabling

69

T. Basten et al. (eds.), Ambient Intelligence: Impact on Embedded System Design, 69-90.
© 2003 *Kluwer Academic Publishers. Printed in the Netherlands.*

infra-structures for Ambient Intelligence. The large number of standards targeted to in-home networking prevents an easy exploitation of home networks.

Well-known communication media standards are IEEE 802.3 (Ethernet) [4] the medium of choice for the computing community, IEEE 802.11 (wireless Ethernet) [6] for the same community but also actively investigated for the home, and IEEE 1394 (Firewire) [5] that supports high throughput guaranteed bandwidth for Audio/Video (A/V) applications. Well-known middleware standards are Home Audio Video interoperability (HAVi) [3] targeted to home entertainment, Universal Plug and Play (UPnP) [16] originallly targeted to PC world and currently extended to the CE world, and Bluetooth [1, 2] targeted to closely spaced audio devices.

For this paper it is assumed that these standards are here to stay, and more standards will follow given evolving technical, political and business interests. Integrating these standards and managing the network resources is a must to reach the Ambient Intelligence goal.

After an introduction to the essential concepts that allow interoperability between devices, the particularities of three standards are described. The standards provide similar facilities. A major difference between them is the reliance of the higher-level facilities on the underlying communication medium facilities. The dependency of the applications on the services provided by the standards is different from standard to standard. After identification of the incompatibilities between standards, existing integration efforts are discussed. Based on these results, the paper defines integration objectives from an application point of view. The paper terminates with a possible design to meet these objectives by employing the High Level Architecture (HLA) IEEE 1516 standard [7].

2.   Underlying Concepts

An in-home digital network consists of a set of interconnected *physical devices* (devices for short). A device is connected to the network via a digital connector that is realized with one or more chips. Examples of digital connectors are interface cards for an Ethernet cable, or a wireless transceiver. A device, A, is connected to another device, B, via the network when:

  a. The digital connector of A can send bit patterns to the digital connector of B such that B can interpret these bit patterns.

  b. A is connected to some device C and C is connected to B.

For example two devices are connected when they are attached to the same ethernet wire, or the signal emitted by one device can be received and decoded by the second device. Two devices are also connected when there exists a path via a set of connected devices. On the network a physical device is identi-

Physical Device

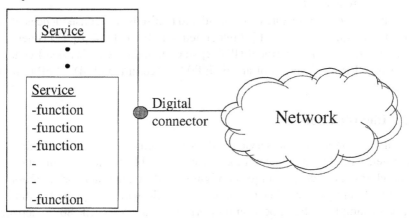

Figure 1. Physical device connected to network.

fied by the identifier of the digital connector. A device has therefore as many identifiers as it has digital connectors. For example a device with two Ethernet cards has two identifiers.

Device *resolution* is making the identifiers of the devices connected to a device, A, available to the applications running on device A. Device *advertisement* is the proclamation by a device of its identifier to a subset of all connected devices. *Discovery* usually covers both aspects.

On a device a set of *services* (see Figure 1) can deliver a 'service' to applications situated on connected physical devices. A service is an instance of a *service type* or service class. An example of a service type is a printing service or a clock service. The meaning and interface of service types is defined by a committee, or by manufacturers, or by individual people. An individual service type may be the returning of video images of the last 5 minutes taken by a home webcam. A service provides a set of functions that can be invoked, and a set of variables that can be read or set. Functions can modify or monitor the software and associated storage or may act on the physical operation of the device. Service *resolution* is making the identifiers of the services available to an application on a connected device. Service *advertisement* is the proclamation by a device of its services to a subset of all connected devices.

3. Heterogeneity

To access services or devices in a network, the applications in the network should be able to identify and address the connected services. Identification takes place at several levels, (user, device, service, ...) that are not the same for all standards. Issues in this context are the scope of uniqueness of the identification, the rigidity of the specification and whether the identification is in a human readable form.

Four network architectures are considered to discuss the consequences of architectural choices: (1) HAVI [3] integrated with IEEE 1394 [6], (2) Bluetooth [1, 2], (3) the Internet Protocol (IP) [8] stack together with Service Location Protocol (SLP) [11] on top of an IEEE 802.x medium, and (4) UPnP [16] on top of IP.

3.1   Identification

The digital connector, the service and its functions can each have their own identifier. The digital connector can have, a possibly world-wide unique network address or a locally unique configuration dependent network address. A world-wide unique identifier necessitates a world-wide recognized authority. A good example is the allocation of world-wide unique IEEE 802.x network interface addresses. Devices can have world-wide unique identifiers and also a world-wide authority is needed to at-least establish a worldwide unique prefix on the identifier. During operation, the digital connector can have an operational identifier. The latter can be shorter than the unique identifier or enables an efficient routing of packets through the network.

Although service-types can be uniquely named, a service is not necessarily uniquely named. For example, instances of the same printer service with the same service name on different network segments are not uniquely named. Often, the service identifier is made unique by aggregating the service type with the unique identifier of the device on which the service executes. Function identifiers are only unique within the context of one service.

3.1.1   HAVi

HAVi is based on the IEEE 1394 communication medium. It is a bus type system. The node id, composed of a bus identifier (*bus_id*) and a physical identifier (*phy_id*), uniquely identifies the digital connector over the local IEEE 1394 network. The *phy_id* can change every time a device is attached or disjoint from a given bus. The *phy_id* is an operational identifier and is used to send messages over a bus to the network connector with the specified *phy_id* value.

The Global Unique Identifier (*GUID*) that uniquely identifies a device, is composed of a unique vendor id and a unique chip id. An IEEE supported authority is responsible for allocating *vendor_id*'s to vendors. A vendor is responsible for uniquely identifying a device through the '*chip_id*'.

The Software Element Identifier (*SEID*) is composed of the *GUID* and a Software Handle (*SWHandle*) that identifies a software element in a device. The *SEID* is used by HAVi applications to address services or other software components.

In every physical device, there is a mapping from *GUID*s to *node_id*s that is updated after every network topology change. Note that HAVi devices always have a single digital connector, which allows a non-ambiguous mapping.

3.1.2 Bluetooth

Bluetooth is based on wireless communication. Bluetooth physical devices are identified by a: *Bluetooth device address (BD_ADDR)*, composed of a company assigned part (LAP) and a company identifier. The identifier is unique over the world.

Within a so-called piconet, composed of a master and a set of slaves, two types of operational identifiers are used to send packets from a master device to the associated slave devices. The *Active member address (AM_ADDR)*, represents one "non parked" connected slave. A master has no AM address. The all zero AM address constitutes a broadcast. Only 8 AM addresses are possible per master, thus limiting a so-called pico-net to 7 members.

A *Parked member address (PM_ADDR)*, represents one parked slave connected to a master. A total of 256 parked addresses are possible in a piconet.

3.1.3 Internet Protocol

The Internet Protocol (IP) stack allocates operational identifiers, called IP address, to devices connected to the Internet. The IP address abstracts from the used communication medium. A digital connector can have one or more IP addresses. Two types of addresses are known: (1) global addresses and (2) private addresses. The global addresses are unique over the total Internet. Private addresses are unique within a network of a given organization. The same private IP addresses can be used in multiple networks administrated by different organizations. The IP address of a device changes over time, given the reorganization of the network. Within each device a mapping between digital connector address and IP address is maintained. The mapping has a limited validity and needs to be refreshed regularly. SLP uses the IP addresses as device identifiers.

3.1.4 UPnP

UPnP devices use the identification provided by the IP addresses.

3.2 Connection

Once services and devices can be identified, it becomes feasible to detect whether they are connected to the network. After device connection, the services can be discovered and made available. To satisfy users of home equipment it is necessary to indicate all services and nothing but the services that can be reached. This means that connection and disconnection of devices should be known to all interested applications within seconds after the physical act. Consequently, the user is always confronted with an up to date list of connected services. Such reaction times are difficult to meet with many technologies. Both identification and connection knowledge are obtained in a network architecture dependent way.

3.2.1 HAVi

Every time a device is connected or disconnected from the IEEE 1394 medium a bus reset is generated. After the bus reset, two things happen: (1) every connected device learns that the other device is disconnected or connected, and (2) the network is reconfigured. Configuration means that a spanning tree is matched over the network. Every device connected to the network and only connected devices compose the spanning tree. The *node_id*s and the mapping from *GUID* to *node_id* are recalculated after every reset.

3.2.2 Bluetooth

Bluetooth is a standard for wireless communication between closely spaced devices. Connection is less straightforward for wireless devices than for wired devices. Devices can come within each other's reach after which they may sense the carrier signal and adjust their clocks to enable data transmission. However once connected, it is difficult to establish whether two devices are disconnected. Especially, mobile devices may drift in and out of reach of each other. A regular sending of "are you present messages" elicits responses of the device to which the message is sent. The absence, or the persisting absence, of an answer is interpreted as disconnection.

Bluetooth knows the concept of pico-net, constituted of one master and a set of slaves. Slaves in a pico-net follow the clock and communication frequency of the master of the pico-net. A device can be master in one pico-net and slave in another. This allows the concept of a scatternet: a set of connected pico-nets. Consequently, a packet can be sent from a device in one pico-net, possibly via various pico-nets, to a device in another pico-net within the same scatternet.

To support disconnection and connection of devices, a Bluetooth device knows a number of states associated with: inquiry, paging, connection and stand-by. In the *inquiry* state the device continuously transmits the enquiry message to discover the presence of other Bluetooth devices. Devices that allow their presence to be known go into an *inquiry scan* sub-state. A slave responds to a master with an inquiry response message containing the slave's device address and other information. A master, in the *page* state, can connect to a slave, in the *page scan* sub-state, by sending a message to the slave's device address. Enough information is exchanged such that master and slave are synchronized and connected. Master and slave have entered the *connection* state. In the connection state there are four modes: *Active, Sniff, Hold* and *Park* mode. The first three differ in the communication channels they can use. Asynchronous capacity can be made free to listen to other devices or a low-power sleep mode can be entered. In park mode the slave remains synchronized with the master but cannot freely communicate with the master.

3.2.3   Internet Protocol

Until recently, the IP address was manually attributed to a network interface or manually introduced from a file server to a Dynamic Host Configuration Protocol (DHCP) [9] server. Originally, the mapping from interface card to IP address was relatively fixed. With the advent of roaming laptops and DHCP servers with private addresses for home networks, IP addresses are allocated to devices when they ask for them. That means that these devices may change their IP address frequently.

Disconnection and connection of devices is not generally communicated to the connected devices. Therefore IP addresses, and the mapping from IP address to digital connector identifier are valid for a limited period of time. A consequence of this approach is that the application is responsible for verifying the disconnection of a device

3.2.4   UPnP

UPnP devices are connected when they are connected according to the IP rules. Connection of UPnP devices does not imply that the associated UPnP services and devices are available and known to other UPnP applications.

4.   Streaming Support over the Network

There are two aspects for streaming support: (1) the efficient, timely and robust sending of the A/V contents over the physical medium, and (2) the middleware to support the control of service to select, start and possibly modify the wanted contents.

4.1   Medium Support

The streaming between two devices makes demands on the network capacities and functionality dependent on the transported Audio Video (A/V) contents. In video streams frames are transported with a frequency of one frame every 20 milliseconds or less. A frame is encoded to reduce the amount of data that needs to be transported. However due to differences in the frame contents, the sizes of the frames differ. Therefore, the required throughput fluctuates in the time. Transporting streams over the same medium may involve the reservation of a given bandwidth to guarantee the throughput. Bandwidth guarantees originated in digital telephony where capacity is reserved for every telephone connection. Reservation mechanisms were used for audio in general and concurrently also for video. Some media do not allow bandwidth reservation such as Ethernet. However, a reservation mechanism can be constructed on top of the Ethernet standard to support bandwidth reservation. Other media such as IEEE 1394 and IEEE 802.11e provide facilities that support bandwidth reservation. These mechanisms are not necessarily compatible. The middleware on top can provide facilities to establish A/V streaming connections between applications, and possibly allocate bandwidth to streams to provide A/V streams of a given quality perceived by the user.

Three media are considered which support bandwidth reservation for A/V streams: IEEE 802.11, Bluetooth, and IEEE 1394.

4.1.1 IEEE 802.11

Wireless media, including IEEE 802.11, are very sensitive to perturbations. To counter this problem, a packet is immediately acknowledged in IEEE 802.11 after correct reception. This results in an efficient retransmission of failed packets.

The IEEE 802.11 standard provides two bandwidth reservation techniques. Neither of these mechanisms is easily quantifiable in bits per second due to the many transmission perturbations. At the time of writing, two mechanisms are used: (1) Extended Distributed Coordination Function (EDCF) is based on conflict sensing delays of varying sizes, and (2) Point Coordination Function (PCF) is based on polling [15].

The EDCF is based on the Collision Avoidance mechanism of IEEE 802.11. Every time a device wants to send, it senses whether the medium is occupied. When the medium is occupied, the device waits a DCF Inter-frame Space DIFS and some chosen back-off time. When the medium is empty during the chosen interval, the device tries to send. The back-off time is selected by taking a value that lies between zero and a specified maximum back-off time that is specified as function of packet priority. By allocating different maximum back-off times to the packet priorities, packets with short maximum back-off times will be sent more frequently and occupy a larger part of the bandwidth than the packets with a larger maximum back-off time.

With PCF, the point coordinator provides contention free access. With regular intervals, contention free periods are made available by sending out packets that delimit the start and end of a contention-free period. During the contention free period the coordination function sends poll packets to pollable devices that can answer with the packet they want to send. Guarantees depend on the number of pollable devices, length of contention free period, etc. Poll packets pass before other packets because the back-off time of pollable packets is always smaller than for normal packets.

4.1.2 Bluetooth

A channel is divided in time slots of 62.5 μs. At the start of a time slot a frequency hop occurs. Most packets occupy one time slot. But there are packets that can extend over 3 or even 5 time slots. In even numbered slots slaves can send a packet and in odd numbered slots masters can send a packet, with this restriction that for a given even numbered slot, only the slave that received a packet from the master in the former slot, can send a packet. A master regularly polls the slaves. A master can support three connections. A slave can support three connections with the same master or two connections with two masters. Bluetooth knows two types of links: Synchronous Connection-Oriented (SCO) and Asynchronous Connection-Less (ACL) links. A master and a slave can agree to set up a SCO link. For SCO links time slots are reserved on a periodic basis. For robustness data is repeated three times in a packets. However, synchronous voice data is not acknowledged and resent.

4.1.3 IEEE 1394

IEEE 1394 provides a possibility to reserve isochronous channels of a guaranteed throughput. A device can ask for channels from the isochronous resource manager. The isochronous resource manager will allocate free channels to reserving devices. Every 125 μsec the cycle master sends a cycle start. At that moment, a node, willing to send, arbitrates for the bus and will send a packet on each channel it has reserved.

4.2 Middleware Support

On top of the connections established over the media, middleware allows applications to discover and manipulate services on a device. Heterogeneity is present at many levels: (1) standard dependent operational identifiers are used to identify services in a given standard. (2) The protocols to discover services over the network differ in the number of messages and the purpose of the message. (3) The syntaxes of the description of the services differ. In one standard, a number describes the function completely and in another standard the service is described with XML. (4) The semantics of the service and the functions differ.

Below, four middleware standards are described. Comparison is done in Section 5.

4.2.1 HAVi

Each device has a local registry in which device- and service-descriptions are stored. At start-up or reset the HAVi registry is locally filled with the local devices and services. A service can be requested by formulating a query over attributes of the services. If no answer is found locally, the local registry forwards the query to the other devices. The protocol uses HAVi messages for communication.

4.2.2 Service Location Protocol

The Service Location Protocol [11] is part of the IP stack to support service advertisement and resolution. Two operation modes exists (1) with directory and (2) without directory. Only mode 1 is discussed. At start-up, a device multicasts a request for the directory identifier. After reception of the identifier, the device sends its service descriptions to the directory. Applications will send requests to the directory to obtain identifiers and descriptions of the specified services. Service queries can specify scopes or attribute values. The protocol uses UDP messages, multicast or broadcast for communication

4.2.3 Bluetooth

After connection, a Service Discovery Protocol (SDP) client in a Bluetooth device is automatically connected to a SDP server. Service discovery passes requests from SDP client to SDP server. In a Bluetooth device there are at most one SDP client and one SDP server. The client can send a request for a list of service classes that is answered by the server. Consecutively the client can ask for a more elaborate description of the service class attributes. These two queries can also be combined in one query. The SDP client can poll on a regular basis to check the connection of the server.

4.2.4 UPnP

From start-up a device multicasts its identifier, and services on a regular basis. Other UPnP devices receive this information and store it locally. Applications can send requests for services with a multicast. A reply message, with identifier and services, is then unicast to the requester by the service provider. Before using the service a more detailed description is retrieved by the client. This detailed description is described according to the rules of the XML standard. A large flexibility is allowed and the description of individual services can be added in a straightforward manner.

5. Comparison

The relation of the middleware with the underlying communication medium determines for a larger part the response times of the middleware with respect to network configuration changes.

In Bluetooth and HAVi, every device knows the identities of all connected devices. The operational identifiers are available and consistent from the moment that two devices are connected. Service discovery exploits the connection knowledge by sending resolution messages to all connected devices. The recognition of disconnection of devices is a problem for wireless communication. In HAVi unconnected devices are not present in the registry because disconnection of a device is signaled to the whole network. In Bluetooth, resolution messages may be sent to (temporarily) unconnected devices.

In the case of IP and consequently UPnP and SLP, the operational IP addresses of connected devices are in general not known. Therefore, at a rather high level in the stack, services are multicast to connected devices to make them known.

One or two-level queries exist. In the one level query all information is returned within one message. In two level queries additional information is asked for after the identification of the service. Although not explicitly mentioned, description formats and grammars are wildly different for all discussed standards. For example, UPnP uses elaborate XML based descriptions while Bluetooth relies on standardized numeric identifiers to identify a service class.

Although all standards know the concept of device and service type, not all standards share the same basic concepts. For example the device type that hosts a fixed number of services, is not known in HAVi and Bluetooth but known to UPnP. This points to the largest heterogeneity problem: concepts in one standard are not necessarily present in another. Semantics of the functions of a given service, known in more than one standard, may differ from standard to standard.

6. Existing Integration Efforts

Two well-known standardization efforts are Open Services Gateway initiative (OSGi) [13] and Salutation [14]. They introduce a new standard to provide a common interface to services in a heterogeneous environment. These two standards do not really integrate middleware standards. A prototype of a special purpose 'bridge' [12] has been constructed to integrate HAVi based and UPnP based functions. The HAVi UPnP bridge is a successful attempt to provide integration at all levels of UPnP and HAVi devices and services.

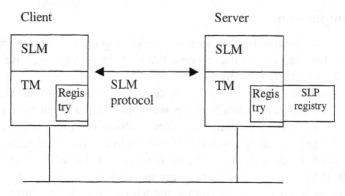

Figure 2. Salutation structure.

6.1 Salutation

The Salutation consortium has developed and distributes a discovery protocol that uses its own standardized service identifiers and descriptions stored in a so-called Salutation Manager (SLM) registry. Salutation managers, present on clients, and servers exchange Salutation manager protocol messages. The Salutation managers are separated from the underlying communication protocol via a Transport Manager (TM) interface (see Figure 2). Salutation device and service description exist independently of other middleware standards. However, possibilities exist in some cases to convert the contents of the Salutation service registry to contents used in other service discovery registries such as SLP of IP (and vice versa). In the Bluetooth standard an extension to Salutation is described. This is also shown in Figure 2, where SLP registry contents coexist with the SLM registry contents. Conversions between SLM and SLP registry contents occur, but only one SLM API is presented to client and server applications.

Salutation does not look at converting service semantics from one protocol to another. Applications are inherently written for Salutation (like HAVi and UPnP) and must use the formats and Application Program Interface (API) as specified for access to the registry and the registry contents.

6.2 OSGi

The Open Services Gateway initiative specifies APIs for a gateway. With the APIs end-users can load network-based services on demand from the service provider. The OSGi specifications are designed to enhance network and middleware standards. Central is the gateway that connects the external service providers to the internal clients. Applications developers combine a set of services in a *bundle* that is downloaded to the gateway. A bundle is a file that contains the JAVA code needed to handle a device or service and also contains references to other needed services. The OSGi framework installs, maintains, and deploys the specified bundles. Applications access external services through one common API. These external accesses are unrelated to accesses to home network under control of the used middleware standard.

6.3   HAVi UPnP Bridging Device

The prototype of a bridging device has been made to connect a UPnP based network and a HAVi based one. Complete transparency is impossible given the semantic difference between the service descriptions that are only available in one of the two standards. The device is proxy-based which means that the bridging device is both a HAVi and a UPnP device and every time a device or service in standard X is added or removed, a proxy to standard Y is added or removed to the bridging device. The following four translations aspects from a set of seven addressed ones are described below:

- *Plug and Play.* When a HAVi device is (dis)connected, the bridging device detects the addition (removal) of devices and services, the corresponding proxies are added (removed) and discovery (bye-bye) messages are broadcast to the UPnP devices. When a UPnP device is added (removed), the registry of the bridging device is modified accordingly, and proper *SEID*s must be allocated. Removal is noticed because bye-bye messages are sent or no alive messages arrive.

- *Device description and look-up.* Lookup can be initiated from UPnP network only. Under HAVi, the registry in the bridging device contains a complete image of the connected UPnP devices. UPnP queries are translated to HAVi queries. The description of HAVi devices can well be translated automatically to UPnP descriptions although some human intervention is necessary to find proper XML character strings that are 'human-readable'. HAVi descriptions require standardized device and service Ids. Remark, UPnP devices that do not exist for HAVi cannot be accessed from HAVi and vice versa.

- *Programmatic control.* Devices that exist in both standards or are specified as manufacturer extension in one of both standards are considered only. Even then applications will need to be UPnP device or HAVi device specific given the different signature of the service's functions. When a HAVi device invokes a UPnP function, messages are sent to the *SEID*, arrive in the bridging devices are translated to required XML syntax and passed on to the UPnP device using the UPnP identifier. When a UPnP device invokes a HAVi device, the XML message is sent to the bridging device, translated to a HAVi message and passed on to the right *SEID*.

- *Streaming mechanism.* Only streaming between HAVi devices is considered. The control of a stream within the HAVi network from a UPnP device is made possible. The source and destinations of the steams are specified from the UPnP device and the streaming between the two HAVi devices is started from the UPnP device. This can be done with the programmatic control concepts described above, provided the streaming end-points of HAVI source and destination can be translated to the appropriate UPnP concepts. This seems to be possible given the last results of the A/V working committee of UPnP.

6.4   Conclusions

From the former sections it is clear that still much work is needed to transport A/V streams seamlessly over heterogeneous middleware standards. Given the different semantics of the service functions and even the different sets of concepts, applications need to be targeted to a given standard, unless the standardization committees address this problem directly and provide descriptions for all devices of all standards. Such an approach, although nice, looks rather improbable.

The streaming over heterogeneous media is an open problem as no translation between bandwidth reservation schemes is provided. A common protocol may be required but translating one protocol into another is also a viable solution. Example of a common protocol is to use RTP protocol [10] and ignore the bandwidth reservation schemes, possibly at the detriment of the video quality. A translation candidate is: the translation of isochronous IEEE 1394 channels to EDCF of IEEE 802.11e. Even when from a communication point of view streaming is possible, additional standardization efforts are needed such that all equipment connected to the network communicate their bandwidth needs to some (possibly distributed) authority to obtain bandwidth reservation that are enforced by the middleware.

Below preliminary results of the FABRIC project are reported, based on the HLA standard, to control streams over heterogeneous networks and to separate applications from the underlying middleware. The project aims at integrating standards, like the UPnP HAVi bridging device, and does not create new standardized device and service descriptions like Salutation.

7.   Objectives

We want to achieve integration such that an application programmer can concentrate on the service semantics and is not unnecessarily hindered by standard's operations and syntax. The best possible integration is the one where the whole heterogeneous network behaves as if communication and middleware differences are completely hidden to the application. Such an objective is

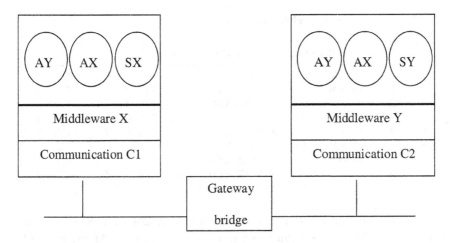

Figure 3.   Compatibility implementation.

completely impossible, already coming from the fact that communication media have different characteristics and are optimised for different applications. The following *compatibility* objectives are specified by us, where X, Y, Z stand for Bluetooth, HAVi, UPnP or some other middleware standard.

1. X application runs on device with middleware standard X and accesses X services on a device of middleware standard X.

2. X application runs on device with middleware standard Y and accesses X services on a device of middleware standard X.

3. X application runs on device with middleware standard X and accesses Y services on a device of middleware standard Y

4. X application runs on device with middleware standard Y and accesses Y services on a device of middleware standard Y

5. X application runs on device with middleware standard Y and accesses Z services on a device of middleware standard Z

Objective 1 is the current situation. Objective 2 means for example that a new application that accesses Bluetooth devices and HAVi devices can execute either on the HAVi or Bluetooth middleware. In objectives 3 and 4 standard X invocations and results must be translated to standard Y invocations and results. Objective 3 is completely fulfilled (for HAVi/UPnP) by the HAVi UPnP bridging devices. Objective 5 assures that anything can be translated to anything and executed on anything.

The objectives can be implemented by putting translators at suitable places. This is illustrated in Figure 3, where two devices, interconnected by a bridge

Table 1.   Realization of compatibility objectives.

Nr.	Obj.	Obj. Rewrite	Application	Bridge/Gateway	Service
1	1	AX→MX→SX	{C1(X)}		>C1; >X
2	2	AX→MY→SX	X|Y; {C2(Y)}	>C2; Y|X; {C1(X)}	>C1; >X
3	2		{C2(Y(X))}	>C2; >Y; {C1(X)}	>C1; >X
4	3	AX→MX→SY	{C1(X)}	>C1; X|Y; {C2(Y)}	>C2; >Y
5	3		X|Y; {C1(X(Y))}	>C1; >X; {C2(Y)}	>C2; >Y
6	4	AX→MY→SY	X|Y; {C2(Y)}		>C2; >Y
7	5	AX→MY→SZ	X|Y; {C2(Y)}	>C2; Y|Z; {C3(Z)}	>C3; >Z
8	5		X|Z; {C2(Y(Z))}	>C2; >Y; {C3(Z)}	>C3; >Z

or gateway, are shown. Applications targeted to middleware X or Y, ovals with AX and AY, access services SX (SY) situated above middleware X (Y). Translations can take place in the application to translate calls to middleware X into calls to middleware Y. The same translation can be done in the gateway. Middleware X is associated with communication C1 and Y with C2. Messages of communication medium C1 can be changed to messages for communication C2 in the bridge. The following notation is used

- {m}   send message m

- X(Y)   Y is tunneled through X

- X|Y   X is translated to Y.

- >X   remove standard X encapsulation

For example C1(Y(X)) means: tunnel a message of middleware X in middleware Y and encapsulate it in medium format C1. Table 1 gives the actions needed at the source, the gateway/bridge, and the destination to realize the compatibility objectives. Two rows are explained in mode detail. In row 1, compatibility objective 1 results in sending a message in format X over communication standard C1. At the service the communication format C1 is removed and the message format X is removed. In row 6, compatibility objective 4 translates message in format X to a message in format Y; the message in format Y is sent over communication standard C2. At the service the communication format C2 is removed and the message format Y is removed. Beware, an application (say AX) is present both in the left (for Obj 1) as in the right device (for Obj 2). The cited middleware Z and associated communication C3 is not shown in Figure 3.

In Table 1, the application always invokes the underlying middleware to transport its message. It is also possible to bypass the middleware and send a message by invoking the underlying communication layer directly.

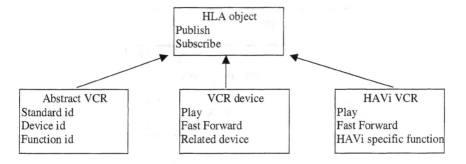

Figure 4.   Possible object example.

8.   HLA Technology Deployment

We achieve integration by providing tools that in a standardized way allow the creation of applications targeted to a known set of standards. The application programmer defines a set of object descriptions to describe the services the application needs. Tools are provided to map the object invocations expressed in the user defined object descriptions to invocations expressed in the format of the standard of the targeted service. At run-time the invocations to the objects are translated to the invocations of the actual services. Users define their own service interface in contrast to Salutation where users need to adhere to the Salutation definitions. Only after a Salutation description has been standardized by the Salutation committee can it be used. In Salutation devices and functions are standardized independent of their standardization elsewhere. In our approach a service can be used the moment a standard publishes its interface and the corresponding objects are defined by the application programmer.

We have used the HLA object model to provide the translation facilities mentioned in Section 7. HLA is based on the Publish/Subscribe concept. Applications or services can publish attributes of objects instantiated from a given class. Applications can subscribe to a set of attributes of a given class. The information is exchanged from publisher to subscriber via the Run-Time Infrastructure (RTI). The conversions from object invocation to services invocation are also part of the RTI. Specific RTIs will be needed for every supported communication medium and middleware.

In Figure 4 an example of a possible object definition is shown. At the root is the HLA object that supports the Publish and Subscribe actions. Three objects representing a Video cassette recorder (VCR) inherit from the HLA object. The abstract VCR is used for service discovery, explained later. The VCR device is the object used by the application. The HAVi VCR provides the link to a VCR device under the HAVi middleware. In the VCR device, the

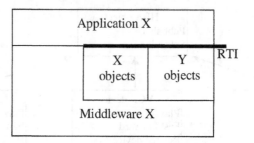

Figure 5.   Application design.

'related device' attribute can point to such a HAVi VCR. The application can use the HAVi VCR object when it wants to do HAVi specific actions.

In short, the application of HLA in the heterogeneous network can bring us the following advantages:

- An abstraction layer to make the applications unaware of the underlying middleware standard.

- The use of Publish/Subscribe to provide a minimal, simple to use, application interface, -under investigation-.

- A (possibly minimal) standardization effort on top of the existing middleware standards, -under development-.

Three aspects are discussed in more detail in the following sections:

1. Heterogeneous access

2. Heterogeneous discovery

3. Heterogeneous streaming control

8.1 Heterogeneous Access

Instead of targeting integration-enabled devices, we aim for integration-enabled applications. We aim at a high degree of backward compatibility to allow access to devices via its standard middleware. An integration-enabled application of standard X in a given device should be able to access services in standard X. All applications of standard X should continue their normal operation after the addition of integration-enabled applications. This requirement allows a gradual introduction of integration-enabled applications within already existing home networking infrastructures.

The design starts from items 7 and 8 of Table 1 and supports compatibility objective 5. A first design decision is to place the RTI on top of the middleware

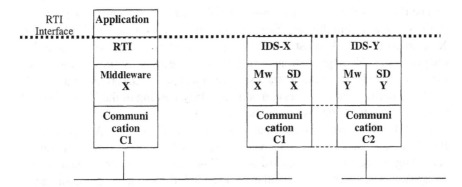

Figure 6.   Integration architecture.

standards such that applications use the API provided by the RTI-interface that includes the objects that represent the controlled services and devices. In

Figure 5, the application accesses the middleware standard X directly and accesses a given set of objects associated with devices of standards X and Y via the RTI-interface. The X-objects and Y-objects are part of the RTI implementation. For a given application an object model needs to be made of the services it will invoke. Using HLA, it is possible to share these descriptions between applications to generate automatically the conversions between the HLA representation and the standard's representation.

In Figure 6 the approach is drawn. An application is separated from the underlying middleware by the RTI. The RTI constructs the appropriate messages of the destination middleware standard. The messages are communicated by invoking the underlying middleware. Via the communication medium, C1 (C2), the messages are directed to the appropriate destination within the network of middleware standard X (Y). This is either the destination within middleware X (Y) or a connecting node (bridge or gateway) to middleware Y (X). At arrival of the message, the service is invoked and returns its result.

8.2   Heterogeneous Discovery

We aim at the following hierarchical and scalable service discovery. The application subscribes to the requested services expressed in the user defined object format. The RTI subscribes to a probably standardized abstract service object. Integration-enabled Directory Services (IDS) publish the abstract service objects. After reception of the abstract service object, the RTI collects the information from the device that contains the description in full conform the middleware X. The RTI presents the received information to the application according to the object definition (see Figure 4).

The IDS uses the service discovery of middleware X to enquire the presence of the services of middleware X. An IDS present on a device with middleware X is called IDS-X. At most one IDS is present on every device that contains integration-enabled applications. An IDS is not necessarily present on all devices. Such an IDS and the abstract service object can be described in the IEEE 1516 standard. Manufacturers can add the IDS according to the HLA standard to the manufacturer extensions of the standard X.

The IDS-X is started as a normal service for middleware X. At start-up, the IDS-X makes itself known. Afterwards, the IDS-X sends a browsing enquiry, according to the rules of standard X, to all connected devices of standard X. Every time a new service of standard X is added, the service is communicated according to standard X, received by the IDS-X, added to its contents, and the associated abstract service is published.

The information stored in the abstract service object can be rather minimal. It needs the device identification, service (and function) identification, and the standard identification.

The question remains how to connect the IDS of different standards (IDS-X and IDS-Y in Figure 6). A simple approach is to situate the IDS in a gateway device that connects to both standards. In the figure, this approach is represented by the dotted lines joining IDS-X and IDS-Y. Consequently all services defined within standard Y are directly accessible to the integration-enabled applications of standards Y and X. Suppose we have three standards X, Y and Z with gateways YX and YZ. All applications in Z know about all services in X, Y and Z. Applications in X only know services in X and Y and application in Z only know services in Y and Z. A protocol is needed to communicate the services of X to applications of Z and vice versa.

When only a router or bridge is used to connect two standards, a similar additional protocol is needed to connect the IDS over the bridge. Each IDS-X publishes its contents over the heterogeneous home network, and each IDS-X subscribes to all IDS-Y. This latter new communication protocol between directories may prohibit the use of this solution. However, it provides a HLA-compatible way to communicate all services to all connected integration-enabled applications.

8.3   Heterogeneous Streaming Control

We envisage the starting, stopping and modification of a stream via a so-called *stream* object. The object is published and stored at the connection source. The HLA standard enables the definition of such an object. Every integration-enabled device will be able to subscribe to stream objects. Each stream object corresponds with a given video stream. Browsing stream objects and selecting the wanted one by any integration-enabled application is made possible. The stream object contains information about the source and destination and the standards to which they belong. The invocations to the stream object are translated to the appropriate middleware invocations. The stream object is special in the sense that the concept is not supported by all middleware standards. The concept can be defined within the HLA standard and added to the manufacturer middleware extensions similar as for the IDS.

Similarly, objects can be defined for other network entities. Good candidates are: user, owner, connection manager and others.

8.4 Meeting Objectives

Our approach allows the realization of objective 5 with an adjustment: Integration-enabled application X runs on top of middleware Y and accesses (legacy) devices of standard Z. In one respect this is more than is realized by the HAVi UPnP bridging device, in another respect it is less. With the bridging device a standard HAVi application can access a standard UPnP service. The latter is not possible with our approach.

In contrast, our approach allows a more flexible stream handling over middleware standards. Users can browse through a list of available streams, select a relevant one and modify its operational characteristics such as, destination, window size, etc.

The major advantage is the definition of service objects independent of the supporting middleware, thus making the application and services independent of the supporting middleware. The use of publish/subscribe paradigm enables easy integration of standards without a tight coupling to the middleware standards from the applications.

9. Conclusions

In conclusion, the Ambient Intelligence vision provides the following challenges: (1) Streaming over heterogeneous media, and (2) translation of function invocations between different middleware services. The paper has described the differences between the more important middleware standards that provide interoperability between devices and services. It is shown that beyond connection and recognition of services, also the heterogeneity of the service semantics needs to be addressed. This paper shows that HLA provides interesting handles to integrate the standards into one conceptually homogeneous home network. Our approach is characterized by creating an abstraction layer between application and middleware. The consequence is an extra discovery repository of abstract objects that connects the service discovery of the connected standards.

Acknowledgements

Tom Suters suggested the contents and format of Table 1. I am grateful to Jean Moonen for many suggestions for improvements. This work has been funded by the FABRIC project in the context of the IST programme of the EC fifth framework. The FABRIC team members have contributed by reviewing and commenting the presented solutions. Special thanks go to Helmut Burklin and Rob Udink for explaining the middleware standards and to Jan-Jelle Boomgaardt and Marco Brasse for their HLA knowledge.

References

[1] Bluetooth SIG, *Specification of the Bluetooth System version 1.1*, volume 1, Core, 2001.

[2] Bluetooth Sig, *Specification of the Bluetooth System version 1.1*, volume 2, profiles, 2001.

[3] HAVi specification, version 1.0, 2000.

[4] IEEE, 802.3: Carrier Sense Multiple Access with Collision Detection, *IEEE 802.3-1985*, 1985.

[5] IEEE, IEEE standard for a High Performance Serial Bus, *IEEE std 1394-1995*.

[6] IEEE, Part 11: Wireless LAN medium Access Control (MAC) and Physical layer (PHY) Specifications, *IEEE 802.11-1999*, 1999.

[7] IEEE, Standard for modeling and simulation High Level Architecture, *IEEE standard 1516-2000*.

[8] IETF, Address Allocation for Private Internets, *RFC 1918*, 1996.

[9] IETF, Dynamic Host Configuration Protocol, *RFC 2131*, 1997.

[10] IETF, Real-Time Transport Protocol, *RFC 1889*, 1996.

[11] IETF, Service location protocol, version 2, *RFC 2165*, 1999.

[12] Jean Moonen, private communication, 2003.

[13] OSGi, *Specification 1.0*, 2000.

[14] Salutation, http://www.salutation.org/.

[15] A.S. Tanenbaum, *Computer Networks*, 4th edition, Prentice Hall, 2003.

[16] Universal Plug and Play Device Architecture, version 1.0, 2000.

The Butt of the Iceberg: Hidden Security Problems of Ubiquitous Systems

Frank Stajano

University of Cambridge, Laboratory for Communication Engineering, Cambridge, U.K.
http://www-lce.eng.cam.ac.uk/~fms27/

Jon Crowcroft

University of Cambridge, Computer Laboratory, Cambridge, U.K.
http://www.cl.cam.ac.uk/~jac22/

Abstract   In Kurt Vonnegut's eponymous novel, there is a moment when an adult demonstrates Cat's Cradle, the art of making patterns with a loop of string held between two hands, to a small child. On dropping the string, the adult says "Look: no cat, no cradle". The child is naturally dismayed.

In a sense, ubiquitous systems are as ephemeral as the Cat's Cradle. The infrastructure to support them is as subtle, and complex, and the trust we have in them as fragile. This paper reveals the hidden problems in store for security in ubiquitous systems. A crucial aspect we wish to stress is that there is a phase change as one moves from classical distributed computing into this new environment: the mere shift in quantity of components is enough to lead to qualitative changes in aspects of security. Moreover, there are inherent qualitative shifts.

Keywords   ubiquitous computing, security, privacy

1.   The Tip of the Iceberg

It is inevitable that the physical manifestations of personal computers will vanish into the surrounding infrastructure and into the very fabric of buildings, clothes, vehicles and so forth. As this happens, computer scientists will need to come to terms with systems that interact in an ad hoc manner with their users, as well as with each other. Concepts of input, output, persistence of information and computation will change. The notion of ownership of information and computation will drastically evolve, along a trajectory that is only partially visible today.

The expression "ubiquitous computing", now often abbreviated to "ubicomp", was made popular by the late Mark Weiser in his well-known 1991 *Scientific American* article [12]. In that early work he candidly admitted that

91

T. Basten et al. (eds.), Ambient Intelligence: Impact on Embedded System Design, 91-101.
© 2003 *Kluwer Academic Publishers. Printed in the Netherlands.*

systems implementing his vision would evolve in ways he could not possibly foresee:

> Neither an explication of the principles of ubiquitous computing nor a list of the technologies involved really gives a sense of what it would be like to live in a world full of invisible widgets. To extrapolate from today's rudimentary fragments of embodied virtuality resembles an attempt to predict the publication of *Finnegan's Wake* after just having invented writing on clay tablets. Nevertheless the effort is probably worthwhile.

In 2003, over a decade later, ubicomp is irrevocably happening. On the computing side, even non-technical people unwittingly interact on a regular basis with invisible microprocessors embedded in vehicles, buildings and entertainment systems. On the communication side, cellular telephony has been growing steadily at an even faster rate than computing; and wireless networking for mobile computing is becoming a commodity item: the availability of 802.11 connectivity at conferences, as happened for VGA projectors a few years back, is soon graduating from a novelty feature to something that attendees naturally expect. Many consequences of ubicomp that were not believable or not obvious back then are now in plain view. But this is still only the tip of the iceberg—more developments will come along and surprise us.

In this paper we explore some of the outstanding long-term challenges for ubicomp, with special attention to the systemic aspects. Embedded computing and wireless networking are no longer in their infancy, but what is still missing is a synergy of the parts: bringing together all these hidden processing nodes into a coherent entity worthy of being called a *system*. Once this is achieved, the complex system will exhibit interesting properties, of which *security* will have special importance. Security is a paradigmatic example of a system property—one that can only be achieved holistically. In these pages we aim to understand the challenges ahead by imagining how ubiquitous systems may develop.

We have been writing on clay tablets for a little while now; perhaps we are already scribbling on papyrus. While it may still be early to envisage *Finnegan's Wake*, for a field that evolves at such speed there is merit in cranking up the imagination dial rather than considering only small incremental developments, even if the result borders on the outrageous. In the next section we are going to do just that.

2.   Application Scenarios for Ubiquitous Systems

Where could ubiquitous systems take us, in the course of the next decade, from the viewpoint of technological feasibility? Let's have a little brainstorming about what might happen, without worrying too much for the moment about whether it actually will.

2.1 Always-on Health-Care

Imagine the entire population of your country being permanently wired for systematic monitoring of various metrics such as heartbeat, skin conductivity, blood sugar, etc. Imagine this data being locally and remotely logged in a secure way via wireless communication. In the short term this is done mainly for people at risk and the data is "phoned in" periodically. In the longer term, monitoring covers everyone, on a 24×7 schedule. One possible justification for this development is long term medical research.

The next stage is to introduce intervention. Firstly, via remote diagnostics and human advice; later, via autonomic responses (e.g. triggering defibrillators for cardiac patients undergoing unattended attack; appropriate other responses for epileptics, diabetics and so on).

The third stage is to synchronize information (telemetry, long term data and local data) between patient, hospital and the paramedics of an arriving emergency service. It may sound optimistic to imagine that the synchronization and intervention will be achieved without problems. However, in any case, the risk levels could be provable properties of the system: an automatically derived explanation would be provided in case of failure, such as "it ran the defibrillation, despite there being a 12% risk that this was not heart failure, because there was a 0.003% risk of this doing harm and a 48% chance of it being beneficial".

It has been observed that in fact deriving this type of explanation is a very hard problem, and that it is all too easy to fool the user into thinking that they have a deeper understanding of a complex system (e.g. the heart) than really possible.

2.2 Surveillance and Evidence-Gathering

Nowadays, in our society, a large proportion of people have mobile phones, enjoy Internet access and exchange electronic messages with greater frequency than they write snail mail. Many commercial transactions, and certainly most of the higher-value ones, are paid for with plastic rather than cash. Furthermore, especially in the UK, most public and many private spaces are under video surveillance. As a consequence, each one of us leaves behind a conspicuous digital trail [5] through which our past movements and actions can be reconstructed by third parties who were not involved in them.

Various agencies will be keen to use more and more of this data in ways that enhance public safety and offer better (in the sense of accurate and provable) evidence for criminal charges. Ideally, this would create disincentives for criminal behaviour without intruding on the privacy and other human rights of the general population.

Two real world examples suggest directions for this.

2.2.1 Locating Mobile Phones

In recent test cases [1, 2], mobile phone call records (which can establish the phone's location quite accurately, to about 100 metres) have been used in defence as "evidence" of a person's absence from a scene.

Strictly speaking, all that such records can prove is the location of the *phone*, not of its owner (short of recording the call and recognizing the voices of the correspondents; but this would require prior warrant to tap the call). For this reason, they cannot be used by the prosecution as evidence of the presence of the accused party at the scene; whereas, in the opposite sense, a defendant is automatically granted the benefit of the doubt, which the hint offered by the location of the phone can corroborate.

The increasing pervasiveness of video surveillance will certainly offer further opportunities to build up "evidence" by combining different kinds of such hints.

2.2.2 Cameras Reading Car Number Plates

In February 2003, the city of London launched a scheme to charge motorists in order to reduce congestion. Cameras along the boundary of the restricted zone check car registration plates automatically and issue fines to the drivers who have not paid the charge.

In the original design, cameras used to monitor vehicles would immediately discard the acquired combination of registration plate, location and time if the database listed the car as having already paid. Under such a policy, there would be no record of the movements of law-abiding people (or, more accurately, vehicles, since the system cannot tell who is driving).

Recently, though, the policy was changed: the police and other authorities will be granted routine access to the location data of law-abiding users, of course within the guarantees of the Data Protection Act.

This poses several threats. Given there are so many people with access to the police computers it is relatively likely that someone can be bribed or blackmailed into giving over data. Drivers may then be embarrassed (e.g. when journalists report them "at their lovers' nest") or fired (when seen at the match instead of home sick) or killed (when terrorists learn the routes taken by their political targets to their offices and constituencies).

2.3 Zero Road Fatalities

More young people die on roads than from any disease. It is amazing that this is socially acceptable. It should be possible to build a system that reduces death for pedestrians, cyclists and drivers to zero. A low-tech solution is to put a 15 cm spike on the centre of the steering wheel—this provides an interesting incentive to drivers. Another is to require an attendant to walk in front of each car with a red flag.

Realistically, it is already possible to get cheaper insurance for sports cars if the system is fitted with a speed limiter. As other mandatory safety features, from seat belts to airbags, gain broad social acceptance, we may imagine the next step to be the automatic override of the vehicle's controls.

With sensors monitoring the presence of humans in the vicinity (on the sidewalk, or nearby and on a collision trajectory) and of obstacles in general, a control system could adjust the speed of the vehicle to a safe limit whenever someone is near. Incentives such as insurance discounts could be introduced to accelerate the deployment. And the system would not necessarily have to force ridiculously low speeds.

The interesting problems with such a system are clear when you consider what a driver would do—people would probably put their foot to the floor at all times. What then if the system fails?

Going further, it is not an enormous leap of the imagination from this to a car equipped with a navigation system, inter-vehicle communication, road traffic monitoring and fully automated driving. Could we then speak a destination, sit back and read the paper? At least in theory, with all cars smoothly cruising along at a constant safe speed without overtaking, this kind of system could not only improve safety but also offer the guaranteed arrival times of a conveyor belt, combining the reliability of a dedicated subway system with the comfort of a private vehicle.

3. The Shape of Ubicomp

Ubiquity of computing and communication poses major technical challenges across a broad range of computer science topics. We will require a coherent approach to networking, operating systems and to the various programming environments for applications, interfaces and user customization. A good Ubiquitous System must accommodate new structures such as combinations of unreliable components and yet appear more reliable and usable than today's PCs.

3.1 Quantitative Changes

The exponential growth predicted by Moore's Law has continued to hold for much longer than originally expected. A rough but essentially accurate summary of the performance evolution of a standard $1000 computer over the past two decades is that we went from "3 Ms" (MB, MHz, Mbps) to "3 Gs" (GB, GHz, Gbps). The trend continues towards the "Ts". This is now far more than enough for most sensible personal computing needs, especially given the ability to provide services within the network. (Except that we keep on inventing new ways to soak up memory, processor cycles and bandwidth—not to mention disk space.)

An apparent trend in computer science is that we will soon produce constant performance at a falling price on the same curve, leading to the facilities in a 2003 desktop being available for under $1 by 2013. There is evidence that this direction is already being followed to some extent, with PDAs, mobile cell phones and hand held gaming computers falling below the $100 mark. However a lot of this is achieved through a reduction in functionality.

Currently, smaller inexpensive computers have user interfaces of limited functionality (displays of lower resolution and smaller physical size, pens instead of keyboards etc.). By contrast, with 3D projection display, voice input and output, and other interaction modes such as gesture, one could consider a wholly different style of use. This would not be a computer for embedded systems work (as in today's phones or PDAs), or a pure tablet (a.k.a. "network computer"), but a fully featured device. The best description of the type of interface we envisage, for science fiction fans, is probably to be found in the 1953 Asimov novel *Second Foundation*, which introduces the Prime Radiant, with full fledged computing, communications, storage, etc. Such a computer may be worn, but it may well also be something that one places in large numbers around one's person and property, as well as in the broader, public environment.

This latter point is critical. When this approach is used, a merely quantitative change in the number of computing devices leads to a qualitative set of changes in the way one deals with information and computation. Key to this is the notion of ownership. The physical instantiation of computing is no longer there to allow one to have even the illusion of ownership and control.

At the same time, comparable computational power is becoming embedded in dedicated objects, as is already happening today, with the difference that ubiquitous networking would bring them together in a system as opposed to isolated parts. Let's briefly review the way this is happening.

3.2 Qualitative Changes

There are already many more computing devices embedded in the environment than associated with individuals in the form of PDAs or wearables. However, to date, these devices are largely independent (e.g. engine management systems), replacing the analog control systems of the past. There is great potential in connecting them into a pervasive information sensing and actuator system.

Already, connectivity such as Bluetooth allows the user to control several objects near their phone. Wireless LAN hotspots and GPRS allow the laptop or PDA to integrate into the world in many places, and to access remote devices.

Low cost home area networks allow domestic appliances and sensors to be accessed and managed by a central computer or from outside the home.

Networking all of these components that were previously autonomous leads to a number of important consequences that we address next.

4. Challenges for Ubiquitous Systems

4.1 Control

Many synonyms have been coined to describe the vision of embedding computing and communication capabilities into everyday objects: we have so far adopted the Weiser locution of "ubiquitous computing" but the reader will have also heard these same ideas variously described as "proactive", "context-aware", "sentient", "calm", "ambient" and "pervasive" computing. This last term, "pervasive", while popular in some circles, is avoided by a number of researchers who prefer not to evoke any sinister overtones of wide-reaching domination of machines over mankind. We, instead, are keen to awaken these fears explicitly in order to help ensure that they won't have a chance to become reality.

At the technology level, the security issues posed by ubiquitous systems can be split into two main categories: the ones that can be solved with the traditional tools developed for distributed systems security, and the new ones that require original solutions. The latter group includes for example secure ad-hoc routing [6], stream authentication for sensor networks [7] and secure transient association [10].

At the highest level of abstraction, though, the application scenarios we presented in Section 2 highlight one major trade-off that ubicomp forces on us: the convenience of self-activating and proactive technology is inescapably intertwined with the risk of losing control. The most fervent proponents of the new way will be happy to delegate away to their car the low-level chore of driving, in the same way in which they gladly switched to automatic transmission years ago; many others, though, are bound to remain sceptical. The trade-off is even more evident in the health-care example.

4.2 Ownership

The problem of control, or lack thereof, is compounded by that of ownership. In a world of systems as opposed to self-contained devices, the service we receive is the result of a synergy between many devices, not all of which will be owned by us.

In one of our ongoing experiments [11] we are building a "sentient car" that, among other features, is capable of seamless handover on the move between heterogeneous wireless networking technologies: it will switch from GSM to

802.11, without dropping connections, as it drives through a "Wi-Fi" hot spot. The connectivity provided by this hot spot, however, is made available by third parties under terms over which the car driver has no say (other than perhaps a boolean "accept or reject"); in this sense it is a rudimentary example of an external dependency. The ubicomp world will present us with many more such dependencies, so the concept of virtual ownership must be addressed.

The move from the mainframe to the PC was a move from central to personal ownership. The systems we envisage are a move towards fully cooperative virtual ownership. Notions of identity and provenance are critical. There are many challenging technical problems here, but also social value: disposable grids; affordable computing for the developing world; collaborative filtering; emergency service and disaster recovery support; location aware services and so on.

Some previous work [10] suggested a strategy through which devices could be securely "lent" to temporary owners. More work is needed, though, on higher level incentives, rewards and accounting.

Ubicomp extends even to "smart dust" or similar intelligent sensor net computing projects: here, a combination of systems co-operate to offer functions previously located in in one small device (personal communicator/digital assistant/games console/toaster). The dynamicity of these systems will be extreme. As users move amongst the computing milieu, they will autonomically create services as easily as one today borrows and uses physical tools such as knives and forks, whether at home, at a friend's or in a restaurant.

4.3   Privacy

The application scenarios of Section 2 also make it evident that ubicomp may become a serious threat to privacy.

At the technical level, some basic countermeasures may include anonymization and brokerage of queries through a layer of trusted middleware. For example, in the surveillance scenario, the middleware would disallow access to the raw data and only permit queries according to a standard set of interfaces. These might enforce rate limitations (limiting the frequency with which a certain kind of question can be repeated) and semantic limitations (allowing for example "were you at location L at time T?" but not "where were you at time T?"). These safeguards would be provable properties of the middleware, which would be open to inspection and verification by the users or their appointed expert advisors. Preventing attackers from inferring higher level information from the allowed queries is, however, a hard theoretical problem [4]. On the practical side, instead, ensuring that the middleware cannot be bypassed to access the data directly is going to be equally arduous.

Going up one level, we ought to be cynically sceptical about the guarantees provided by policy assurances such as the ones that the above middleware layer would be charged with enforcing. The London congestion charging scheme mentioned above provides a neat example of a significant change in policy introduced after the deployment of the surveillance infrastructure.

As appropriately noted by Zimmermann [13] well before 9/11,

> while technology infrastructures tend to persist for generations, laws and policies can change overnight. Once a communications infrastructure optimized for surveillance becomes entrenched, a shift in political conditions may lead to abuse of this new-found power. Political conditions may shift with the election of a new government, or perhaps more abruptly from the bombing of a Federal building.

Responsible architects of ubiquitous systems will take this into account and examine in detail the ways in which their envisaged systems could be abused.

There is often a conflict of interests for researchers working on new and exciting technologies: being too lucid and honest about the dangers for society of one's new creation may hamper or block its development. It is therefore easy to self-justify the decision not to spend time and effort trying to limit the potentially dangerous capabilities of the invention by arguing that other less scrupulous competing researchers would otherwise be first to market, of course with the unrestricted version. It is also frequently argued that, except for extreme cases, it is not the technology itself that is inherently evil but only some of the uses to which it may be put. While these arguments have merit, it would be shortsighted to use them as excuses to avoid a critical assessment of the risks that the new technology might pose to society in case it were intentionally misused once deployed.

Let us revisit some of the problems associated with just one aspect of security, namely the normal non-technical notion of privacy, to see some of the effects of ubiquity. Spärck Jones [9] envisages the following properties of the technology that create novel pressures on the perception of privacy.

Permanence Information in a ubicomp world attains a persistence hitherto unforeseen.

Volume The range of types of data available in the infosphere is massively increased.

Neutrality The data may easily be divorced from associated meta-data (or real world context, such as location, ownership) and as such used in ways that are unexpected.

Accessibility The availability of data on a network is much wider.

Assembly Data may be combined and mined.

Remoteness Aliens may see data.

This adds up to another vivid representation of the fact that ubicomp is no longer just a quantitative change but a significant *qualitative* change, bringing about new issues, opportunities, problems and threats that simply didn't exist before.

5.   Mistrust

As computers disappear into the environment they become less threatening to non-technical users but at the same time even more mystifying, especially when they don't work as expected. It is hard enough for a regular human being to understand why a regular computer misbehaves; it may be a lot harder when there appears to be no computer. Witness digital cameras (or cars!) that lock up and can't even be switched off with their "off" button—the only escape route being a hard reboot via removal of the battery pack.

Some current research [8] uses augmented reality to visualize the state of ubicomp devices that lack an obvious user interface. The effort needs to scale up to higher levels of abstractions too, in order to help non-technical users build correct and reliable mental models of the behaviour of the ubiquitous systems around them.

Earlier on we discussed a health-care system that could trace back and justify its choices—particularly the ones that later turn out to be inappropriate. More still needs to be done on this aspect, especially with regard to incorporating human intuitions about the expected behaviour. The more ubicomp systems behave counterintuitively and unpredictably, the less they will be trusted by their users.

6.   Conclusions

We have examined some consequences for security for ubiquitous computing at the levels of systems and of users. The assertion is not that ubicomp is bad. Rather, we want to draw attention to the qualitative nature of the changes brought about by the quantity and type of data and processing available in such an environment. We could picture a major crisis for the introduction of new services if awareness of this big picture is not taken on board. The types of reactions might fall into the following categories.

- Principle of unintended consequences

- Confounded expectations

- Surprise

- Frustration

■ Rejection

These reactions might not be irrational.

Acknowledgements

Some of this material, including the three example scenarios, was originally developed by a brainstorming group sponsored by the BCS and UK CRC [3] in the context of the UK Grand Challenges in Computer Science initiative. We also acknowledge a useful discussion of novel problems in privacy with Karen Spärck Jones.

References

[1] BBC News. Evidence 'proves Hamiltons innocent', 2001. http://news.bbc.co.uk/1/hi/uk/1488859.stm.

[2] BBC News. At a glance: Damilola CPS report', 2002. http://news.bbc.co.uk/1/hi/uk/2558777.stm.

[3] Jon Crowcroft, editor. Scalable Ubiquitous Computing Systems or just Ubiquitous Systems. http://www.nesc.ac.uk/esi/events/Grand_Challenges/proposals/US.pdf.

[4] Dorothy Denning. *Cryptography and Data Security*. Addison-Wesley, 1982.

[5] Simson Garfinkel. *Database Nation*. O'Reilly, 2000.

[6] Yih-Chun Hu, Adrian Perrig, and David B. Johnson. Ariadne: a secure on-demand routing protocol for ad hoc networks. In *Proc. MOBICOM-02*, pages 12–23, September 2002.

[7] Adrian Perrig and J. D. Tygar. *Secure Broadcast Communication In Wired and Wireless Networks*. Kluwer, 2002.

[8] Kasim Rehman, Frank Stajano, and George Coulouris. Interfacing with the Invisible Computer. In *Proc. NordiCHI 2002*, October 2002.

[9] Karen Spärck Jones. Privacy: What's Different now?, October 2002. British Academy Conversazione, http://www.cl.cam.ac.uk/users/ksj/privksjtext1.pdf.

[10] Frank Stajano. *Security for Ubiquitous Computing*. John Wiley and Sons, February 2002.

[11] Pablo Vidales and Frank Stajano. The Sentient Car: Context-Aware Automotive Telematics. In *Proc. LBS-2002*, September 2002.

[12] Mark Weiser. The Computer for the Twenty-First Century. *Scientific American*, 265(3):94–104, September 1991.

[13] Philip R. Zimmermann. Testimony of Philip R. Zimmermann to the Subcommittee on Science, Technology, and Space of the US Senate Committee on Commerce, Science, and Transportation, 26 June 1996.

Emerging Challenges in Designing Secure Mobile Appliances

Srivaths Ravi and Anand Raghunathan
NEC Laboratories America, Princeton, NJ, USA
{ sravi, anand } @nec-labs.com

Jean-Jacques Quisquater
Université Catholique de Louvain, Louvain-la-Neuve, Belgium
jjq@dice.ucl.ac.be

Sunil Hattangady
Texas Instruments Inc., Dallas, TX, USA
sunil@ti.com

Abstract   Realizing the visions of ubiquitous computing and communications, and ambient intelligence will require modern electronic and computing systems to pervade all aspects of our everyday lives. These systems are used to capture, manipulate, store and communicate a wide range of sensitive and personal information. It is, therefore, not surprising that security is emerging as a critical concern that must be addressed in order to enable these trends. Mobile appliances, which will play a critical role in ambient intelligence, are perhaps the most challenging to secure. They often rely on a public medium for (wireless) communications, are easily lost or stolen due to their small form factors and mobility, and are highly constrained in cost and size, as well as computing and battery resources.

This paper presents an introduction to security concerns in mobile appliances, and translates them into challenges that confront system architects, HW engineers, and SW developers. These challenges include the need to bridge the mismatch in security processing requirements and processing capabilities (processing gap), the need to address the burden of security processing on battery life, the need for flexible security processing architectures to keep up with evolving and diverse security standards, and, lastly, a need for providing countermeasures against various kinds of attacks and threats. We also survey recent innovations and emerging commercial technologies that address these issues.

Keywords   ambient intelligence, mobile appliances, security, cryptography, system design, security processing, security platforms, attacks, tamper resistance, battery life

T. Basten et al. (eds.), Ambient Intelligence: Impact on Embedded System Design, 103-127.

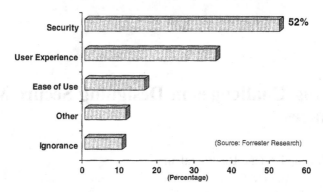

Figure 1.   Obstacles preventing consumers from adopting m-commerce.

1. Introduction

Mobile appliances, including cell phones, PDAs, and smart cards, account for a large segment of the electronics and semiconductor industries. Due to their convenience and ubiquity, it is widely accepted that such mobile appliances will evolve into "personal trusted devices" that pack our identity and purchasing power, benefiting various aspects of our daily lives [54, 32]. On the other hand, mobile appliances are also likely to play an important role in enabling the vision of an intelligent ambience, by collecting and communicating various personal habits and preferences, and enabling our environments to sense and react to us.

Due to the aforementioned trends, the usage of mobile appliances will frequently involve the storage of, access to, and communication of sensitive information, making security a serious concern. Indeed, the success and adoption of several emerging applications and services are predicated on the ability of mobile appliance manufacturers and service providers to ensure adequate security, thereby gaining the trust and confidence of consumers and other parties involved. For example, 2.5G and 3G wireless applications, including mobile commerce (m-commerce), multimedia messaging, mobile content delivery, and mobile office deployment, require high levels of security. In fact, security is cited as the single largest concern among surveys of prospective m-commerce users. Figure 1 shows a survey from Forrester Research that lists security as the primary bottleneck to m-commerce adoption, according to nearly 52% of mobile appliance users [30].

Thanks to the evolution of the Internet, information and communications security has already attracted significant attention [86, 91, 84, 77]. While the knowledge and experience gained from the wired Internet, including cryptographic algorithms, security protocols, and standards, give us a head start in the quest to secure mobile appliances, there are several challenges unique to mobile appliances that must still be addressed.

- Mobile appliances often use a public transmission medium for (wireless) communication, which implies that the physical signal is easily accessible to eavesdroppers and hackers. Wireless security is a challenging problem, even more so than wired security in many respects [55, 9, 90, 38, 73], that must be addressed by mobile appliances.

- Unlike desktop computers, which operate in physically secure environments and have fixed or limited "points of access", mobile appliances are free to operate in far more hostile environments (due to the potentially unlimited points of access), and over a wide range of bearer technologies such as cellular networks (*e.g.*, GSM / GPRS), wireless local area networks (*e.g.*, 802.11a/b), and personal area networks (*e.g.*, Bluetooth).

- Due to their small form factors, mobile appliances are quite vulnerable to theft, loss, and corruptibility. Security solutions for mobile appliances must, therefore, provide for security under these challenging scenarios.

- Constraints on cost and weight, and the need to operate mobile appliances off batteries, imply that they are quite constrained in their processing capabilities and energy supplies. The processing and energy overhead required to provide sufficient security can be significant, and overwhelm the modest capabilities of mobile appliances [73, 15].

The challenges of securing mobile appliances can be adequately addressed only through measures that span virtually every aspect of their design — hardware circuits and micro-architecture, system architecture, system and application software, and design methodologies. This paper describes the challenges that security poses to mobile appliance designers, and briefly surveys emerging technologies that can be used to address them. Despite significant recent interest and notable innovations in this area, many challenges remain that will require further attention and awareness of security among hardware, software, and system designers.

The rest of this paper is organized as follows. Section 2 highlights the various security concerns in a mobile appliance. Section 3 translates these concerns into system design challenges. Section 4 focuses on the various HW/SW solutions that have been proposed to enhance the security of mobile appliances. Section 5 concludes and outlines directions for further work in this area.

2. Security Concerns in Mobile Appliances

The role of security mechanisms is to ensure the privacy and integrity of data, and the authenticity of parties involved in a transaction. In addition, it is also

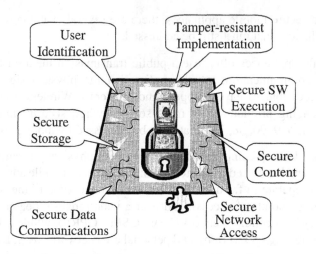

Figure 2.   Security concerns in a mobile appliance.

desirable to provide functionality such as non-repudiation, copy protection, preventing denial-of-service attacks, filtering of viruses and malicious code, and in some cases, anonymous communication [84, 77].

Figure 2 illustrates some of the major security concerns from the perspective of a mobile appliance.

- *User identification* attempts to ensure that only authorized entities can use the appliance.

- *Secure storage* addresses the security of sensitive information such as passwords, PINs, keys, certificates, *etc.*, that may reside in the secondary storage (*e.g.*, flash memory) of the mobile appliance.

- A *secure software execution environment* is necessary to ensure that attacks from malicious software such as viruses and trojan horses are prevented or do not compromise the security of the system.

- A *tamper-resistant system implementation* is required to ensure security of the underlying implementation from various physical and electrical attacks.

- *Secure network access* ensures that only authorized devices can connect to a network or service.

- *Secure data communications* considers the privacy and integrity of data communicated to and obtained from the mobile appliance.

- *Content security* refers to the problem of ensuring that any content that is downloaded or stored in the appliance is used in accordance with the terms set forth by the content provider (*e.g.*, read only, no copying, *etc.*).

We illustrate mobile appliance security concerns through the example of secure data communications. Wireless data communications can be secured by employing security protocols that are added to various layers of the protocol stack, or within the application itself. Security protocols utilize cryptographic algorithms (asymmetric or public-key ciphers, symmetric ciphers, hashing functions, *etc.*) as building blocks in a suitable manner to achieve the desired objectives (peer authentication, privacy, data integrity, *etc.*). In the wired Internet, the most popular approach is to use security protocols at the network or IP layer (IPSec), and at the transport or TCP layer (SSL/TLS) [84, 77].

In the wireless world, however, the range of security protocols is much broader. Different security protocols have been developed and employed in cellular technologies such as CDPD [21] and GSM [31, 19], wireless local area network (WLAN) technologies such as IEEE 802.11 [39], and wireless personal area network technologies such as Bluetooth [13]. Many of these protocols address only network access domain security, *i.e.*, securing the link between a wireless client and the access point, base station, or gateway. Several studies have shown that the level of security provided by most of the above security protocols is insufficient, and that they can be easily broken or compromised by serious hackers [33, 65, 88, 18, 7, 53, 44]. While some of these drawbacks are being addressed in newer wireless standards such as 3GPP [1, 12] and 802.11 enhancements [39], it is generally accepted that they need to be complemented through the use of security mechanisms at higher protocol layers such as WTLS [89], IPSec [84, 77] or SSL/TLS [84, 77].

The need to support security protocols and mechanisms, such as those described above, translates to various challenges in the design of the mobile appliance. The rest of the paper focuses specifically on these system design challenges and solutions that address them.

3.   Secure Mobile Appliances: System Design Challenges

In this section, we describe the various challenges and considerations involved in designing secure mobile appliances. Section 3.1 first discusses the diversity and evolutionary nature of security protocols and cryptographic algorithms, and the consequent need for flexibility in the security processing architecture of a mobile appliance. Section 3.2 analyzes the computational requirements of security processing, while Section 3.3 examines the impact of security processing on battery life. Section 3.4 addresses the important problem of securing the system implementation and the resultant need for building in attack resistance

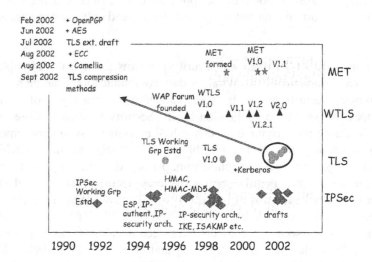

features. Finally, Section 3.5 concludes with a brief note on other factors which must be considered in the mobile appliance design cycle.

3.1 Flexibility

A fundamental requirement of a mobile appliance is the ability to cater to a wide variety of security protocol standards in order to facilitate inter-operability in different environments. For example, an appliance that needs to work in both 3G cellular and wireless LAN environments would need to execute security algorithms specified by 3GPP [1, 12] as well as the Wired Equivalent Privacy (WEP) algorithm specified by the 802.11 standard [39]. Additionally, a device is often required to support distinct security process-ing standards at different layers of the network protocol stack. For example, a wireless LAN enabled PDA that supports secure web browsing may need to execute both WEP (Link Layer) and SSL (Transport Layer), while the same PDA, if required to connect to a virtual private network (VPN), may also need to support IPSec (Network Layer).

Complicating the above picture is the fact that any security protocol stan-dard typically allows for a wide range of cryptographic algorithms. To illus-trate this scenario, let us consider the SSL protocol [82], which supports the use of different ciphers for its operations (authenticating the server and client, transmitting certificates, establishing session keys, *etc.*). For key exchange, cryptographic algorithms such as RSA and KEA are possible choices. For symmetric encryption, an RSA key exchange based SSL cipher suite would need to support 3-DES, RC4, RC2 or DES, along with the appropriate message authentication algorithm (SHA-1 or MD5). Since the mobile appliance may have to communicate with a server/client that uses any specific combination of cipher suite and key exchange algorithm, it is desirable to support all the allowed combinations so as to inter-operate with the widest possible range of peers.

Finally, security protocols are not only diverse, but are also continuously evolving over time. This has been and is still witnessed in the wired domain, wherein, protocol standards are revised to enable new security services, add new cryptographic algorithms or drop weaker ciphers. Figure 3 tracks the evolution of popular security protocols in the wired domain (IPSec [43] and TLS [85]). We can see that even a well-established protocol such as TLS is subject to constant modifications (*e.g.*, in June 2002, TLS was revised to accommodate the new symmetric encryption standard, AES).

The evolutionary trend is much more pronounced today in the wireless domain, where security protocols can be termed to be still in their infancy. Figure 3 also outlines the evolution of the wireless security protocols, WTLS [62] and MET [54]. Many of the security protocols used in the wireless domain are adaptations of the wired security protocols. For example, WTLS bears a close resemblance to the SSL/TLS standards. However, it is possible that future security protocols would be specifically tailored from scratch for the wireless environment. This presents a formidable challenge to the design of a security processing architecture, since flexibility and ease-of-adaptation to new standards become equally important design considerations as traditional objectives such as power, performance, *etc.*

3.2 Processing Requirements

The computational power available in a mobile appliance is significantly limited compared to the processing capabilities of a desktop computer. To illustrate the difference, compare the MIPS ratings of a 2.6 GHz Pentium 4 processor powered desktop and a state-of-the-art PDA featuring the Intel StrongARM 1100 processor. While the former is capable of delivering roughly 2890 MIPS, the latter can supply only 235 MIPS at its fastest clock speed (206MHz) [41]. The above scenario actually represents the higher end of the embedded processor spectrum. At the other end of this spectrum, we have processors such as the Motorola 68EC000 DragonBall processor used in Palm OS products rated at approximately 2.7 MIPS [29], while the ARM7/ARM9 CPU used in cell phones typically deliver 15 to 20 MIPS running at speeds of 30 to 40 MHz.

While power dissipation and size requirements of mobile appliances continue to restrict the processor architectures and, hence, their MIPS ratings, the level of security desired in data communicated by the mobile appliance remains unchanged or even increases. As a consequence, the computational

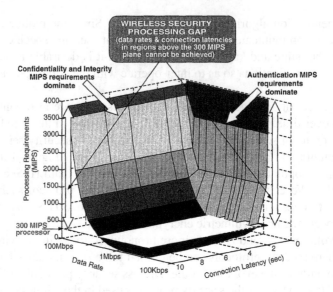

Figure 4. The wireless security processing gap (Source: [73]).

requirements of standard security protocols tend to be significantly higher than the embedded processor capabilities [73, 15]. Data presented in [73] reveal that the total processing requirements for software implementations of a security protocol executing on an embedded processor is around 651.3 MIPS, when the protocol uses 3DES for encryption/decryption and SHA for message authentication, at 10 Mbps (current and emerging data rates for wireless LAN technologies are in the range of 2-60 Mbps).

A similar trend has also been observed for RSA based connection set-ups performed in the client/server handshake phase of the SSL protocol. A 235 MIPS embedded processor running 1024-bit RSA can be used to establish connections at latencies of $1sec$, but not at $0.1sec$. Thus, there exists a clear mismatch between the security processing requirements and the available processor capabilities, even if the processor is assumed to be completely dedicated to security processing. In other words, this mismatch is likely to be worse in reality, since the processor is typically burdened by a workload that also includes application software, network protocol and operating system execution.

The effective computational requirements of a typical security protocol that performs RSA based connection set-up, 3DES-based data encryption and SHA-based integrity are shown in Figure 4 for various combinations of connection latencies and data rates. The processing capabilities of an embedded processor can be represented as a plane in the 3-dimensional space (see, for example, the 300 MIPS plane). Clearly, the processing requirements above the plane (and, hence, the corresponding combinations of connection latencies

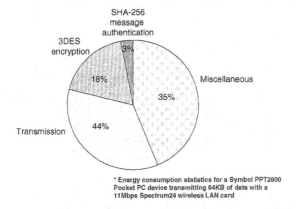

* Energy consumption statistics for a Symbol PPT2800
Pocket PC device transmitting 64KB of data with a
11Mbps Spectrum24 wireless LAN card

Figure 5.   The energy consumption profile of a sample secure wireless session (Source:[46]).

and data rates) can not be supplied by the embedded processor, leading to a *security processing gap*. While embedded processor performance can be expected to increase due to improvements in fabrication technologies and innovations in processor architectures (please refer to Section 4.2 for security-specific hardware customizations), the increase in data rates (due to advances in wireless communication technologies), and the use of stronger cryptographic algorithms (to stay beyond the extending reach of malicious entities) threaten to further widen the security processing gap.

3.3   Battery Life

The computational requirements of security protocols stemming from the inherent complexity of cryptographic algorithms suggest that the energy consumption of these algorithms can be very high. For battery powered mobile appliances, the energy drawn from the battery directly impacts the system's battery life, and, consequently, the duration and extent of its mobility, and its overall utility. To illustrate the impact of security processing on battery life, consider the energy consumption characteristics of a wireless handheld (Symbol PPT2800 PocketPC) conducting a secure wireless session that employs 3DES for bulk data encryption and SHA for message authentication. When the handheld is securely transmitting 64 KB of data, Figure 5 (source: [46]) shows that a considerable part (nearly 21%) of the overall energy consumption is spent on security processing.

The energy overheads of security processing clearly imply that the designer must clearly decide on the security services (privacy, message integrity, authentication *etc.*) that the mobile appliance must support. Correspondingly, he/she must select the "best" cryptographic algorithm used to perform a given security service as well as the determine the "right" settings for various cipher

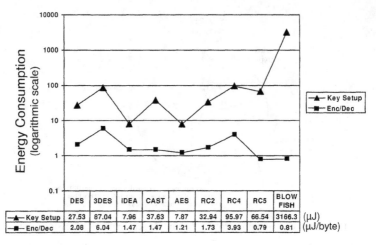

	DES	3DES	IDEA	CAST	AES	RC2	RC4	RC5	BLOW FISH	
▲ Key Setup	27.53	87.04	7.96	37.63	7.87	32.94	95.97	66.54	3166.3	(µJ)
■ Enc/Dec	2.08	6.04	1.47	1.47	1.21	1.73	3.93	0.79	0.81	(µJ/byte)

Figure 6.   Energy consumption data for various symmetric ciphers (Source:[67]).

parameters (key size, cipher mode, *etc.*) so as to minimize energy consumption, while achieving satisfactory levels of security.

Figure 6 (source: [67]), for example, shows significant variations in the energy consumption profile of various symmetric ciphers (measured on an iPAQ 3870 PDA running the OpenSSL crytographic suite [63]). Energy numbers for the key setup phase and energy-per-byte numbers for encryption/decryption phases are shown for each cipher. The results are reported for one specific mode of each block cipher - ECB or electronic code book, where a given plain-text block always encrypts to the same cipher-text block for the same key. The only exception is RC4, which is a stream cipher. Taking into account both the key setup and encryption/decryption costs, we can see from Figure 6 that AES has the least energy cost and BLOWFISH the greatest. The large cost of BLOWFISH is primarily due to its very high key setup cost, since the expanded key in BLOWFISH consists of sub-keys totaling 4168 bytes, which delivers very robust security. The per-byte cost of BLOWFISH encryption/decryption is quite small. In the case of sufficiently large data transactions, one would expect the cost of key setup to be amortized by the low encryption cost. It is interesting to note that the energy costs of IDEA for both encryption/decryption and key-setup compare favorably to those of AES. However, the crypt-analytical strength of AES is superior to that of IDEA, making the former an attractive option. A detailed energy analysis of other cryptographic algorithms and the SSL protocol can be found in [67]. The study also illustrates that significant "energy versus security" trade-offs can be explored by identifying and varying cipher parameters in a cryptographic algorithm.

While the above examples illustrate that the energy requirements for security must be reduced, improvements on the supply side (battery) can also benefit the so-called battery gap. It must be noted here that there has only been a slow growth (5-8% per year) in the battery capacities [50]. However, recent successes with alternative technologies such as fuel cells [78, 66] promise to significantly lengthen the battery lifetimes of mobile appliances.

3.4 Attacks

Most security protocols and mechanisms address security of a mobile appliance without regard to the specifics of the implementation. Theoretical analyses of the strength of cryptographic algorithms assume that malicious entities do not have access to the implementation (classical cryptanalysis). Here, a cryptographic primitive is viewed as an abstract mathematical object, that is, a mapping of some inputs into some outputs parameterized by a secret value, called the key. An alternative view of the cryptographic primitive comes from its implementation. Here, the cryptographic primitive is embodied in a hardware circuit or as a program that will run on a given embedded processor, and will thus present very specific characteristics. Such a view implies that security protocols and cryptographic algorithms can be broken by observing properties of the implementation (for example, "side-channel information", such as timing, power, behavior in the presence of faults, *etc.*). Sensitive data can also be compromised, while it is being communicated between various components of the system through the on-chip communication architecture, or, even when simply stored in the mobile appliance (in secondary storage like Flash memory, main memory, cache, or even CPU registers). Thus, secure design of the the HW/SW system architecture becomes as important as the theoertical strength of the cryptographic primitives employed.

The cost of not addressing various aspects of security in any electronic system can be very high, as illustrated by data from several sources including Computer Security Institute [23] and Counterpane Internet Security, Inc. [25]. While various threats to a mobile appliance are possible, it is important to safeguard the appliance against attacks that are easily deployable, require low-cost infra-structure to launch and can potentially inflict costly losses. Therefore, it is critical to have a clear understanding of the different ways in which a mobile appliance can be "attacked". Figure 7 lists the various categories of techniques used to attack a mobile appliance. At the top-level, mobile appliance attacks are classified into two broad categories: *physical and side-channel attacks*, and, *software attacks*.

3.4.1 Physical and Side-Channel Attacks

Physical and side-channel attacks refer to attacks that exploit the system implementation and/or identifying properties of the implementation. It is not surprising that the first target of these attacks [5, 6, 49, 72] are mobile devices such as

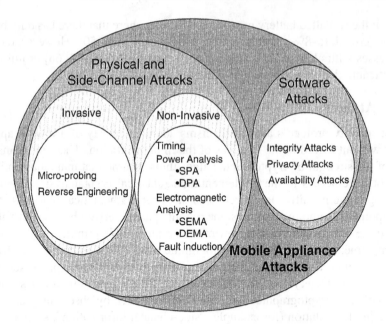

Figure 7.   Taxonomy of mobile appliance security attacks.

smart cards. For concreteness, the discussion here will be put in that context, although most of it applies to other (cryptographic) devices as well. Physical and side-channel attacks are generally classified into *invasive* and *non-invasive* attacks.

- Invasive attacks involve getting access to the appliance to observe, manipulate and interfere with the system internals. Since invasive attacks typically require relatively expensive infra-structure, they are much harder to deploy. Examples of invasive attacks include the following.

  Micro-probing: This technique uses a micro-probing workstation to remove the passivation layer (external layer) of a chip. Subsequently, an attacker can use probes to establish direct contact with the integrated circuit (usually the system bus). An illustration of such an attack to directly eavesdrop on the (secret) data during the execution of sensitive algorithms can be found in [37].

  Reverse engineering: Several attack techniques target particular parts of a device (buses, memory, CPU, co-processor, sensors, *etc.*). Such attacks typically require an initial invasive effort to access the layout of the chip, in order to locate and distinguish its internals. Image processing and form recognition systems are used to retrieve the hardware structure from simple microscope pictures (*e.g.*, optical microscope with a CCD camera). Recent techniques such as [76] illuminate the unplugged chip with a focused laser spot and probe the variation of current between power and ground. Shining light on a transistor makes it generate a micro-current depending on its state. Thus, this technique can be used to reveal the mapping of the integrated circuit as well as the (private or secret) data stored.

- Non-invasive attacks, as the name indicates, do not require the device to be opened. While these attacks require knowledge of the system, they tend to be cheap and scalable (compared to invasive attacks). There are many forms of non-invasive attacks.

Timing attacks: The timing attack [47, 28] exploits the observation that the computations performed in some of the cryptographic algorithms often take different amounts of time on different inputs. For example, the use of Chinese Remainder Theorem (CRT) in the RSA public-key cryptosystem makes it vulnerable to timing attacks.

Power analysis: These attacks [48] record and analyze the power traces leaked by a device. Two kinds of power analysis attacks are typically used: *Simple Power Analysis* (SPA) and *Differential Power Analysis* (DPA). SPA techniques perform a simple inspection of the power consumption traces and rely on the identification of the Hamming weight of the data during encryption/decryption. DPA methods allow sensitive information to be uncovered by performing a statistical analysis (difference in magnitude of average) of the power consumption records. Differential analysis techniques are very powerful since analysis is architecture-independent and relies on the fact that noise effects tend to cancel out.

Electromagmetic analysis (EMA): EMA attacks involve analyzing the electromagnetic radiation around the device to recover sensitive information [87]. More recent demonstrations of these attacks include [70, 34], wherein the equivalents of SPA and DPA were introduced. These techniques were termed SEMA and DEMA (*Simple* and *Differential ElectroMagnetic Analysis*), respectively. Using such techniques, it becomes possible to monitor only the signal from one particular location of the device (*e.g.*, cryptoprocessor, buses, oscillators, *etc.*), without being significantly affected by the noise produced from the rest of the chip.

Fault induction: These techniques manipulate the environmental conditions of the system (voltage, clock, temperature, radiation, light, eddy current, *etc.*) to generate faults and to observe the related behavior [16, 11]. Most of these attacks target data being computed by a sensitive algorithm, but some attempt to corrupt the data directly in the memory. While there are many ways of producing a fault in a mobile device, these attacks can be termed "semi-invasive" as knowledge of the architecture is often required.

Such attacks can be engineered by simply illuminating a transistor with a laser beam, which causes it to conduct. In this way, a transient corrupted bit can be generated [6, 79]. Glitch attacks generate a malfunction by changing the working frequency during computation, thereby, causing some bits to assume the wrong value. The notion of using an error induced during a computation to guess the secret key has been practically observed in implementations of the RSA that use the CRT [6, 45, 17]. Fault induction techniques can lead to both transient and non-reversible faults [6, 71].

3.4.2   Software Attacks

Software attacks are based on malicious software being run on the mobile appliance, that exploits weaknesses in security schemes and the system implementation. The likelihood of software attacks tends to be high in systems such as mobile terminals, where application software is frequently downloaded from the Internet. The down-loaded software may originate from a non-trusted source and, hence, can be be used to implement attacks. Compared to physical attacks, software attacks typically require infrastructure that is substantially cheaper and easily available to most hackers, making them a serious immediate challenge to secure system design.

Popular examples of software attacks include viruses and trojan-horse applications that exploit OS weaknesses and procure access to the system internals. There are many categories of software attacks. While *integrity attacks* can manipulate sensitive data or processes, *privacy attacks* lead to disclosure of private or confidential information. *Availability attacks* (also refered to as denial of service attacks), on the other hand, can prevent the secure functioning of the system by denying access to system resources.

3.5   Other Factors

In an ambient intelligence setup, a challenge stems from most of today's security mechanisms relying on the authentication of the client device. The lack of end-user authentication is thus a weak link in the overall chain of trust. Biometric technologies such as finger print recognition and voice recognition are, therefore, needed for enabling a secure wireless environment, while maintaining ease-of-use for end-users.

Another area of major interest in the industry is content security or digital rights management, where the content providers attempt to ensure that the application content remains protected against any unauthorized use. Techno-

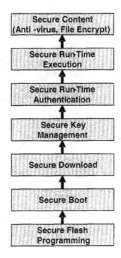

Figure 8. A layered hierarchical approach to security.

logical questions remain on determining the correct framework for protecting the content providers, while ensuring minimum hindrance to the end-user. Several on-going efforts in this regard have come from consortiums such as OpenIPMP [64], MPEG [59], ISMA [42], MOSES [60] *etc.*

4. Secure HW/SW Platform Architectures

In this section, we describe approaches to design secure mobile appliance architectures that address some of the challenges described in Section 3.

4.1 Elements of a Secure Mobile Appliance Architecture

Security challenges are usually complex even when viewed in a limited perspective of a mobile appliance. Thus, from a systems perspective, it is imperative to take a hierarchical approach, where each layer of security provides a foundation for the one above it (Figure 8).

The base architecture must be flexible and scalable to meet the needs of each stratum in the marketplace. Flexible and scalable base architectures simplify the development and deployment of new applications and services and the associated security requirements. Figure 9 shows an example of such a base architecture. At the core is a powerful crypto engine surrounded by firmware and an application-programming interface (API) which speeds the integration of various security applications and peripherals.

Combining hardware and software crypto components plays a significant role in providing a strong crypto foundation that meets basic security requirements such as authentication, confidentiality, integrity, and non-repudiation.

Figure 9.   A modular base architecture for secure mobile appliances.

The foundation of secure crypto operations includes true random number generation, which may be provided for with a HW-based random number generator. Crypto HW accelerators are one method to provide significant performance and power efficiencies to critically used algorithms such as DES/3DES, AES, SHA1/MD5, and public key operations (RSA/DH) necessary in the mobile environment. A low-power DSP in a dual-core processor, such as TI's OMAP1510 processor [61], accelerates critical and performance intensive crypto operations, freeing up much-needed headroom on the main applications processor.

Additionally, HW components such as secure RAM and secure ROM in conjunction with HW-based key storage and appropriate firmware can enable an optimized 'secure execution' environment where only trusted code can execute. A secure execution mode can be used for critical security operations such as key storage/management and run-time security, to provide a strong security foundation for applications and services.

4.2   Security Processing Architectures

Security processing refers to computations that need to be performed specifically for the purpose of security. For example, in a secure wireless data transaction, security processing includes the execution of any security protocols that are utilized at all layers of the protocol stack (and the cryptographic algorithms employed therein). As demonstrated in Section 3, the computational requirements of security processing place a significant burden on the embedded processors used in mobile appliances, and can lead to significant degradations in battery life.

Recognizing these issues, various technologies have been developed in order to enable efficient security processing in mobile appliances. Figure 10

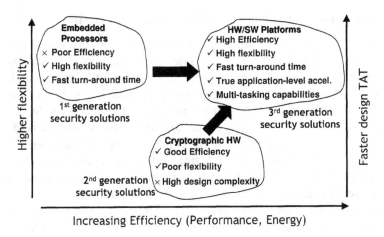

Figure 10.   Taxonomy of security processing architectures.

shows three generations of security processing architectures, and compares them in terms of performance and energy consumption efficiencies, flexibility as well as design turn-around times. The first generation solutions perform security processing as software support running on the processors embedded in the mobile appliance. While good flexibility and fast design times are possible with software-based solutions, these solutions are not efficient in terms of their performance and energy consumption characteristics. Custom cryptographic hardware, on the other hand, can be designed to deliver high performance and consume lower energy. These constitute the second generation solutions. Since these solutions sacrifice the flexibility and design times of the first generation solutions, third generation solutions have been proposed that can capture the benefits of both the first and second generation solutions. These solutions have a programmable component in their architecture. Examples include enhancements of embedded processors for security and programmable security protocol engines. We now provide a brief overview of various second and third generation security processing architectures.

Cryptographic hardware accelerators

Highest levels of efficiency in processing are often obtained through custom hardware implementations. Since cryptographic (asymmetric, symmetric, hash) algorithms form a significant portion of security processing workloads, various companies offer custom hardware implementations of these cryptographic algorithms suitable for mobile appliances including smart cards and wireless handsets [26, 75]. Several vendors also offer integrated microcontrollers that contain embedded processors, cryptographic accelerators, and other peripherals [80, 83].

Embedded processor enhancements for security processing

There have been several attempts to improve the security processing capabilities of general purpose processors. Since most microprocessors today are word-oriented, researchers have targeted accelerating bit-level arithmetic operations such as the permutations performed in DES/3DES. Multimedia instruction set architecture (ISA) extensions such as those in PA-RISC's Max-2 [51] or IA-64 [40] already incorporate instructions for permutations of 8-bit or larger sub-words. For arbitrary bit-level permutations, efficient instructions have been recently proposed [52]. Instruction set extensions have also been proposed for other operations such as substitutions, rotates and modular arithmetic present in different private-key algorithms [20].

Many such extensions have already been applied to embedded processors used in the wireless handset domain. For example, the SmartMIPS [81] cryptographic enhancements extend the basic 32-bit MIPS ISA to speed up security processing. Similar features are also found in the ARM SecureCore family [8]. The security processing capabilities of SecureCore processors can also be further extended by adding custom-designed cryptographic processing units through a co-processor interface. This is useful for delivering high performance, without having to re-design the basic processor core.

Programmable security protocol engines

While cryptographic accelerators alleviate the performance and energy bottlenecks of security processing to some extent, achieving very high data rates or extreme energy efficiency requires a holistic view of the entire security processing workload. In addition to cryptographic algorithms, security protocols often contain a significant protocol processing component, including packet header/trailer parsing, classification *etc.* Security protocol engines (*e.g.*, the IPSec packet engine from Safenet Inc. [75]) accelerate all or most of the functionality present in a security protocol, resulting in higher efficiency than cryptographic accelerators.

As mentioned in Section 3, security standards for mobile appliances are in their infancy, and are expected to evolve significantly [62, 54, 56, 57, 24]. Hence, it is desirable to provide sufficient flexibility in security processing architectures so that they can be re-used, or adapted with minimal effort, to support new standards or enhancements of existing standards. Programmable security protocol engines, such as the MOSES platform developed at NEC [74, 69, 68] combine the benefits of flexibility and efficiency for security processing.

4.3   Design Techniques for Attack Resistance

In this section, we provide a brief overview of various countermeasures that can provide resistance against mobile appliance attacks.

One way to protect a device from EM-analysis is to break the principle of locality [70]. The idea is to spread out the design of main blocks (crytopro-cessor, ROM, RAM, buses, *etc.*) over the whole surface of the chip. It is an important subject of research for the next generation of secure chips. Another related idea is to spread out computations and data in a random manner so that it becomes difficult to predict where exactly the data is located and which operations are related.

Differential (power or electromagnetic) analysis cannot be successful if data, computed (loaded or stored) during the algorithm, are not easily corre-lated to a few key bits and the recorded trace. Data transformation and masking techniques such as the duplication method [36], the masking method [35] and the approaches of [3, 4] have been shown to be effective countermeasures.

Attacks based on DPA rely heavily on the correlation between power con-sumption traces average, the key bits and the inputs/outputs. Exploiting this correlation requires the identification of a synchronizing event that arises at a certain point of the computation. For a naive implementation, critical events always occur at the same time in the trace. So, desynchronization is often used to limit the effectiveness of DPA. Infineon Technologies has deployed such techniques by introducing some fake clock cycles during a cryptographic computation. Such techniques are also described in [49]. More recently, ST Microelectronics proposed a desynchronization scheme based on a weak jitter that is completely controlled. A way to defeat such counter-measures lies in classic resynchronisation techniques based on jitter killer, phase locked loop (PLL) or Costas loop (used for satellite signal tracking).

Another way to defeat DPA attacks is to increase the noise so as to drown the signal. Some smartcard manufacturers have tried to take advantage of other blocks on the card (apart from the cryptographic hardware) to mix the power consumption pattern. The idea is to connect analog or even numeric systems to some blocks so that they could randomly modify the current drawn. A practical way to obtain counter-measures that allow a sufficient security level involves the use of advanced noise makers. Another idea proposed in the European G3card project [2] involves exploring the use of asynchronous logic in smart-card processors to increase their attack resistance. Design techniques that use balanced logic (dual rail logic), self-timed circuits [58], *etc.*, have also been shown to result in significantly higher levels of physical security.

Innovations in other fields have also benefited resistance against attacks. For example, research in micro-electronics related to the reduction of power consumption have led to developments such as the Silicon-On-Insulator (SOI).

SOI allows for decreasing the current consumed by the processor, thereby re-ducing the radiated electromagnetic field [27].

Fault insertion counter measures are required to guarantee the integrity of the chip. For this reason, a designer may include some sensors (sensible to temperature, pressure, light, radiation, *etc.*) to protect the chip against any unusual change in the environment [76]. The chip can react by locking itself or by erasing its memory. In addition, some hardware and software protection schemes (*e.g.*, checking of program results [17]) are needed to prevent transient corruption of the circuit.

Building attack resistance especially into software [10, 14, 22] would neces-sitate one or more of the following measures: (i) finding a means to ascertain the operational correctness of protected code and data, before and at run-time, (ii) providing protection against trojan horse applications trying to steal data (*e.g.*, cryptographic keys) from a security application that is run on behalf of the user, and (iii) protecting against probing (looking at the memory used by secure applications) and software reverse engineering (de-compilation, flow analysis, profiling *etc.*).

5.   Conclusions and Directions for Future Work

Security is critical to enabling a wide range of applications involving mobile appliances. While security has been addressed in the context of traditional computing systems and the wired Internet, mobile appliances usher in many new challenges. This paper highlighted the problems faced by designers of mobile appliances, and outlined recent technological developments and inno-vations to address them. Several issues, however, remain open at the intersec-tion of security and embedded system design.

The interplay of flexibility, cost, performance and power consumption makes choosing the "right" security HW/SW platform a complicated process. In many design scenarios today, it becomes hard to evaluate the effectiveness of a given solution due to the absence of complete system-level simulation and performance evaluation tools. The selection process can be facilitated by the development of enabling design automation technologies.

In addition to the metrics mentioned above, cost and design turn-around times play a crucial role in deciding the underlying hardware fabric and hence, the security architecture. Developments in the semi-conductor fabrication in-dustry can alter the choice of security hardware used. For example, the increas-ing success of technologies such as field programmable logic devices (PLDs) in meeting good performance and lower design turn around times is prompting designers to examine (or re-examine) their usage as the underlying HW fab-ric. Consequently, their impact on performance and power consumption of any cryptographic architecture need to be carefully studied.

Further efforts are also needed in designing the cryptographic algorithms and security protocols that must be supported in these devices (more so in an ambient intelligence setup, where, devices need to support only specific security services). Scalability in algorithms and protocols makes it easier for a security scheme to be effective in a wide range of devices — from low- to medium- to high- end devices.

The attacks described in this chapter are applicable to a wide range of electronic systems. A clear cost/risk analysis is necessary to determine the levels of security that a device must support. Since improvements in attack sophistication continue to happen, development of countermeasures remains a challenging and on-going exercise. It is also important to note that countermeasures applicable to smartcards may not be able to applicable to other embedded systems such as PDAs or Smartphones. Thus, system-specific attack-resistance measures are essential.

Security will thus increasingly impact various aspects of the system design process, including hardware circuits and micro-architecture, software, system architecture, and design methodologies.

Acknowledgements

The authors would like to thank Sylvie Baudine and Eric Peeters for their significant contributions.

References

[1] *3GPP Draft Technical Specification 33.102, 3G Security Architecture.* http://www.3gpp.org.

[2] *3rd Generation Smart Card Project (G3Card).* http:/www.g3card.org.

[3] M.-L. Akkar and C. Giraud, "An implementation of DES and AES, secure against some attacks," *Cryptographic Hardware and Embedded Systems*, pp. 309–318, 2001.

[4] M.-L. Akkar and L. Goubin, "A generic protection against high order differential power analysis," in *Proc. ACM Symposium Foundations of Software Engineering*, pp. 201–216, Sept. 2003.

[5] R. Anderson and M. Kuhn, "Tamper resistance - a cautionary note," 1996.

[6] R. Anderson and M. Kuhn, "Low cost attacks on tamper resistant devices," in *IWSP: Intl. Wkshp. on Security Protocols, Lecture Notes on Computer Science*, pp. 125–136, 1997.

[7] W. A. Arbaugh, *An inductive chosen plaintext attack against WEP/WEP2*. IEEE document 802.11-01/230, May 2001. http://grouper.ieee.org/groups/802/11/Documents/

[8] *ARM SecureCore*. http://www.arm.com.

[9] P. Ashley, H. Hinton, and M. Vandenwauver, "Wired versus wireless security - The Internet, WAP and iMode for e-commerce," in *Proc. 17th Annual Computer Security Applications Conf.*, Dec. 2001.

[10] D. Aucsmith, "Tamper resistant software: An implementation," *Information Hiding, Springer Lecture Notes in Computer Science*, vol. 1174, pp. 317–333, 1986.

[11] E. Biham and A. Shamir, "Differential fault analysis of secret key cryptosystems," *Lecture Notes in Computer Science*, vol. 1294, pp. 513–525, 1997.

[12] C. W. Blanchard, "Wireless security," *BT Technology Journal*, vol. 19, pp. 67–75, July 2001. http://www.bt.com/bttj/

[13] *Bluetooth security white paper*. Bluetooth SIG Security Expert Group, Apr. 2002. http://www.bluetooth.com/

[14] M. Blum and S. Kannan, "Designing programs that check their work," in *Proc. ACM Symposium on Theory of Computing*, pp. 86–97, 1989.

[15] D. Boneh and N. Daswani, "Experimenting with electronic commerce on the PalmPilot," in *Proc. Financial Cryptography*, pp. 1–16, Feb. 1999.

[16] D. Boneh, R. DeMillo, and R. Lipton, "On the importance of checking cryptographic protocols for faults," *Springer-Verlag Lecture Notes in Computer Science, Proceedings of Eurocrypt'97*, vol. 1233, pp. 37–51, 1997.

[17] D. Boneh, R. DeMillo, and R. Lipton, "On the importance of eliminating errors in cryptographic computations," *Cryptology*, vol. 14, no. 2, pp. 158–172, 1999.

[18] N. Borisov, I. Goldberg, and D. Wagner, "Intercepting mobile communications: The insecurity of 802.11," in *Proc. ACM Int. Conf. Mobile Computing and Networking*, pp. 180–189, July 2001.

[19] C. Brookson, "GSM security: A description of the reasons for security and the techniques," in *Proc. IEE Colloqium on Security and Cryptography Applications to Radio Systems*, pp. 2/1–2/4, June 1994.

[20] J. Burke, J. McDonald, and T. Austin, "Architectural support for fast symmetric-key cryptography," in *Proc. Intl. Conf. ASPLOS*, pp. 178–189, Nov. 2000.

[21] *Cellular Digital Packet Data System Specification, Release 1.1*. CDPD Forum, Jan. 1995.

[22] C. S. Collberg and C. Thomborson, "Watermarking, tamper-proofing, and obfuscation - tools for software protection," *IEEE Transactions on Software Engineering*, vol. 28, pp. 735–746, August 2002.

[23] Computer Security Institute, *2002 Computer Crime and Security Survey*. http://www.gocsi.com/press/20020407.html.

[24] *Consortium for efficient embedded security*. http://www.ceesstandards.org/.

[25] *Counterpane Internet Security, Inc.* http://www.counterpane.com.

[26] *Cryptocell^{TM}*. Discretix Technologies Ltd. http://www.discretix.com.

[27] N. de Mevergnies, D. Flandre, and J.-J. Quisquater, "Feasibility of smart cards in silicon-On-Insulator (SOI) technology," in *Proc. USENIX Wkshp on Smartcard Technology*, pp. 1–7, May 1999.

[28] J.-F. Dhem and F. Koeune and P.-A. Leroux and P. Mestre and J.-J. Quisquater and J.-L. Willems, "A practical implementation of the timing attack," in *Proc. Third Working Conf. Smart Card Research and Advanced Applications*, pp. 167–182, Sept. 1998.

[29] *DragonBall processor family*. http://www.motorola.com.

[30] *ePaynews - Mobile Commerce Statistics*. http://www.epaynews.com/statistics/mcommstats.html.

[31] *European Telecommunication Standard GSM 02.09*. Digital Cellular Telecommunications System (Phase 2+): Security Aspects.

[32] P. Flavin, *Who needs a credit card when you have a mobile?*
http://www.btignitesolutions.com/insights/visionaries/
flavin_mobile.htm.

[33] Y. Frankel, A. Herzberg, P. A. Karger, H. Krawczyk, C. A. Kunzinger, and M. Yung, "Security issues in a CDPD wireless network," *IEEE Personal Communications*, vol. 2, pp. 16–27, August 1995.

[34] K. Gandolfi, C. Mourtel, and F. Olivier, "Electromagnetic analysis: Concrete results," *Cryptographic Hardware and Embedded Systems*, pp. 251–261, 2001.

[35] L. Goubin, "A sound method for switching between Boolean and arithmetic masking," *Cryptographic Hardware and Embedded Systems*, pp. 3–15, 2001.

[36] L. Goubin and J. Patarin, "DES and differential power analysis," *Cryptographic Hardware and Embedded Systems*, pp. 158–172, 1999.

[37] H. Handschuh, P. Paillier, and J. Stern, "Probing attacks on tamper-resistant devices," *Cryptographic Hardware and Embedded Systems*, pp. 303–315, 1999.

[38] S. Hattangady and C. Davis, *Reducing the Security Threats to 2.5G and 3G Wireless Applications*. Texas Instruments Inc. http://focus.ti.com/pdfs/vf/wireless/securitywhitepaper.pdf.

[39] *IEEE 802.11 Wireless LAN Standards*. IEEE 802.11 Working Group
http://grouper.ieee.org/groups/802/11/.

[40] Intel Corp., *Enhancing Security Performance through IA-64 Architecture*. http://developer.intel.com/design/security/rsa2000/itanium.pdf, 2000.

[41] *Intel StrongARM SA-1110 Microprocessor Brief DataSheet*.
http://www.intel.com/.

[42] *Internet Streaming Media Alliance*. http://www.isma.tv/home.

[43] *IPSec Working Group*.
http://www.ietf.org/html.charters/ipsec-charter.html.

[44] ISAAC group, U. C. Berkeley, *GSM cloning*.
http://www.isaac.cs.berkeley.edu/isaac/gsm.html.

[45] M. Joye, A. K. Lenstra, and J.-J. Quisquater, "Chinese remaindering based cryptosystems in the presence of faults," *Cryptology*, vol. 12, no. 4, pp. 241–245, 1999.

[46] R. Karri and P. Mishra, "Minimizing Energy Consumption of Secure Wireless Session with QOS constraints," in *Proc. Int. Conf. Communications*, pp. 2053–2057, 2002.

[47] P. C. Kocher, "Timing attacks on implementations of Diffie-Hellman, RSA, DSS, and other systems," *Advances in Cryptology – CRYPTO'96, Springer-Verlag Lecture Notes in Computer Science*, vol. 1109, pp. 104–113, 1996.

[48] P. Kocher, J. Jaffe, and B. Jun, "Differential Power Analysis," *Advances in Cryptology – CRYPTO'99, Springer-Verlag Lecture Notes in Computer Science*, vol. 1666, pp. 388–397, 1999.

[49] O. Kommerling and M. G. Kuhn, "Design principles for tamper-resistant smartcard processors," in *Proc. USENIX Wkshp. on Smartcard Technology (Smartcard '99)*, pp. 9–20, May 1999.

[50] K. Lahiri, A. Raghunathan, and S. Dey, "Battery-driven system design: A new frontier in low power design," in *Proc. Joint Asia and South Pacific Design Automation Conf. / Int. Conf. VLSI Design*, pp. 261–267, Jan. 2002.

[51] R. B. Lee, "Subword parallelism with Max-2," *IEEE Micro*, vol. 16, pp. 51–59, Aug. 1996.

[52] R. B. Lee, Z. Shi, and X. Yang, "Efficient permutations for fast software cryptography," *IEEE Micro*, vol. 21, pp. 56–69, Dec. 2001.

[53] A. Mehrotra and L. S. Golding, "Mobility and security management in the GSM system and some proposed future improvements," *Proceedings of the IEEE*, vol. 86, pp. 1480–1497, July 1998.

[54] *MeT PTD definition (version 1.1)*. Mobile Electronic Transactions Ltd., Feb. 2001. http://www.mobiletransaction.org/

[55] S. K. Miller, "Facing the Challenges of Wireless Security," *IEEE Computer*, vol. 34, pp. 46–48, July 2001.

[56] *Mobey Forum*. http://www.mobeyforum.org/.

[57] *Mobile Payment*. http://www.mobilepaymentforum.org/.

[58] S. Moore, R. Anderson, P. Cunningham, R. Mullins, and G. Taylor, "Improving smart card security using self-timed circuits," in *Proc. of Eighth Intl Symposium on Asynchronous Circuits and Systems*, pp. 193–200, Apr. 2002.

[59] *Moving Picture Experts Group*. http://mpeg.telecomitalialab.com.

[60] *MPEG Open Security for Embedded Systems (MOSES)*. http://www.crl.co.uk/projects/moses/.

[61] *OMAP Platform - Overview*. Texas Instruments Inc. http://www.ti.com/sc/omap.

[62] *Open Mobile Alliance*. http://www.wapforum.org/what/technical.htm.

[63] *Open SSL Project*. http://www.openssl.org.

[64] *OpenIPMP*. http://www.openipmp.org.

[65] S. Patel, "Weaknesses of North American wireless authentication protocol," *IEEE Personal Communications*, vol. 4, pp. 40–44, june 1997.

[66] *Poly Fuel, Inc.* http://www.polyfuel.com.

[67] N. Potlapally, S. Ravi, A. Raghunathan, and N. K. Jha, "Analyzing the energy consumption of security protocols," in *Proc. Int. Symp. Low Power Electronics & Design*, Aug. 2003.

[68] N. Potlapally, S. Ravi, A. Raghunathan, and G. Lakshminarayana, "Algorithm exploration for efficient public-key security processing on wireless handsets," in *Proc. Design, Automation, and Test in Europe (DATE) Designers Forum*, pp. 42–46, Mar. 2002.

[69] N. Potlapally, S. Ravi, A. Raghunathan, and G. Lakshminarayana, "Optimizing public-key encryption for wireless clients," in *Proc. IEEE Int. Conf. Communications*, pp. 1050–1056, May 2002.

[70] J. J. Quisquater and D. Samyde, "ElectroMagnetic Analysis (EMA): Measures and Counter-Measures for Smart Cards," *Lecture Notes in Computer Science (Smartcard Programming and Security)*, vol. 2140, pp. 200–210, 2001.

[71] J. J. Quisquater and D. Samyde, "Eddy current for magnetic analysis with active sensor," in *Proc. Esmart*, pp. 185–192, 2002.

[72] J. J. Quisquater and D. Samyde, "Side channel cryptanalysis," in *Proc. of the SECI*, pp. 179–184, 2002.

[73] S. Ravi, A. Raghunathan, and N. Potlapally, "Securing wireless data: System architecture challenges," in *Proc. Intl. Symp. System Synthesis*, pp. 195–200, October 2002.

[74] S. Ravi, A. Raghunathan, N. Potlapally, and M. Sankaradass, "System design methodologies for a wireless security processing platform," in *Proc. ACM/IEEE Design Automation Conf.*, pp. 777–782, June 2002.

[75] *Safenet EmbeddedIP^{TM}*. Safenet Inc. http://www.safenet-inc.com.

[76] D. Samyde, S. Skorobogatov, R. Anderson, and J.-J. Quisquater, "On a new way to read data from memory," in *Proc. First Intl. IEEE Security in Storage Wkshp*, pp. 65–69, Dec. 2002.

[77] B. Schneier, *Applied Cryptography: Protocols, Algorithms and Source Code in C*. John Wiley and Sons, 1996.

[78] *SFC Smart Fuel Cell AG*. http://www.smartfuelcell.com.

[79] S. P. Skorobogatov and R. J. Anderson, "Optical fault induction attacks," *Cryptographic Hardware and Embedded Systems*, pp. 2–12, 2002.

[80] *SLE 88 family*. Infineon Technologies http://www.infineon.com.

[81] *SmartMIPS*. http://www.mips.com.

[82] *SSL 3.0 Specification*. http://wp.netscape.com/eng/ssl3/.

[83] *ST19 smart card platform family*. STMicroelectronics Inc. http://www.st.com.

[84] W. Stallings, *Cryptography and Network Security: Principles and Practice*. Prentice Hall, 1998.

[85] *TLS Working Group*.
http://www.ietf.org/html.charters/tls-charter.html.

[86] U. S. Department of Commerce, *The Emerging Digital Economy II*, 1999. http://www.esa.doc.gov/508/esa/TheEmergingDigitalEconomyII.htm

[87] W. van Eck, "Electromagnetic radiation from video display units: an eavesdropping risk?," *Computers and Security*, vol. 4, no. 4, pp. 269–286, 1985.

[88] J. R. Walker, *Unsafe at any key size: An analysis of the WEP encapsulation*. IEEE document 802.11-00/362, Oct. 2000.
http://grouper.ieee.org/groups/802/11/Documents/

[89] *Wireless Application Protocol 2.0 - Technical White Paper*, Jan. 2002.
WAP Forum http://www.wapforum.org/

[90] *Wireless Security Basics*. Certicom
http://www.certicom.com/about/pr/wireless_basics.html.

[91] World Wide Web Consortium, *The World Wide Web Security FAQ*, 1998.
http://www.w3.org/Security/faq/www-security-faq.html

Part II

Developments

"What a shame," Kalten sighed. "So many good ideas have to be discarded simply because they won't work."

David Eddings, The Sapphire Rose, 1991

The Domestic Robot - A Friendly Cognitive System Takes Care of your Home

Fiora Pirri, Ivo Mentuccia and Sandro Storri

"La Sapienza" University of Rome, DIS, ALCOR group, Rome, Italy
{ pirri, ivomen } @dis.uniroma1.it, s.storri@email.it

Abstract   The system we describe in this paper is an indoor intelligent system, that is still under development. Our goal is to evolve the system with abilities suitable for a domestic robot, that has to cope with a continuously evolving ambient, the house. The interest of the system relies in its architecture that has been studied and designed for the purpose of integrating some specific cognitive behavior like perception, symbolic map construction and making decisions in a whole hierarchy of intercommunicating levels.

Keywords   autonomous robots, cognitive robotics architecture, perception and action in agents, embedded systems, ambient intelligence, household appliance, personal assistants

1.   Fiction and Reality

Do we really believe that intelligent robots in the near future will be part of our home life and, as other old and familiar devices like the refrigerator or the washing machine, that they would help us in the daily home maintenance? Do we really need an intelligent machine at home? We are surrounded by hundreds of pseudo intelligent devices, helping and entertaining us, but also needing assistance and knowledge and a huge amount of time for management. The key technology of ambient intelligence is to build an intelligent environment and everyday objects will be endowed with abilities like in the Merlin castle. How do we expect them to operate correctly every time without tedious training or updates and management? What if another device, a truly candid intelligent machine, would take care of the less intelligent (e.g. an active badge or a transponder) and the less autonomous (a curtain going up and down in the presence of sun or shade, but not able to detaching from the window and walking around) companions? Such a device would talk to them and listen to their problems and complaints, and be able to learn devices' updates and their

T. Basten et al. (eds.), Ambient Intelligence: Impact on Embedded System Design, 131-158.
© 2003 *Kluwer Academic Publishers. Printed in the Netherlands.*

management, quickly and silently. This device could be our personal robot, a *dom*estic ro*bot* or "domobot"[1].

For a long time now, at least since the 19th century, human beings have been dreaming about artificial machines able to work, play, think, and interact side by side with everyday people. From the Allan Poe "Maelzel's Chess Player" (the very first automaton), the science fiction literature is huge in this field endowing real science with terms like *robot*, from the Czech author Karel Capek, *Rossum's Universal Robot*, to the term *robotics* coined in [2, 3]. Science fiction has been transforming the vague imagination about robots in names and shapes like in the movie *2001 A Space Odyssey* , with "HAL" having full control over the entire spacecraft, interacting with astronauts, making decisions, and seeming to be aware of its own role and existence. Likewise in the *Star Wars* saga with the *R2D2* android and *C3PO* robot. *R2D2* could control and monitor every actuator and sensor in a building or on the space station. Far from the mind's eye of writers and movie makers the *real robotics*, the one everybody is taking benefits from today in a hidden fashion and that will spread amongst common people sooner than expected, originated around the middle of the last century. A starting point was the tele-operated arm, and it was the era of industrial robots and robots built for specific tasks to be repeated thousands times a day. Around 1969, soon after the Artificial Intelligence dawn [39, 37], Shakey [42], the first autonomous robot as we now regard it, saw the light of day.

Nowadays almost every robotic research center has a robot (at least a manipulator), and cognitive robotics laboratories are either building or buying mobile bases to work on, and to settle upon them as much intelligence as they can, tangling them with a wide set of sensors and actuators. Some researchers have gone into business, selling any kind of robots from expensive toys (Sony Aibo) to research platforms (iRobot), from experimental custom-made personal assistants (NEC with PaPeRo) to industrial vacuum-cleaners (e.g. the Auto VacC 6).

Intelligent stand-alone or multi-connected devices, able to exchange informations and instructions, are becoming reality too. Intelligent devices are meeting people's needs, providing comfort and convenience in several human activities. The upshot is the proliferation of a huge amount of complex devices both at home and in the social environment, mostly factories. Domotics (*Dom*estic Rob*otics*), is slowly increasing its impact on society and in a few years we will see an explosion of applications in this field: heating and ventilation, clothes care, safety and security, food preparation and conservation,

---

[1]We shall call the kind of research dealing with the automatization of a domestic ambient with the introduction of autonomous intelligent systems like cognitive robots, "domotics", from the contraction of the terms "domestic robotics".

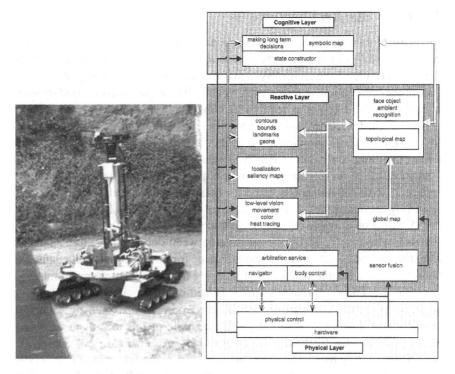

Figure 1.   ArmHandX (on the left) and (on the right) the Internal Data Flow (Dark-Gray=Real Time Events, White-Dashed=API Interface Commands, Light-Gray=Asynchronous Data Update).

health care, communication, plant care, tele-working, banking, shopping, entertainment and thousands of other applications that we can hardly foresee. The role of a domestic robot in the future of ambient intelligence seems to be even more crucial, because it would deal with this new technological life that is developing around us, and whose management will not be easy at all; it will also be able to face the generational gap, by adapting, learning the user needs and teaching how to deal with new devices.

The paper is organized as follows. In the next section the concept of Cognitive Architecture is introduced; the description of a specific aspect of the cognitive component, concerned with the formalization of decision making in the Situation Calculus, is discussed in Sections 3 to 6. Then some working component of the reactive level of the robot ArmHandX, namely the navigation component, is discussed, and finally the hardware and software architecture of the robot are explained.

2. Cognitive Architectures

Cognitive Robotics is a discipline born from Artificial Intelligence whose concern is devising *embodied artificial intelligences*, usually accomplished with the design of the whole control system influenced by high-level cognitive skills (reasoning, perception, decision making etc.). The requirements raised by such an embodiment, have been leading to a thorough analysis of the interaction between high-level and low-level processes (providing more detailed models for them, e.g. [23]), and yielding a deeper understanding of the relationships among symbolic and numerical approaches.

As a result of these research efforts, high level control systems have been enhanced with capabilities suitable for managing tasks traditionally [21, 20] reactively accomplished. Earliest examples of agents of this kind can be found in [55, 10]. For example [32, 16] emphasize the role of high level monitoring systems, allowing for a strict interaction between deliberation and execution. Here the on-line behavior of the system can be formally verified and adapted to new events in a flexible way without affecting the system basic constraint (e.g. in [25, 62]). These approaches aim at providing alternatives to rigid layered architectures where the supervision and execution are managed by separated modules (e.g. [10, 21]). A paradigmatic example of this effort can be found in the context of the Remote Agent project [41]. Another example can be observed considering the evolution of the supervision module from RHINO [10] to MINERVA (see [58]). High-level control traditionally has taken care of all the cognitive functionalities of the system, like reasoning, learning, planning making decisions; since all these human-like abilities are deliberative, i.e. are "thinking abilities", not directly connected to sensing functionalities. In fact, the problem of navigation in an unknown environment has been traditionally decomposed into the SLAM (Simultaneously Localization And Mapping problem) and the decision-making about choices, when the robot needs either to reach a desired location or to explore the environment, the most meaningful approaches being those of ([57, 58]); here the knowledge about the robot position and the map is represented by probability distributions, playing the role of beliefs, and are continuously improved through successive Bayesian updates that exploit current data from sensors. Nowadays cognitive architectures support navigation and (topological) map building tasks in several ways: checking consistency, driving the exploration, making decisions in case of incorrect or suspicious behavior and all the levels in which sensory inputs are needed are strictly interlaced. The construction of topological maps are faced also in the Spatial Semantic Hierarchy (SSH) [29], a framework addressing the management of large-scale space knowledge. Topological maps are expected to be computed by a progressive abduction from the navigation experience and according to the layers of the SSH. Abduction, see [45], is a proof methodology allowing one to form hypotheses as a consequence of observations.

In this paper we are mostly concerned with the agent cognitive behavior (Section 3 to 6), its reactive control based on fuzzy logic (Section 7) and finally hardware and basic communication/distribution system (Section 8). Given the complexity of the project we have to leave aside the basic sensor fusion for simultaneous localization and mapping, and the processes concerning focus of attention and feature extraction, together with the whole perception and recognition system leading to the construction of a cognitive map. These are rather articulated topics (see also [49]) needing a dedicated presentation. The robot is, in fact, endowed with a rich sensory system made up of sonar (a dynamic neural network is used to build a basic local map outputting a probability distribution of space occupancy), encoders, range finders, two CCD color cameras mounted on a pan tilt head and coupled with a telemeter and an infrared camera, a line generator and an inertial system. The control of these devices, together with the entire perceptual system is roughly described in Figure 1. It is important to notice that each of these aspects is faced according to mainstream research in its area. In particular, we have been extending the approaches introduced in Itti and Tsotos (see [26, 59]) to integrate the focus attention with the learning phase and, finally, for the recognition process we have been using the recognition by component approach (see [44, 12, 60, 35]) to recognize Symgeons (see e.g. [6, 63, 43], which we use as basic primitives to individuate either landmarks (like doors or any other simple geon-like shapes) or simple human-made artifacts. Furthermore we are experimenting with infrared to detect temperatures. A picture of ArmHandX is presented in Figure 1.

3. A Cognitive House Keeper

To discuss the basic cognitive components of an intelligent agent, we shall introduce a simple example concerning some tasks that should be accomplished in the house. Suppose that our house keeper has to execute, in a continuous process, the following tasks:

- Turn down the heating if nobody is at home;

- open the door whenever someone rings;

- raise the temperature as soon as someone is at home (if winter, or lower the temperature if summer), and

- re-charge the batteries, if their level goes under 27%.

We can easily add, modularly, several more tasks, like watering the garden when none is using water and if it is not raining; starting the washing machine if it is fully filled of dresses, and control, through the transponders, if dresses require the same washing program; control – always using transponders on the food – whether the refrigerator is well equipped, and whether some food has expired. The difficulty in all these tasks, lies in their correct execution, in the integration of all the architectural layers involved in the execution, and in the choice of a unified formalization to deal with all the cognitive aspects, including making decisions. We focus only on the above mentioned tasks, to simplify the example, as it could be useful to understand the logic underlying its decision making. The above tasks require the robot to reason about what is going on in the house (people coming in and out) and what to do, by saving resources, specifically the heating and its battery. The agent can choose what to do according to a rational behavior inducing it to select actions on the basis of the maximum expected utility. This view can be quite profitable. In fact the agent can choose a task or another depending on the effects of its own actions, a sort of self-referential behavior, where the goals consist of reaching an optimal state.

This is a very simple but interesting case, in which making decisions is compelled by the need to save resources. We do not get into details about the formalization, since we try to give a hint of the agent behavior. The cognitive level is concerned with the following tasks:

- Definition of the symbolic language, used by the robot to specify the domain.

- Representation of the a priori (certain) knowledge. This concerns both non dynamic properties of the environment (e.g. for an indoor robot the position of the house entry or the position of the doors), and causal laws ruling the effects of actions and their preconditions.

- Characterization of uncertain knowledge, the probability distribution of stochastic actions, the probability distribution for sensors, the probability distribution associated with some properties of the environment which are not known a priori.

- Definition of the perception process, concerning the way knowledge is updated, that is, the way uncertainty about the environment is dynamically transformed. The *symbolic map construction and maintenance*.

- Transformation of both the language and the causal laws, according to the learning process, where learning is achieved both supervised (e.g. for updating the language) and unsupervised, e.g. through reinforcement.

- Monitoring the system behavior and making decisions about what to do next according to the main tasks to be executed.

We use the Situation Calculus [38, 51, 47] as our formal tool to describe the agent's ontology and both Ecl^{i}ps^{e}Prolog (of IC·parc) and Golog (alGOl in

LOGic) [33], wrapped into C++, to implement the agent's reasoning. To describe our simple example we introduce some preliminaries about the chosen formal language, i.e. the Situation Calculus.

3.1 Preliminaries

The Situation Calculus (SC) is a many sorted first order logic with equality defined by a signature $SC = \langle Sort, Fun, Pred, Var \rangle$, a language $\mathcal{L}_{SC}$ (the language of Situation Calculus) and an interpretation $\mathcal{I} = (D, \_^{Fun}, \_^{Pred})$. *Sort* is a set, each element of which is a sort. The core *Sort* consists of three sorts: *action, situation* and *object* = *worldEntities* $\cup$ *genericObjects*. Where the sort *genericObject* is a catch-all that can include any possible additional sorts needed in an application, like the reals or sets. In [16] the core has been extended with four new sorts: *perceptible, outcome, stochasticAction, event*. The sort event consists of two sub-sorts, *event* = *condEvent* $\cup$ *seqEvent*. *Fun* is a family of sets including, for example, functions of sort genericObject e.g $\{+, -, /, length, \ldots\}$, functions of sort action e.g. $\{grasp, goTo, follow, approach, slide, \ldots\}$, functions of sort perceptibles for instance $\{isOn, isOpen, isAtHome, \ldots\}$; functional fluents, that is, dynamic functions taking as arguments situations like, for instance *BatteryLevel*, and so forth: each element in a set of *Fun* is a function symbol whose sort is in *Sort* and having as arguments terms which, in turn, belong to *Sort*. *Pred* is a family of sets including, for example, relations taking as arguments terms of sort object and terms of sort situation, which are called relational fluents (i.e. dynamic relations, because their truth values depend on their situation argument) like $\{Heating, On, At\}$; the relation $\{Percept\}$, which takes as argument perceptibles and terms of sort situation; non-dynamic relations like $\{Table, Block, Set, Number, \ldots\}$. Each element in a set of *Pred* is a predicate symbol having as arguments terms, which belong to *Sort*. *Var* is a family of sets. Finally D is a sorted domain. Observe that D could be infinite or uncountable even if some of its subset is finite. For example we would be interested to impose a finite domain for world entities, while we would require probabilities to range over the reals. We shall use the following alphabet, with subscripts and superscripts, to denote elements of *Var*: a, for variables of sort action; s, for variables of sort situation; p, for probabilities; pr, for variables of sort outcome and $x, y, z...$, for variables of sort object. Each of the above sets *Sort, Fun, Pred* and *Var* are pairwise disjoint. The family of terms over SC and the set of formulae $\mathcal{L}_{SC}$ over SC are inductively defined as usual.

The Situation Calculus is a formal theory of actions and its core ontology is given by three sets of axioms. The first set is defined by the set of properties and facts that holds about the initial situation (called S_0), this set is $\mathcal{D}_{S_0}$. The second set is formed by the *Successor State Axioms* (SSA), that is, a set of axioms, one for each fluent, stating what becomes true about the domain, once a give action occurred. The set of SSA determine the Markovian property of SC. Notice that there are different kind of actions, and for example perceptual actions do not directly affect the world but only the internal state of an agent, therefore the SSA describing the effects of perceptual actions (e.g. $Percept(isOpen(x), pr, do(sense(x), s))$ are all bounded by the predicate $Percept$. The third set of axioms is the set of *Action Precondition Axioms* (APA), one for each action specified in the language. So, through the APA, each action has its precondition. E.g.:

$$Poss(goTo(x, t), s) \equiv$$
$$\exists y \, At(y, s) \wedge \exists z \, FreePath(z, y, x, s) \wedge BatteryLevel(s) > 18\%. \quad (1)$$

While the SSA stipulate the effects, on the domain, of the execution of an action, the APA stipulate the preconditions for an action that have to be satisfied for executing it. Other axioms are needed to specify the constraints on the state transition: these constraints are independent of the domain.

In all sentences, if not otherwise specified, all the terms are universally quantified. For example

$$On(x, do(a, s)) \equiv a = switchOn(x) \wedge CloseTo(x, s) \vee On(x, s)$$

is a shorthand for the following sentence:

$$\forall x \, y \, a \, s. \, On(x, do(a, s)) \equiv a = switchOn(x) \wedge CloseTo(x, s) \vee On(x, s) \quad (2)$$

A crucial characteristic of the SC is the concept of situation. A situation is a history of actions, and when mentioned in a fluent it denotes the history of actions that make this fluent hold or not. For example, after the sequence

$$s = do(pickUp(glass), do(throwTo(glas, floor), s'))$$
$$\text{(deterministic situation)}$$

then the fluent $Broken(glass, s)$, has to hold. In classical SC a situation term is recursively built using a situation, an action, and the function *do*; since we treat non deterministic actions, having an outcome, success or failure, then we slightly modify this notation introducing the ∘ operator, therefore

$$\langle 1, pickUp(glass) \rangle \circ \langle 1, throwTo(glass, floor) \rangle \circ s'$$
$$\text{(non deterministic situation)}$$

is analogous as the above situation, but it is an event. A single action is denoted with the pair $\langle v, a \rangle$, where v is the outcome of action a, and they have associated a probability of success. Given that E is the current knowledge, the probability of an action is represented as follows (see [16]):

$$P(\langle 1, a \rangle \mid s, E) = p \text{ and}$$

$$P(\langle 0, a \rangle \mid s, E) = 1 - P(\langle 1, a \rangle \mid s, E) = 1 - p \qquad (3)$$

Therefore a deterministic action is just one whose outcome and probability, is 1. Under this view, a situation is a history of actions with outcome 1, e.g. a sequence of the form (from left to right):

goTo(batteryLocation, t),	*follow(walls, t)*,	*approach(plug, t)*,
goTo(doorLocation, t),	*grasp(handle, t)*,	*approach(door, t)*,
goTo(heatingLocation, t)	*insert(finger, t)*,	*waitUntil(charged, t)*,
open(door, t),	. . .	. . .

is a successful history, while a sequence in which some actions were not successful would be:

$$\langle 0, goTo(batteryLocation, t) \rangle, \langle 1, follow(walls, t) \rangle,$$
$$\langle 0, approach(plug, t) \rangle, \ldots$$

The above implicitly describes an event, that is a tree of actions, with associated their outcome. Given a situation, there is a function $tr : situation \mapsto event$, transforming a situation into a tree of events, there exists also an inverse, which is defined as an explanation, given an event tree and an observation.

Time can be added as a new function symbol $time : action \mapsto real$, and $time(a)$ denotes the time of occurrence of action a. For an extension of SC to include time see [51].

Finally observe that probability is a term, when applied to actions, and it is denoted by P, on the other hand when probability takes as argument a sentence, and it is denoted by Pr, then it is a macro whose definition is given according to the probability value computed in the initial situation, where the sentence is regressed. We refer the reader to [47, 16, 51] for a detailed description of the above arguments concerning probability of sentences and the concept of regression.

4. The Initial Language and the Initial State

In our formalization in the Situation Calculus, we take states as class of situations in which a set of properties hold, therefore several histories of actions can lead to the same state. In principle there can be an infinite class of situations identifying the same state. Therefore a state is the class of situations satisfying a given set of requirements. Let σ be a sequence of actions (e.g. $[\alpha_1, \ldots, \alpha_k]$), and $\phi(\sigma)$ be a sentence, e.g. $\phi(\sigma) = \exists x At(home, x, \sigma) \wedge$

$BatteryLevel(\sigma) < 27\%$. Then, any history of actions that induced the consumption of batteries and the presence of someone at home, can be considered to belong to the same class:

$$state(\phi(\sigma)) = \{\sigma' \mid \mathcal{D} \models \phi(\sigma')\}$$

Here $\mathcal{D} = \mathcal{D}_{S_0} \cup \mathcal{D}_{SSA} \cup \mathcal{D}_{AP} \cup \Sigma \cup \mathcal{D}_{UNA}$ is the basic theory of actions, that is, the model of the state transition. Observe that in the above set $\mathcal{D}_{UNA}$ is the set of unique name axioms, imposing that each action name is different from all the others, and Σ is the set of axioms modeling the transition.

The initial theory of actions is the agent's initial a priori knowledge of the world which implicitly define its symbolic language. The a priori knowledge is the agent core information about its domain, and includes all the laws ruling the domain that are not subject to uncertainty, for example to open a door either it has to turn the handle or to push it, to get into a room the robot has to open the door unless the door is already open, no object can be on itself, any object is subject, on the earth, to gravitational force, if a person is seen in a room and another person is seen in another room the persons are the same person if they look the same, i.e. they have the same set of features, and so on. The need of a set of laws depends on the fact that, if ever possible, it would take a huge amount of time to teach the common sense laws to the robot. Actions are functions taking as argument also time, and are divided into three main sets, perceptual actions ruling observations, exogenous actions induced by nature (see [48]), and control actions, that is, actions performed by the robot, that affect the domain. The agent a priori knowledge will be formed by a set of axioms, for each fluent and action in the language, of the form:

$$AtHome(x, \langle v, a \rangle \circ s) \equiv$$
$$\exists t.a = openDoor(t) \wedge v = 1 \wedge t \leq timeLimit(a) \vee$$
$$AtHome(x, s) \tag{SSA}$$

$$Percept(isAtHome(x, t), pr, \langle v, a \rangle \circ s) \equiv$$
$$pr = 1 \rightarrow (\exists z.a = observe(z, t) \wedge v = 1 \wedge is(x, z, s) \in Image(z)) \wedge$$
$$pr = 0 \rightarrow ((\forall t. \, t \leq timeLimit(a) \rightarrow (a \neq observe(z, t))) \vee$$
$$\exists t' \, z.a = observe(z, t') \wedge t' \leq timeLimit(a) \wedge v = 1 \wedge$$
$$is(z, y, s) \notin Image(y) \wedge Percept(isAtHome(x, t), 0, s)) \tag{SSAP}$$

$$Poss(openDoor(t), s) \equiv$$
$$BatteryLevel(t, s) \geq 27\% \wedge closeTo(door, s) \tag{APA}$$

The above axioms specify the contextual knowledge for the agent to get aware that someone is at home. In the first axiom (a successor state axiom) is specified that a person is at home if either the agent has opened successfully the door, at a given time in which the operation was possible, or she was already at home. The second axiom specifies the condition for perceiving that someone is either at home or not. Other axioms will be concerned for finding free paths, see for example axiom (1). In the third axioms (an action precondition axiom) it is specified that to open a door the agent has to be close to the door and in the condition to do an action, i.e. its battery charge is greater than the allowed limit. Observe that preconditions can also be given in terms of probabilities. Note, also, that in principle the robot can either move any "movable objects" (chairs, balls, papers, pencils, pillows, cats, dogs, people...) or wait until they move, it follows that any obvious path is free for it. In other words any path free for a person should be free for the robot.

Each time the robot wake up, it find itself in an initial situation which is the same as the last situation, left before "falling asleep". To achieve this the knowledge base needs to be *progressed*, likewise it needs to be updated each time a new observation is performed. However, observations are not directly managed by the cognitive component (which is considered to be pure reasoning) of the agent, as it can be seen in the design of its structure (see Figure 1), but it is mediated by the reactive control, that assembles and process all the sensory information and delivers an *observation state* which is a concise array of data collecting and recording all the crucial informations about the current operations: an observation state is not just a snapshot of the world, as it embodies positions, dimensions, colors, shapes, interesting objects in the foreground, etc. and a history of the low level actions expanding the control and perceptual actions. These last, in fact, are instructions to the reactive level to gather information where and when needed to disambiguate the observation state.

As we argued above, the cognitive level is implemented in Ecl^{i}ps^{e}Prolog (of IC·parc) and Golog (alGO1 in LOGic) [33], wrapped into C++, and as soon an observation state can be delivered by the reactive level, an event asks to the prolog program to access these data, which are read and transformed into percepts, to which a given probability is associated, depending on the sensors having collected the information. The probability of the current knowledge base is then suitably updated, exploiting the connection between perceptibles and fluents (see [4, 46], following a Bayes filtering, the outcome is that some facts become more reliable and some less. In this way the current knowledge base is updated by the invocation of the observation state.

5.   Making Decisions

A crucial component of the robot's ability of doing the correct sequences of actions, leading to tasks completion, is its skillfulness to reason about its tasks, and thus to get all the information needed to keep the house well managed. For example, it has to monitor whether there is someone at home or not, and to do this it has to patrol all rooms. Since nobody would tell the robot that she/he is leaving the house.

Suppose that at time t the agent has concluded that P is in room R_i: it is expected that the heating is on. Then, if the robot can conclude that "Someone is at home at time t in the situation in which room R_i has been visited by the robot, and in such a situation both the heating is on and the battery level is high", that is:

$$\exists x AtHome(x, \langle 1, visiting(R_i, t) \rangle \circ s)) \wedge$$
$$Heating(on, \langle 1, visiting(R_i, t) \rangle \circ s)) \wedge$$
$$BatteryLevel(\langle 1, visiting(R_i, t) \rangle \circ s)) > 27\%.$$

Then the robot is in an optimal state, because it has achieved its task. Now it could rest, but for how long? This is the crucial aspect of its skill in executing the tasks, that is, the momentum of its actions. In the sequel we shall explain better the above considerations.

Given the tasks it has to execute to manage the house, the states in which the work is done can be clustered into two optimal states. The first state, say $state_N$, is the one in which the robot has its batteries charged, nobody is at home and the heating is turned off. In the next optimal state, say $state_P$, the robot has its batteries charged, someone is at home and the heating is on.

Furthermore, once a set of requirements have been defined, an optimal state is any situation in which the requirements hold. Therefore we give a weight to the optimal state with a reward function. Let s be any situation, we define a real valued rewarding function **Rew** : $situations \mapsto [0, 1]$. Then a situation is said to be optimal, denoted by s_{opt}, whenever its value is greater than a given threshold. The optimal states of the house keeper robot are described by the following equations:

$$[HeatingOn(s_{opt}) \wedge BatteryLevel(s_{opt}) \geq 27\% \wedge$$
$$P(\exists x AtHome(x, s_{opt})) > P(\forall x \neg AtHome(x, s_{opt}))] \vee$$
$$[HeatingOff(s_{opt}) \wedge BatteryLevel(s_{opt}) \geq 27\% \wedge$$
$$P(\forall x \neg AtHome(x, s_{opt})) > P(\exists x AtHome(x, s_{opt}))] \equiv \mathbf{Rew}(s_{opt}) = 0.9$$

and

$$[(HeatingOff(s') \wedge P(\exists x AtHome(x, s')) > P(\forall x \neg AtHome(x, s'))] \vee$$
$$BatteryLevel(s_{opt}) < 27\% \vee$$
$$[HeatingOn(s') \wedge P(\forall x \neg AtHome(x, s')) > P(\exists x AtHome(x, s'))]$$
$$\equiv \mathbf{Rew}(s') = strategy(s', \mathbf{Rew}(s_{opt}))$$

The above sentences say that whenever the robot reaches an optimal state, identified by the above specified set of properties, it gets a reward of 0.9, and in any other state it has a reward depending on the strategy it chooses to reach an optimal state, from the current one. Therefore the reward, in this case, is a quantity used *ex ante* to evaluate choices under risk and uncertainty. Given that the transition model of the Situation Calculus is Markovian, the strategy is chosen by assembling a sequence of actions maximizing the expected utility. Here we intend subjective expected utility in terms of state-preference (for choice under conditions of uncertainty, where probabilities for events are subjectively determined or perhaps undetermined).

Observe that actions are non deterministic and the perception not necessarily exact. A strategy thus has always to take care of the outcome of actions and the outcome of observations, and since an observation, at the cognitive level, is the percept action activated by the observation state, through *observe*, and a situation is the outcome of an action, then a strategy is tuned by the utility of the situation the robot can end up into.

For example, given that the robot, in situation s, is in the kitchen, nobody is at home, and the battery level is high, and someone rings at the door, what is the probability that the sequence

$$goTo(doorLocation, t), \ldots, open(door, t + m),$$
$$goTo(heatingLocation, t + m + k), \ldots,$$
$$turnOff(heating, t + m + k + ..), \ldots$$

will lead the robot into a state with high reward? Or, in particular, what is the probability that the above sequence will lead the robot into an optimal state? According to [5, 11], the strategy is chosen as a function of the utility of the outcome situation (viz. state) $\langle v, \alpha_i \rangle \circ s$ which, in turn, is computed by choosing the action α_j that maximizes the probability to end up in such a situation, given that the action is executed at the situation s, where the robot believes to be, plus the reward of the outcome situation:

$$\mathbf{U}(\langle v, \alpha_i \rangle \circ s) = \mathbf{Rew}(\langle v, \alpha_i \rangle \circ s) + \arg \max_{\alpha_i} \sum_{s \in S} P(\langle v, \alpha_i \rangle \circ s \mid E, s) \times \mathbf{U}(s)$$

Here S is the state space, E is the current knowledge, which depends on the observation. In other words, the probability that the robot, doing action α_i, with outcome v will end up in situation $\langle v, \alpha_i \rangle \circ s$, depends on its knowledge about the current situation, and the outcome v of the action. Given the above equation, a strategy is chosen according to the action that maximizes the utility of the outcome state. Recently this problem in the Situation Calculus has been faced, in a similar way in (see [51]), suitably adapting the Bellman value iteration algorithm [5, 11].

As we outlined above, the critical problem here consists not only in reaching an optimal state, but in shuttling between two optimal states, in so forcing the robot to act even when it thinks that it could rest, where acting depends on resource consumption. In fact, to reach an optimal state the agent has to derive that either there is someone at home, or nobody. Since it cannot logically (monotonically) derive a knowledge variation, it has to update its knowledge through the *observation state*. To get this information it has to monitor the house, but by monitoring the house it might end up in a state with negative reward, e.g. one in which the battery level is lower than 27%. Therefore it has to evaluate the advantage of getting the required information, against the advantage of doing nothing, waiting until some external event happens. Let E be the current knowledge of the agent about the environment, and $\delta(E, \langle v, \alpha \rangle)$ the information that results after executing action α and the update knowledge, and let s be the current situation. If the agent is in situation s, and $E = e_s$, resting in the current state has payoff equivalent to the utility of the current state times the probability that the previous information is still correct (observe that resting is obviously always successful):

$$\mathbf{U}(\langle 1, rest(t) \rangle \circ s) = P(\delta(E, \langle 1, rest(t) \rangle) = e_s \mid E, \langle 1, rest(t) \rangle \circ s) \times \mathbf{U}(s)$$

Therefore the agent has to evaluate the payoff of patrolling the house (acquiring information) against resting, in so not consuming its own resources. If the robot is resting, the probability of observing any variation of the environment diminishes with time (despite once in a blue moon it could increase, because of exogenous events). Therefore the sentence $\forall x \neg AtHome(x, s)$ (resp. $\exists x AtHome(x, s)$) becomes less and less reliable, i.e.:

$$\lim_{t \mapsto \infty} \sum_{s \in S} P(\delta(E, \langle 1, rest(t) \rangle)) = e_s | E, \langle 1, rest(t) \rangle \circ s) = 0$$

We claim that the expected utility of a state, as a result of executing an action α, depends also on the knowledge (information) acquired by the agent in that state:

$$\mathbf{U}(\langle v, \alpha \rangle \circ s) =$$
$$\mathbf{Rew}(\langle v, \alpha \rangle \circ s) +$$
$$\arg \max_{\alpha} \sum_{s \in S} P(\delta(E, \langle v, \alpha \rangle) = e_s \mid E, \langle v, \alpha \rangle \circ s) \times VI(\delta(E, \langle v, \alpha \rangle)) +$$
$$P(\langle v, \alpha \rangle \circ s \mid E, s) \times \mathbf{U}(s) - P(BatteryLevel(\langle v, \alpha \rangle \circ s) < 27\% \wedge \Psi) \}$$

Here Ψ is the sentence saying that either someone is at home and the heating is off, or nobody is at home and the heating is on. In other words the term $P(BatteryLevel(\langle v, \alpha \rangle \circ s) < 27\% \wedge \Psi)$ is completely concerned with resource waste.

On the other hand VI denotes the value of the delta information it gets by patrolling the house. Let us explain the above notion of expected utility of a state (see also Figure 2). Let $state_P$ be the optimal state in which someone is

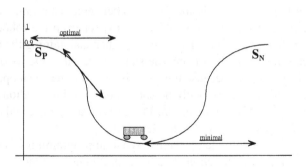

Figure 2.   Oscillation between two optimal states.

at home, the heating is on and the battery are at the right level. And let *state$_N$* be the state in which nobody is at home, the heating is off and the battery level is the required one.

Let us suppose that the agent is at *state$_P$*. If the robot would know that nobody is at home, then it would no more be in *state$_P$*, and since it would lose point, in a non optimal state, it is eager to change state and reach *state$_N$*. In order to move to *state$_N$* it should turn off the heating. Analogously if it is in *state$_N$*, and it would know that somebody just came in. How much would it pay to get this information? to get the information "nobody is at home" or "somebody is at home" it would pay the amount necessary to discount the risk of sliding into a non optimal state, unwarned about its current state (while the robot is in a non optimal state, being unaware of that, it loses points). On the other hand, since to get the information it has to visit the house, the information is uncertain. This happens because it can never get the certainty that "nobody is at home", i.e. it can never conclude with probability 1, that $\forall x \neg AtHome(x, s)$. In fact if it believes that somebody is at home at which point it will start to believe that nobody is at home? I.e. when:

$$P(\forall x \neg AtHome(x, \langle v, \alpha \rangle \circ s))) > P(\exists x AtHome(x, \langle v, \alpha \rangle \circ s))$$

The value of information $VI(\delta(E, \langle v, a \rangle))$ depends on the action performed, and the whole expected utility of the state depends on the equilibrium between awareness of the current state (acquired with the information about the environment) and the resource consumption, affected by the battery charge consumed. Therefore the arbitration is done by the battery level. This means that the agent is motivated to observe, and monitor the house, whenever its battery level allows it for exploration.

In other words, the robot is not allowed to rest whenever its battery level is greater than the minimum tolerated. On the other hand, given that an optimal state is independent of the physical position of the agent, the optimal actions will be those that position the robot in the same room of the people at home, and those shuttling it backwards and forwards between that room, the entrance door and the re-charging location. What we expect from the robot behavior is that it passes from one optimal state to the other maintaining the maximum utility. We call this oscillation from one state to the other the robot unstable equilibrium. It is unstable because the robot is forced to leave an optimal state to make observations. To cope with the above notion of information, the agent receives from the low levels (see Figure 1) the observation state and eventually a stream of interrupts in case of urgent decisions.

To come to more pragmatic problems, dynamic programming is in general used for computing value functions like the expected utility function of a state. These computations then allow the agent to find the optimal policy, that is the optimal sequence of actions that would allow it to shuttle between these two states. Optimal here is meant in the sense of a strategy, that maximizes the expected utility of the agent (see [50]). In particular we are using EPLEX: the Ecl^{i}ps^{e}Prolog interface to CIPLEX.

6.　High Level Behaviors

Given the discussion in the previous section, three items are now ascertained:

1. Even if the agent is reward driven, there can be opposite requirements that put it in a stalmate situation, e.g. keeping the battery level high and exploring the house.

2. The agent's actions need to have a momentum that pushes it out from a lazy state, that can be determined by the mere absence of information due to logical monotonicity.

3. Observations, even if causing risk, are necessary to update the knowledge, thus the value of information is a crucial component of the agent's expected utility, i.e. the domain of consequences is not state independent ([54]).

Clearly classical planning cannot solve the above problems. On the other hand there are tasks that are both unfailing and not determined by risk or subjective probability, like *opening the door*, walking through a room, plugging in the end effector into the battery recharger, etc.. However these tasks might have parameters depending on the context, therefore they might require to solve a planning problem. When this is the case, a solution is found by an *open world planning problem*, that is a problem whose solution requires to deductively find, even under the condition of current uncertain knowledge (e.g. the agent has no definite knowledge about its position, before stretching out the end effector to open the door), a sequence of actions leading the agents to a state in which the goal is satisfied. A plan in the situation calculus is computed deductively (for planning in open world, in the SC, see [18, 48]).

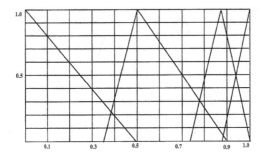

Figure 3. Input variable "front".

However a plan is just a special case of a policy, in which the expected utility of each state is known, and to maximize its utility the agent has to reach the goal state.

7. ArmHandX Reactive Control

Once the cognitive system has decided which action the robot must execute, the action is taken under control by the reactive module that has to perform it actively: managing all the information that are not considered in a cognitive decision, constantly monitoring the dynamic real world that is spinning around, introducing minor but essential adjustments to the planned action, thus, safely sailing the robot in every single step of a plan, from the initial state to the goal state.

Focusing on the reactive navigation problem, the challenge is to perform the required routing in an hostile world like it is one with static and dynamic, small and huge obstacles, where wheels slippage and skidding are the everyday bread.

ArmHandX reactive navigation is based on a Fuzzy Logic Controller (FLC) [64, 34, 31, 52]. The FLC shapes, in real-time, high-level commands, obtained reasoning on a virtual model of the world, and built in low-level motor directives forged on the real world perception. Differently from classical control methods like artificial potential field[28], edge detection and PID controllers, the FLC guarantees robust navigation in a dynamic and unstructured real world, such as the home world, and it naturally copes with data uncertainty and doesn't require an accurate model of the environment.

In everyday situations ArmHandX FLC instinctively shuns obstacles while the robot achieves actions proposed by the cognitive level; there are different behaviors that can be performed by the FLC, the main two are:

- Obstacle-avoidance.

- Point-to-point navigation.

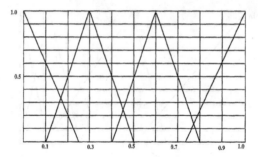

Figure 4.   Input variable "side".

Obstacle-avoidance behavior is intrinsically endowed in every action; for each high level command the fuzzy inference system (FIS) includes a specific rule-base that is able to manage the compromise between avoiding impacts and reaching the given goal. A collision-free navigation is always guaranteed.

Fuzzy linguistic rules map fuzzy inputs into fuzzy outputs, ArmHandX inputs, pinning upon the control, come mainly from a sonar system mounted all around the robot and from encoders assembled in the wheels. The obstacle-avoidance behavior takes five input values that represent the obstacle perception in each of the five areas in which the nearby space, in front of the robot, is considered:

- *side-front-left, front-left, front, front-right*, and *side-front-right*.

Each region is screened by three sonars. The minimum values obtained from one of these sonars is taken as the crisp input for the associated and homonymous fuzzy variable. Every fuzzy variable has four fuzzy terms that share the same labels but not the same membership functions for all the fuzzy variables (see Figures 4 and 3), the linguistic fuzzy terms are:

- *very near, near, safe*, and *far*

The outputs, with the related linguistic terms, that guide the robot in the obstacle-avoidance behavior are:

- speed: *very low (VL), low (L), high (H), very high (VH)*.

- steer: *negative big (NB), negative medium (NM), negative small (NS), zero (Z), positive small (PS), positive medium (PM), positive big (PB)*.

Rules are of the type *IF (x is A) and/or (y is B) THEN (z is C)*; if an obstacle is on the way almost on the right (left) and there are impediment on the left (right) try to avoid the obstacle turning on the left (right), slowing the speed to better control the manoeuvre. The point-to-point behavior has the same outputs of the obstacle-avoidance behavior (speed and steer) but it has different inputs, what is important for the ArmHandX navigation in this case are the following two variables:

- *distance* and *orientation error.*

Both are computed with respect to the actual robot configuration in the work space (robot position and orientation) and the spot to reach. The basic rules are like: if goal is far (close) and orientation error is big positive then speed is low (very low) and steering angle is positive (big positive). ArmHandX FLC applies Mamdani's fuzzy inference method [34], conjunction and disjunction operators are defined to be the min and max functions respectively, the same functions characterize the implication (min) and aggregation (max) method. Defuzzification is performed applying the popular centroid technique. To merge different commands [53, 52, 65] for the same output variable (speed or steering angle) coming from two or more behaviors the fuzzy outputs of each behavior are aggregated together and weighted by a parameter computed upon ArmHandX state of affair, for instance if an obstacle is near to the robot then obstacle-avoidance weights more than point-to-point behavior and the whole fuzzy set for the output variable is then normally defuzzified with the centroid method.

To map the discovered world, increasing knowledge and augmenting reliability of future plans and navigations, ArmHandx concurrently keeps track of its position and acquires and updates the workspace representation, in doing so the robot manages the simultaneous localization and mapping (SLAM) problem [56, 14, 15, 13].

The bottom commission of the reactive layer and, in particular, of the reactive navigation, is to translate single ideal commands that see the robot structure as a whole in commands accepted by each servo controller involved in the final effect; for instance navigation draws in eight servo controllers, in fact to have maximum flexibility, slippage reduced to minimum and no mechanical differential devices, ArmHandX wheels are independent each others in speed and steer [61], hence when the robot turn right (left) in a zero side slip manoeuvre the front wheels turn right (left) but rear wheels turn left (right), in the meanwhile right (left) wheels speed is reduced and left (right) wheels speed is accelerated.

8.   ArmHandX Hardware Architecture

The hardware and software architectures that allow one to control each single servo and sensor on the robot, and that permit layers communication and data exchange, is quite complex. All the architecture is based upon asyncronous communication between different modules and components participating to the system. The goal is to obtain real time data commmunication between the different layers allowing for:

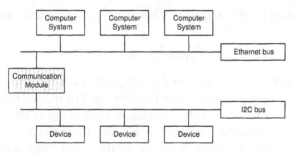

Figure 5.   ArmHandX General Harware Architecture.

- a parallel and independent control of actuators. Each actuator has its own control circuitry connected to the network.

- A parallel data communication. All computers and controllers are wired to the network and provide a data flow about their status and operations.

- customized configurations describing the system configuration via software modules. This gives an abstract devices description to the higher levels of the system yielding a true separation between logical and phisical layers.

- Distributing and dispatching system status data. Therefore event-driven software development can be based upon low level hardware and device events.

All the devices can be operated in parallel, in fact, the main focus of the hardware project is to exploit the concurrent processing capabilities of two communication buses, Ethernet and I2C:

- Ethernet provides wiring and connectivity for several x86 computers that will host all the high-level processes such as reactive layer and cognitive layer. All the system, at this level, is based on a LAN (Local Area Network) installed on the robot, consenting all the subsystems to communicate and exchange informations and commands.

- I2C provides wiring and connectivity for all low-level modules operating actuators and sensors. This part of the network is based on a Multi Master Architecture I2C and every actuator or sensor has its own address on the network bus, granting a complete independence of each device.

Since there is an independent control module for every actuator or sensor and every module contains its own control program that adheres to robot framework specifications, it is possible to provide functionalities similar to Plug 'n Play on PCs, ensuring reconfigurability of the entire robot, at needs. As a result we get a system that can represent and operate each actuator with its own specialized logic. One of the modules wired to I2C bus is the so-called *Communication Module*. This kind of control module provides a communication bridge between the I2C layer and the Ethernet layer. In fact, this module purpose is to provide a routing mechanism for data packets between the I2C bus and the Ethernet bus. At a higher level the ethernet bus provides wiring for several Intel Motherboards (so far we have three motherboards). On the Ethernet bus the Communication Module is just an Ethernet device such as a Network Card. The final result of this configuration is as follows: at a lower level, the control hardware is placed, together with the very low-level control logic, as the feedback control from encoders, and at a higher level a LAN is placed, where a multilayered and multithreaded client-server control system is running. The software modules runninng on the PCs wired to the LAN can access the data on modules on the I2C bus using the bridging capabilities of the Communication Module. The Communication Module acts like a multithreaded TCP Server, routes data packets from/to the devices (actuators/sensors), allowing for a parallel access and control to all the devices on the robot.

9. ArmHandX Software Architecture

Figure 5 depicts the general architecture. A more accurate schematics will depict the complete architecture with the vision subsystem. This subsystem is directly connected to the computer network via PCI connections available on framegrabbers, thus the video services are provided only by software and not by hardware architecture (see Figure 6). The software project is split into several components.The first component is constituted by embedded software running on all the PICs connected to I2C bus. This software contains a very low level logic and it is focused to control actuators via PWM amplifiers, to read data from sensors and to manage sensors polling to get input data from sonars, IR range finders and encoders. A second component is focused on the *Communication Module* that provides a routing mechanism to/from control modules, maintaining a map of that devices to correctly address them. The third component is a multithreaded software router, that provides an enqueuing/scheduling mechanism to give access to the hardware resources available on the robot.

This last software component is located on one of the x86 motherboards connected to the Ethernet bus and forms the other end-point of software connection to the Communication Module. Over this software layer an API interface providing an abstract description of all the devices residing on the robot, has been built. One of the latest versions of this software component contains some logic to process data input from an Inertial Platform that gives attitude and acceleration data.

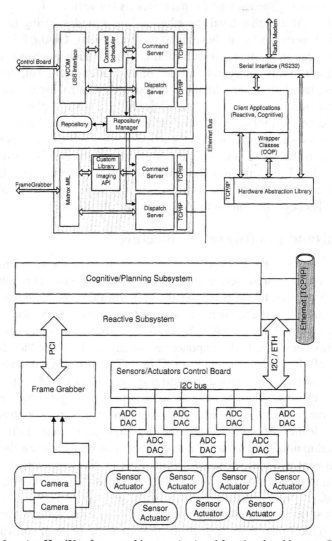

Figure 6.   ArmHandX software architecture (top) and functional architecture (bottom).

The final result is that the real robot modules such as direct and inverse kinematics, fuzzy navigation, sensor fusion, reactive system and cognitive system interact with an abstract robot represented by the API interface published by the so-called *Dispatcher* (the software interface). This architecure ensures a multiporcessing / multithreading approach plus a high level system reconfigurability, since all the hardware parameters of the robot are contained in the lower levels. Furthermore the interface provides some communication functionalities to build sophisticated software. In fact the API does not publish robot functions only, but, through a streaming mechanism, provides a continuous data stream to give the chance to build reactive and cognitive processing software that could be interruptable and event driven.

10.   Conclusions

The robot assistant, the personal robot, the home care-robot (see [24, 7, 22]), the domobot, is a figure that is becoming part of our language, and our real scientific projects, even before a real and complete experiment has been produced. A real experiment would require a robot to behave in complete autonomy in a non-engineered environment, like a house, for at least one week, and we are probably still far from a similar successful experiment. As such an experiment would require the robot to really be able to autonomously re-charge her batteries. However a deep understanding of the requirements for a cognitive architecture, and its complex interactions, together with a more concrete discernment about the role of each component, control, localization, mapping, navigation, reasoning, planning, making decisions, perception etc. and the sensory system indispensable for each component, is now achieved. The early AI approaches (e.g. Shakey [42], the Maoravec CART [40] and SOAR [30]) were based on a SPA (sense-plan-act) architecture, purely deliberative and top down, in which perception, planning and execution are strictly executed in this order. SPA architectures require searching over a large state-space, leading to slow, resource-hungry operation. In the mid 1980s Brooks developed an alternative approach to robot control in response to these problems (see [8, 9]) introducing the Subsumption Architecture, a new conception towards a behavior based architecture with a bottom up approach. This represents a break from the SAS cycle. Brooks' paradigm largely does away with a central representation of the world and uses many simple, high-speed (reactive) processes coupling simple sensory systems directly to action, operating in a highly parallel manner. Behaviors are distributed processes having direct connections to each others or to the sensors and actuators on the robot. They have been applied to a diverse collection of tasks such as navigation, terrain mapping, distributed group foraging, etc. (see [36]). Brooks' paradigm became increasingly studied and several computation formalisms for connecting sensors to actuators in agents, based on the synergy between sensation and actuation in lower animals such as insects, followed in the trail of this concept. Brooks' concept of limited representation, not requiring complete information about the environment seemed to be applicable in real world implementations. Brooks' approach has shown the relevance of low-level behaviors like object avoidance and wall following. Behavior-based systems have been extensively reported in the literature [1]. However, despite the importance of the behavior-based architectures, it soon became apparent that, in its pure form, Brooks' reactive behavior paradigm is difficult to program as rather complex problem solving patterns are attempted, as in practical applications the lack of any ability to carry out high-level planning and problem solving is a real concern. A second wave of architectural paradigms aimed at formalizing more complex structures in which both reactive behaviors and problem solving could take place. Most significant representatives of this new wave are Saphira [27] and Xavier [55]; they have shown how several layers are needed to get a scalable autonomous behavior, likewise the early cognitive robots Rhino and Minerva [58]. New hybrid architectures (see also [21, 19, 17]) are now being developed but the most crucial aspect to be developed is that of integration, which goes far beyond a simple architectural concept of layers and connections, as it involves all the semantics aspects of the combination of formal approaches to cognition. ArmHandX' cognitive architecture has exploited this vast experience and even if it prefers a utility driven behavior to planning, it can still make use of a library of precompiled plans (within GOLOG), and further features, such as a sort of interruptible reasoning process, have been investigated and added. Furthermore a State Constructor Component, the Reactive to Cognitive interface, is continuously aware of all events in the system. This feature is fundamental to build an always up-to-date state for the cognitive layer that can adapt itself to the new conditions. We have not been able to describe the crucial perceptual structure which is embedded both into the reactive and cognitive layers; however the state constructor gives a hint on the way reasoning is updated, according to perceptions.

References

[1] R.C. Arkin. *Behavior-Based Robotics*. MIT Press, 1998.

[2] I. Asimov. *Runaround*. Faucett Crest, New York, 1942.

[3] I. Asimov. *I Robot*. Doubleday, 1950.

[4] F. Bacchus, J. Halpern and H. Levesqueč. "Reasoning about noisy sensors in the situation calculus." *Artificial Intelligence*, 111:171–208, 1999.

[5] R.E. Bellman. *An Introduction to Artificial Intelligence: Can Computers Think?* Boyd and Fraser Publishing Company, 1978.

[6] I. Biederman. "Recognition by components - a theory of human image understanding." *Psychological Review*, 94(2):115–147, 1987.

[7]   R. Bischoff. "Towards the development of 'plug-and-play' personal robots." In *Proceedings of 1st IEEE-RAS International Conference on Humanoid Robot*, 2000.

[8]   R. Brooks. "A Robust Layered Control System for a Mobile Robot." *IEEE Journal of Rototics and Automation*, 1(2):14–23, 1986.

[9]   R. Brooks. "Intelligence without representation." *Artificial Intelligence*, 47:139–129, 1991.

[10] W. Burgard, A. Cremers, D. Fox, D. Hahnel, G. Lakemeyer, D. Schulz, W. Steiner and S. Thrun. "The interactive museum tourguide robot." In *Proceedings of AAAI'98, Madison, Wi*, pages 11–18, 1998.

[11] A.R. Cassandra. "Optimal policies for partially observable markov decision processes." Technical Report CS-94-14, CS, 1994.

[12] S. Dickinson and D. Metaxas. "Integrating qualitative and quantitative shape recovery." *IJCV*, 13(3):311–330, 1994.

[13] G. Dissanayake, P. Newman, S. Clark, H. Durrant-Whyte and M. Csorba. "A solution to the simultaneous localisation and map building (SLAM) problem." *IEEE Transaction of Robotics and Autonation*, 17(3):229–241, 2001.

[14] G. Dissanayake, P. Newman, H. Durrant-Whyte, S. Clark and M. Csorba. "An experimental and theoretical investigation into simultaneous localisation and map building (SLAM)". In *Proceedings of 6th International Symposium on Experimental Robotics*, pages 171–180. Sydney, Australia, March 1999.

[15] G. Dissanayake, P. Newman, H. Durrant-Whyte, S. Clark and M. Csorba. "An experimental and theoretical investigation into simultaneous localisation and map building (SLAM)". In P. Corke and J. Trevelyan, eds, *Experimental Robotics VI*, pages 265–274. Springer, London, 2000.

[16] A. Finzi and F. Pirri. "Combining probabilities, failures and safety in robot control". In *Proceedings of IJCAI-01*, volume 2, pages 1331–1336. Seattle (WA), USA, 2001.

[17] A. Finzi, F. Pirri, M. Pirrone, M. Romano and M. Vaccaro. "Autonomous mobile manipulators managing perception and failures." In *AGENTS'01, The Fifth International Conference on Autonomous Agents*, pages 196–203, 2001.

[18] A. Finzi, F. Pirri and R. Reiter. "Open world planning in the situation calculus." In *Proceedings of AAAI-2000*, pages 754–760, 2000.

[19] A. Finzi, F., Pirri and M. Schaerf. "A system integrating high level and low level planning with a 3d-visualizer." In *International Symposium on Artificial Intelligence, Robotics and Automation in Space (ISAIRAS)*, pages 521–528, 1999.

[20] R. Firby, P. Propopowicz and M. Swain. "The animate agent architecture." In Kortenkamp, D., Bonasso, R., and Murphy, R., editors, *Artificial Intelligence and Mobile Robots: Case Studies of Successful Robot Systems*, chapter 10. AAAI Press/The MIT Press, ISBN 0262611376, 1998.

[21] E. Gat. "Three layered architectures." In Kortenkamp, D., Bonasso, R., and Murphy, R., editors, *Artificial Intelligence and Mobile Robots: Case Studies of Successful Robot Systems*, chapter 8. AAAI Press/The MIT Press, ISBN 0262611376, 1998.

[22] B. Graf, R. Schraft and J. Neugebauer. "A mobile robot platform for assistance and entertainment." In *Proceedings of 31st Int. Symposium on Robotics, Montreal*, pages 252–253, 2000.

[23] H. Grosskreutz and G. Lakemeyer. "On-line execution of cc-golog plans." In *Proceedings of IJCAI 2001*, pages 12–18, 2001.

[24] M. Haegele, J. Neugebauer and R. Schraft. "From robots to robot assistants." In *Proceedings of the 32nd International Symposium on Robotics ISR 2001*, volume 1, pages 404–409. Seoul, Korea, 2001.

[25] F. Ingrand and F. Py. "Online execution control checking for autonomous systems." In *7th International Conference on Autonomous Systems, IAS-7*, pages 273–280. Marina del Rey, USA, 23–25 March 2002.

[26] L. Itti, C. Koch and E. Niebur. "A model of saliency-based visual attention for rapid scene analysis." *IEEE Transactions on Pattern Analysis and Machine Intelligence*, 20(11):1254–1259, 1998.

[27] K. Konolige, K.L., Myers, E.H., Ruspini and A. Saffiotti. "The Saphira architecture: A design for autonomy." *Journal of experimental & theoretical artificial intelligence: JETAI*, 9(1):215–235, 1997.

[28] Y. Koren and J. Borestein. "Potential field methods and their limitations for mobile robot navigation". In *Proceedings of IEEE International Conference on Robotics and Automation*, pages 1398–1404, 1991.

[29] B. Kuipers. "The spatial sematic hierarchy." *Artificial Intelligence*, 119:191–233, 2000.

[30] J.E. Laird, A. Newell and P.S. Rosenbloom. "Soar: An architecture for general intelligence." *Artificial Intelligence*, 33:1–64, 1987.

[31] C. Lee. "Fuzzy logic in control systems: Fuzzy Logic Controller". *IEEE Trans. Sys. Man Cyber.*, 2(20):404–435, 1990.

[32] Y. Lespérance, K. Tam and M. Jenkin. "Reactivity in a logic-based robot programming framework (extended version)." In Levesque, H. J. and Pirri, F., editors, *Logical Foundation for Cognitive Agents: Contributions in Honor of Ray Reiter*, pages 190–207. Springer, 1999.

[33] H. Levesque, R. Reiter, Y. Lesperance, F. Lin and R. Scherl. "Golog: A logic programming language for dynamic domains." *Journal of Logic Programming*, 31:59–84, 1997.

[34] E. Mamdami and S. Assilian. "An Experiment in Linguistic Synthesis with a Fuzzy Logic Controller". *Int. J. Man Machine Studies*, 7(1):1–13, 1975.

[35] D. Marr. *Vision: A Computational Investigation into the Human Representation and Processing of Visual Information*. W.H. Freeman, ISBN 0716715678, San Francisco, 1982.

[36] M.J. Mataric. "Behavior-based control: Examples from navigation, learning, and group behavior." *Journal of Experimental and Theoretical Artificial Intelligence*, Vol. 9(2-3):323–336, 1997.

[37] J. McCarthy. "Programs with common sense". In *Proceedings of the Symposium on Mechanisation of Thought Processes, Vol 1*, pages 77–84, 1958.

[38] J. McCarthy. *Situations, actions and causal laws*. MIT Press, Cambridge, MA, 1969.

[39] W. McCulloch and W. Pitts. "A logical calculus of the ideas immanent in nervous activity". *Bulletin of Mathematical Biophysics*, 5:115–137, 1943.

[40] H.P. Moravec. "The cmu rover." In *Proceedings of AAAI 1982*, pages 377–380, 1982.

[41] N. Muscettola, P.P. Nayak, B. Pell and B.C. Williams. "Remote agent: To boldly go where no AI system has gone before." *Artificial Intelligence*, 103(1-2):5–47, 1998.

[42] N. Nilsson. "Shakey the robot." Technical report, A.I. Center Technical Note 323, SRI International, 1984.

[43] L.N. Patel and P.O. Holt. "Modelling visual complexity using geometric primitives: Implications for visual control tasks." In *Proceedings of the 19th European Annual Conference on Human Decision Making and Manual Control*, pages 3–8, 2000.

[44] A. Pentland. "Perceptual organization and the representation of natural form." *Artificial Intelligence*, 28(2):293–331, 1986.

[45] C.S. Pierce. "Abduction and induction." In J. Buchler, editor, *Philosophical writings of Pierce*, pages 150–156. New York, Dover Publications, Inc, 1903.

[46] F. Pirri and A. Finzi. "An approach to perception in theory of actions: Part i." *ETAI*, 4, 1999.

[47] F. Pirri and R. Reiter. "Some contributions to the metatheory of the situation calculus." *Journal of the ACM*, 46(3):325–362, 1999.

[48] F. Pirri and R. Reiter. "Planning with natural actions in the situation calculus." In J. Minker, editor, *Logic-Based Artificial Intelligence*. Kluwer, 2000.

[49] F. Pirri and M. Romano. "A situation-bayes view of object recognition based on symgenons." In *The Third International Cognitive Robotics Workshop*, pages 25–34, 2002.

[50] R.S. Sutton and A.G. Barto. *Reinforcement Learning: An Introduction*. A Bradford book. MIT press, 1998.

[51] R. Reiter. *Knowledge in Action: Logical Foundations for Describing and Implementing Dynamical Systems*. MIT press, 2001.

[52] A. Saffiotti. "Fuzzy Logic in Autonomous Robotics: behavior coordination". In *Proceedings of the 6th IEEE International Conference on Fuzzy System*, pages 573–578, 1997.

[53] A. Saffiotti, E. Ruspini and K. Konolige. "Blending reactivity and goal-directness in a fuzzy controller". In *Proceedings of IEEE International Conference on Fuzzy Systems*, pages 134–139, 1993.

[54] L. Savage. *The Foundation of Statistics*. Wiley, 1954.

[55] R. Simmons, R. Goodwin, K. Haigh, S. Koenig, J. O'Sullivan and M. Veloso. "Xavier: Experience with layered robot architecture." *ACM SIGART Bulletin Intelligence*, pages 1–4, 1997.

[56] R. Smith, M. Self and P. Cheeseman. "Estimating uncertain spatial relationships in robotics." In I.G. Cox and G.T. Wilfong, eds, *Autonomous Robot Vehicles*, pages 167–193. Springer, 1990.

[57] S. Thrun. "Probabilistic algorithms in robotics." *AI Magazine*, 21(4):93–109, 2000.

[58] S. Thrun, M. Beetz, M. Bennewitz, W. Burgard, A. Cremers, F. Dellaert, D. Fox, D. Hahnel, C. Rosenberg, N. Roy, J. Schulte and D. Schulz. "Probabilistic algorithms and the interactive museum tour-guide robot minerva." *Journal of Robotics Research*, 19(11):972–999, 2000.

[59] J. Tsotsos. "Analyzing vision at the complexity level." *Behavioral and Brain Sci.*, 13:423–469, 1990.

[60] J. Tsotsos, G. Verghese, S. Dickinson, M. Jenkin, A. Jepson, E. Milios, F. Nuflot, S. Stevenson, M. Black, D. Metaxas, S. Culhane, Y. Ye and R. Mann. "Playbot: A visually-guided robot for physically disabled children." *Image and Vision Computing*, 16(4):275–292, April 1998.

[61] D. Wang and F. Qi. "Trajectory Planning for a Four-Wheel-Steering Vehicle". In *Proceedings of the IEEE International Conference on Robotics and Autonamtion*, pages 3320–3325. Seoul, Korea, May 21-26, 2001.

[62] B.C. Williams, S. Chung, and V. Gupta. "Mode estimation of model-based programs: Monitoring systems with complex behaviour." In *Proceedings of IJCAI 2001*, pages 579–585, 2001.

[63] K. Wu and M. Levin. "3D object representation using parametric geons." Technical Report CIM-93-13, CIM, 1993.

[64] L. Zadeh. "Fuzzy sets." *Information and Control*, 8:338–353, 1965.

[65] J. Zhang and A. Knoll. *Integrating deliberative and reactive strategies via fuzzy modular control.* In D. Driankov and A. Saffiotti, eds, *Fuzzy Logic for Autonomous Vehicle Navigation*, Springer, 2001.

QoS-based Resource Management for Ambient Intelligence

Clara M. Otero Pérez, Liesbeth Steffens, Peter van der Stok and Sjir van Loo
Philips Research Laboratories Eindhoven, Eindhoven, The Netherlands
{ clara.otero.perez, liesbeth.steffens, peter.van.der.stok, sjir.van.loo } @philips.com

Alejandro Alonso and José F. Ruíz
Universidad Politécnica de Madrid, Madrid, Spain
{ aalonso, jfruiz } @dit.upm.es

Reinder J. Bril
Philips Research Laboratories Eindhoven, Eindhoven, The Netherlands, and
Eindhoven University of Technology, Eindhoven, The Netherlands
r.j.bril@tue.nl

Marisol García Valls
Universidad Carlos III de Madrid, Madrid, Spain
mvalls@it.uc3m.es

Abstract   Future homes will probably be equipped with in-home networks, combining a backbone of wired networks and a large number of devices connected through wireless links, to provide a responsive and supportive environment, known as Ambient Intelligence. Many of the applications provided to the user lean heavily on media processing and streaming data. Therefore it is expected that consumer terminals will play an important role in providing new experiences to the users. Most of the devices have to be very cost and power effective, while digital media processing is able to consume all the resources a device can offer, and more. Typically, the number of applications and the resource needs of the applications change over time. To adapt to these variations, applications have the ability to trade resource usage for quality of service (QoS). QoS based resource management enables these tradeoffs in resource-constrained systems. In this paper we present our QoS approach, and we explore an integrated approach that addresses terminal and network resources, and takes power issues into account.

Keywords   QoS, QoS resource management, ambient intelligence, in-home networks

T. Basten et al. (eds.), Ambient Intelligence: Impact on Embedded System Design, 159-182.

1.   Introduction

Ambient Intelligence (AmI) refers to a vision where people will be surrounded by intelligent and intuitive interfaces embedded in everyday objects, and an environment recognizing and responding to the presence of individuals in an invisible way [1].

With respect to their power consumption, AmI devices can be classified in three categories, as described in [2]: power autonomous microwatt nodes, personal milliwatt nodes, and static watt nodes. The work presented in this paper focuses on the latter two categories. We use the generic term 'consumer terminal' or just 'terminal' to refer to these devices. These terminals range from small portable screens with very simple display and compute functionality to very powerful media processing and storage servers.

AmI applications will combine a high degree of interactivity with media processing, such as audio and video, 3D graphics, and speech recognition, in all possible combinations. The system requirements imposed by the user on the applications (responsiveness, low latency, quality) and on the platforms (robustness, predictability, stability), together with those imposed by the manufacturers (cost-effectiveness, low power) are highly conflicting. The design and architecting of such systems becomes a complex task, especially when taking into account application, system, and user dynamics.

Fortunately, ambient intelligence is not about computationally exact results, but about providing experiences and serving people. The ultimate judges of such systems are the human users, whose perception of media applications is highly subjective and content dependent. Therefore, we can fruitfully apply the notion of Quality of Service (QoS), which is defined as the "collective effect of service performances that determine the degree of satisfaction of the user of that service" (*ITU-T Recommendation E.800 - Geneva 1994*). Systems architects can exploit the QoS concept to design systems that make run-time tradeoffs between delivered QoS and consumed resources.

In a combined research effort, Philips Research and DiT/UPM have developed a software framework for QoS-based resource management in consumer terminals, see [4, 10]. This effort, which constitutes the basis of the approach presented in this paper, aimed at easing system design, enabling run time tradeoffs, and meeting the system requirements. AmI networks contain both elements: network and terminal. It is our belief that an integrated approach for both networks and terminals is needed. In this paper we present our QoS-based resource management framework for a multimedia terminal, an AmI scenario in the home environment, and the initial ideas and motivation towards an extended framework that integrates terminal and network QoS in an extended home environment.

Our approach combines resource reservation and application adaptation in a multi-layer QoS architecture. QoS-based systems have been extensively researched, especially in the context of large and open networks [8, 20]. Work on ubiquitous QoS delivery is presented in [6]. Terminal QoS management based on resource reservation ([16], [19]) has been pioneered in [15]. Multi-layer QoS representations are proposed in [7], and finally, the combination of resource reservation and application adaptation is presented in [9].

2. Framework for QoS-based Resource Management in Consumer Terminals

Our framework for QoS-based resource management should lead to cost-effective systems that are robust, predictable, and stable, enabling QoS trade-offs. Cost-effectiveness requires that the resources are allocated, provided, and used effectively, towards the goal of maximizing overall QoS. It does not make sense to spend resources on applications that are of no interest to the user (not effective), to allocate resources that will not be needed (not efficient), or to under-utilize a resource when it can be used to increase the overall QoS.

Sections 2.1 and 2.2 introduce the basic concepts, section 2.3 presents the framework architecture. The framework consists of a reservation-based resource manager and a multi-layer control hierarchy, which are described in Sections 2.4 and 2.5.

2.1 Application Execution Model

The applications we are dealing with are *media applications* (mainly audio and animated or live video). Such applications are also known as streaming applications, because they process streams of data. A *stream* is a sequence of data objects of a particular type (audio samples, video pictures, video lines, or even pixels). For example, a video stream is a sequence of pictures, with a given picture rate: the number of pictures to be displayed per second. A *composite* multi-media stream consists of multiple, temporally related streams. For instance, a movie consists of a video stream and an audio stream. The streams are temporally related. Certain audio samples and certain video pictures belong together in time: if their rendering times are too far apart, the perceived quality of the movie is greatly reduced. In this paper we will mainly focus on video streams.

A stream is typically produced by one streaming task and consumed by some other concurrent asynchronous streaming task. The part of the stream that has been produced but not yet consumed, is temporarily stored in a buffer, or is being transferred, from producer to buffer, or from buffer to consumer.

Our execution model for streaming applications on a terminal consists of a connected graph in which the nodes represent either a *task* (an independent,

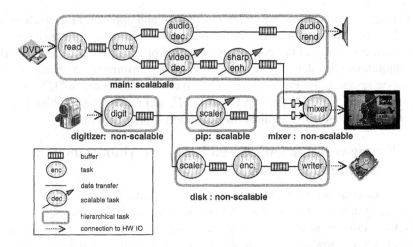

Figure 1.   Execution model of a terminal application in an A/V terminal.

asynchronous, active component that uses processing and memory resources) or a *buffer* (a passive component that uses memory resources). The interconnections represent the data transfer, using transportation resources, such as buses or networks on chip. Tasks can be scalable, i.e. they can trade QoS for resources. The execution model is hierarchical. At higher levels of abstraction, a connected graph can again be viewed as a single task in a connected graph. In the model, we can therefore have applications at different levels. *User application* denotes a functionality the user can start or stop; *terminal application* denotes the union of all connected graphs that exists within the terminal.

Figure 1 depicts an example of a terminal-application, taken from [17], and first presented in [13], showing two levels of abstraction. There are three user-applications, *main*, *pip*, and *disk*, and two independent streaming tasks, mixer and digitizer. The *mixer* and *digitizer* belong to the non-scalable supporting infrastructure of the terminal. Hierarchical tasks are denoted with the rounded rectangles.

2.2   Modes and Quality Levels

A (user) application running on a terminal can be in a number of different modes. Each *application mode* is determined by a set of *mode parameters*. Very often, mode parameters are characteristics of the input streams or the output streams of the application, and determine the *stream modes* of these streams. A typical mode parameter is the video resolution required at the output, e.g. QCIF, SD or HD. Modes and mode parameters have been introduced in [5] [11]. The application mode determines the *application graph* and the *task modes* of the tasks within the application graph. As an example, we take the application *disk* from the previous section. Depending on the available disk space, the user may want to store the video image as CIF or SD. The selection determines the local streaming graph within the application. Assuming that the input from the camera is SD, the HW scaler is needed for a CIF output and not needed in case of an SD output. Similarly, the user may decide to have the image encoded using MPEG1 or MPEG2. In this case, the selection determines the *task mode* of the *encoder* task. Note that task modes and application modes are strongly related with stream modes. In the example, the application parameters are also stream parameters of the output stream of *disk* and *encoder*. Buffer modes were not considered in our work so far, and will not be discussed in this paper.

Within a task mode, the task can operate at different *quality levels*, allowing run-time tradeoffs between output quality and resource usage, by controlling the *operational quality* of the task. A task can operate at one or more quality levels within each mode. Each quality level leads to a different quality of the output stream.

For the resource consumption by a task at a particular quality level within a particular mode, we define a *resource estimate*. Resource estimates can be computed at run time, or predetermined in the laboratory. The former approach is used for scalable video [4, 12], the latter for 3D graphics [14, 21].

2.3 Architectural Design

We designed our framework as a combination of a reservation-based resource manager and a multi-layer control hierarchy. The resource manager is responsible for efficient resource provision and for the immediate allocation of otherwise wasted, volatile, resources such as CPU cycles. The control hierarchy is responsible for effective, dynamically adjusted, resource allocation, and for effective use of the allocated resources. The combination of resource reservation and adaptation has also been reported in [9].

Figure 2 shows the basic elements of our framework. At the bottom of the control hierarchy are the system-application, which consists of several operational parts, and the resource manager. Optionally, an operational part has an application-specific controller.

Robustness and predictability are a joint responsibility of the resource manager and the application, whereas the control hierarchy addresses stability by choosing appropriate time scales for the control layers, and appropriate, non-interfering, control algorithms.

Separation of concerns has been a leading principle in the design of our framework. The separation of resource allocation and resource provision is one example. Resource allocation is driven by effectiveness criteria whereas resource provision is driven more by efficiency concerns.

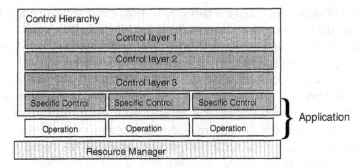

Figure 2.   QoS resource management framework.

2.4   Reservation-based Resource Manager

The sharing of resources is a potential source of interference that leads to un-predictability, and jeopardizes overall system robustness. Resource reservation is a well-known technique in operating system research [16, 19], and is recognized as a basis for QoS management [15], to improve robustness and predictability. It is based on four components, admission control, scheduling, accounting, and enforcement. When properly combined they provide guaranteed reservations.

Based on this approach, our resource manager provides resource budgets to the resource consuming entities. A *resource budget* is a guaranteed resource allowance. A *resource consuming entity (RCE)* is an active component to which a limited resource budget is allocated, and that is able to run with acceptable results on this limited resource budget. In Figure 1, the hierarchical tasks correspond to RCEs. For simplicity, we assume a one-to-one relation between applications and RCEs.

Guaranteed CPU budgets provide temporal isolation, which contributes to robustness. However, applications have to contribute as well. Two issues must be addressed at the application level to really obtain robustness at system level. First, the cost-effectiveness requirement leads to average-case resource allocation. Given the dynamic load fluctuations depicted in Figure 3, RCEs will be faced with temporal (or stochastic) and structural (or systematic) overloads. The resulting robustness problems are to be resolved by the RCEs themselves: the RCEs have to *get by* with their budget. The getting-by requires adaptive behavior and is therefore embedded in the control hierarchy. Secondly, the RCEs may be connected in the terminal flow graph, which leads to temporal dependencies, and propagation of temporal errors. Time-outs are used to prevent propagation of temporal errors. For example, in Figure 1, if *main* or *pip* fail to provide an input on time, the *mixer* will not wait, but will redisplay the previous picture. Similarly, the *digitizer* follows the timing of the camera. If

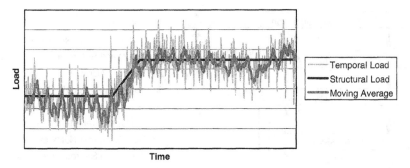

Figure 3. Load changes in MPEG decoding for DVD.

pip or *disk* fail to consume its output on time, digitizer will overwrite the un-consumed output. In both cases, time-outs confine the temporal errors to the RCEs that cause them.

Our current implementations provide CPU budgets only [10, 17]. The resource manager is built on top of a commercial RTOS, and the CPU resource reservation is based on the fixed-priority scheduling model used by that RTOS. The general expectation is that a commercial RTOS will provide this functionality in the near future.

2.5 Multi-layer Control Hierarchy

In a running terminal, the following load changes can be observed: application changes (start/stop), mode changes (aspect ratio of a stream changes from 3:4 to 9:16), structural load changes, and temporal load changes. A structural load change occurs when, for instance, the complexity of the incoming stream changes considerably, i.e. at a scene change. In contrast, temporal load changes are continuously happening. Temporal load changes are partly stochastic in nature, but there may also be significant contributions from other factors, such as different picture types in the encoded input stream of a decoder. Structural and temporal load changes are depicted in Figure 3.

Going from application changes to temporal load changes, the frequency of the changes increases with a step size between one and two orders of magnitude for each step.

Figure 4 depicts our layered control hierarchy, which addresses these multi-level changes. Application changes are addressed by the Application Manager, mode changes by the Mode Manager, structural load changes by the Quality Manager, and finally, the RCE controller primarily deals with temporal load changes.

Each layer has its primary responsibility, but the layers cooperate towards the common goal of optimizing overall QoS for all the application in the sys-

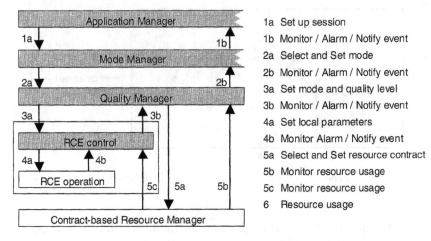

1a Set up session

1b Monitor / Alarm / Notify event

2a Select and Set mode

2b Monitor / Alarm / Notify event

3a Set mode and quality level

3b Monitor / Alarm / Notify event

4a Set local parameters

4b Monitor Alarm / Notify event

5a Select and Set resource contract

5b Monitor resource usage

5c Monitor resource usage

6   Resource usage

Figure 4.   Control architecture.

tem. If the user requests to start a new application, the Application Manager requests a feasibility test from the Mode Manager. The request may ripple down all the way to the Resource Manager. On the other hand, if an RCE controller cannot control its operation to satisfaction, it will send an alarm to the Quality Manager. If the terminal load is very high, this alarm may even reach the Application Manager, which will then decide to stop one application (possibly with user intervention). In general, the effects of decisions taken at a higher level will ripple down to the lower levels. Inversely, the inability of solving a problem at a lower level implies a change at the higher levels. Control decisions are much less frequent in the higher layers than in the lower layers, and the corresponding control algorithms may therefore be more elaborate.

Although their responsibilities differ, the controllers have many commonalities from a structural point of view. Therefore, we have designed a reusable control component, to unify the architecture. This control component is presented in the next subsection. The remaining subsections provide more detail on the RCE controller, the Quality Manager, and the Mode Manager.

2.5.1   Reusable Control Component

A control layer in a hierarchy serves as a *controller* for the layer below, *and* as a *controlled system* for the layer above. The functions of a controller are to select and set operating conditions for, monitor the behavior of, and react to alarms from the controlled system. A controlled system operates according to the settings determined by the controller, allows the controller to monitor its behavior, and sends alarms to the controller if it cannot maintain the desired behavior. The separate select function allows a multi-level selection process,

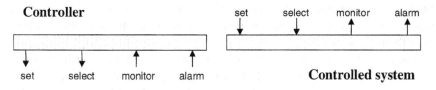

Figure 5.   Template for control component.

in which the control component tentatively sends a selection down the hierarchy to check for feasibility. Figure 5 depicts a template for a reusable control component. As a controller, it requires a complete interface (*select, set, monitor,* and *alarm*) from its controlled system(s). As a controlled system, it provides a complete set of interfaces (*select, set, monitor,* and *alarm*) to its controller(s). The arrows in the picture indicate the information flow. The direction of function calls of the interfaces is always downwards. Since alarms have to be initiated by the controlled system, the alarm interface has to be a subscription interface that allows a callback function.

Every control component is responsible for making and effectuating a control decision. This decision is based on a *selection range* and on *selection criteria*, and results in a *selection result*. According to the selection criteria, the control decision selects one control setting from the set of candidates specified by the selection range. The control settings and the selection range are different for each control component, and determine to a large extent *what* the specific control component has to do. At any time, the *selection range* is determined by the parameters of select or set functions that have been called by the controller over time. In general, a control component performs the following steps:

1. Determine the selection range and selection criteria.

2. Select a feasible control setting.

3. Effectuate the selected control setting.

4. Monitor the controlled system to verify the adequacy of the control settings.

5. Accept notifications or alarms from the controlled system(s).

6. Whenever needed as a result of step 4 or 5, adjust the control setting(s) by going back to step 2.

7. When step 2 cannot find an adequate control setting, notify the controller.

Steps 1-3 can be triggered by select and set calls from the controller, or by step 6. Step 4 determines the autonomous behavior of the controller, with its characteristic frequency.

The ultimate feasibility check is done by the resource manager (in its function as controlled system), which determines the availability of resources. Before a set call is made at the higher levels, the select call has to go down the hierarchy for feasibility testing. However, this approach may require a lot of up-and-down communication. As an alternative, the control component can prune the search tree by using a local feasibility test based on a local model of the controlled system.

2.5.2 Application Manager

The Application Manager determines which applications can run concurrently in the system, and starts and stops user applications. Since the Application Manager is at the top of the hierarchy, it does not need to provide a control interface. However, being a controller, it does require a control interface from the Mode Manager. The Application Manager interacts with the user(s) and with the applications.

When a user requests to start an application, the Application Manager performs an admission test, based on the application description for that application. An *application description* is a full description of an application, in terms of modes, quality levels, and resource estimates. We assume that it is accessible in the terminal.

The Application Manager follows the general steps of a control component. Its scope is the set of user applications. Its *selection range* is the set of applications currently requested by the user. The user determines the *selection criteria*, for instance the relative importance of the applications. Selection is only needed when the terminal is, or threatens to be, overloaded. The *selection result* is the set of selected applications. The Application Manager monitors the state of the Mode Manager to observe the overall match of the current application mix to the platform.

2.5.3 Mode Manager

The Mode Manager determines the appropriate application mode for each application, and takes care of smooth mode-change transitions. It provides a full control interface to the Application Manager, and requires a full control interface from the Quality Manager. The parameters of the select and set functions provided to the Application Manager are (access pointers to) the relevant application descriptions. The Mode Manager interacts with the user(s).

Each application description consists of a set of application mode descriptions. An *application mode description* provides a connected graph of RCEs, and the corresponding RCE modes for the scalable RCEs. These RCE modes must be mutually consistent. For example, if a producer RCE produces a QCIF stream, the consumer RCE cannot consume an SD stream.

The Mode Manager follows the general steps of a control component. Its scope is the set of user applications. Its *selection range* is a set of application descriptions. The user determines the *selection criteria*; one of them will be the application mode parameters described in Section 4.1.2. If the selection criteria allow several application modes, the Mode Manager selects the most appropriate one. The *selection result* is the set of selected application mode descriptions, one for each application. The Mode Manager monitors the state of the Quality Manager to observe the overall match of the current application mode mix to the platform.

2.5.4   Quality Manager

The Quality Manager determines the quality levels and associated budgets for the RCEs in the terminal, and orchestrates the transitions in case of quality-level and budget changes. The Quality Manager provides a full control interface to the Mode Manager, and requires partial control interfaces from the Resource Manager and the RCE controllers. The parameters of the select and set functions provided to the Mode Manager are (access pointers to) the relevant RCE mode descriptions. The Quality Manager does not interact with the user(s): the user has no knowledge of RCEs, and the Quality Manager has no knowledge of user applications.

The scope of the Quality Manager is the terminal: it is aware of the terminal resources and translates the estimated resource needs into guaranteed budgets. For a given RCE in a given RCE mode, the *quality mapping* is the set of possible quality levels, each with the corresponding resource needs.

The Quality Manager follows the general steps of a control component. Its *selection range* consists of the current quality mappings for the RCEs in the terminal. A major selection criterion is that the resource budgets of all RCEs match the available resources in the terminal. From the selection range and the selection criteria, the Quality manager determines the *selection result*: a quality level for each RCE, and the corresponding resource budgets. The Quality Manager monitors the state of the RCEs to observe the overall match of the current RCE loads to the allocated budgets. An example of a QM selection algorithm can be found in [15].

2.5.5   Resource Consuming Entity (RCE)

As shown in Figure 3 the RCE consist of two parts: an operational part and an (optional) controller. If the operational part is not controllable, as is often the case for legacy code, a controller does not make sense. The *operational part* of a controllable RCE is at the bottom of the control hierarchy; it is the controlled system for the RCE controller. The RCE controller provides a full control interface to the Quality Manager, and requires a proprietary intra-RCE interface from the operational part. In addition, the RCE controller requires a monitor function from the Resource Manager. The parameter of the provided select and set functions is (an access pointer to) the RCE quality level (which includes the RCE mode).

The scope of the RCE controller is, of course, the RCE. An *RCE description* provides the local connected graph that makes up the RCE, and the possible operational settings of the scalable tasks. These settings must be mutually consistent. For example, if a producer task produces a QCIF stream, the consumer task cannot consume an SD stream.

The *RCE controller* follows the general steps of a control component. The *selection range* consists of the local task set, the operational settings of the tasks, and the quality levels of the scalable tasks. The *selection criteria* determine which combinations of operational settings and task quality levels are allowed. They are an integral part of the RCE design. For example, if a producer task produces a low quality or QCIF stream, the consumer should not expect a high quality/SD stream, respectively. The *selection result* is the set of selected tasks, and the operational settings and/or quality levels for the selected tasks. The RCE controller monitors the state of the budget provision, which is maintained by the Resource Manager, to observe the current match of RCE load with the available budget. Examples of RCE control algorithms can be found in [22].

Some operational parts are able to detect changes in the incoming stream that affect the mode of this and possibly other RCEs. These events will be signaled to the Mode Manager via the Quality Manager, by the alarm callback.

3.   AmI Scenario

The previous sections explained our existing QoS-based resource management framework for tasks that consume CPU resources in a single terminal. However, as described in the introduction, AmI applications are distributed among heterogeneous terminals, which communicate along different types of networks. We use the following home-based scenario to explore how our framework can be extended towards AmI systems in the extended home environment.

> Mary is filling out a tax form in her home office. At the same time, she watches a movie on the screen in the office. A small window in the corner of the screen shows the baby room. She decides to call it a day, and goes to the living room to join her husband John, who is watching the news. As she leaves the home office, the movie she was watching is being recorded on a central recorder server for later view. As she enters the living room, a small window pops up in the corner of the large living room screen, showing the baby room again.

Three (potentially) distributed applications are active initially:

A1. Mary is watching a movie from a broadcast channel on the home office screen;

A2. John is watching the news from a second broadcast channel on the living room screen;

A3. Mary is supervising the baby room in a small window on the home office screen.

> When Mary leaves the home office, application A1 continues in a different mode: instead of being shown on the office screen, the movie is now being recorded on the central recorder. At the same time application A3 goes on hold, sensing Mary's whereabouts. When Mary enters the living room, A3 is reactivated, allowing Mary to continue supervising the baby room on the living room screen.

Now the same 3 applications are active in a different configuration:

A1. The movie from the broadcast channel is being recorded on the central recorder;

A2. John and Mary are watching the news from a second broadcast channel on the living room screen;

A3. Mary is supervising the baby room in a window on the living room screen.

The broadcast streams (50 Hz, SD resolution) are received in a home gateway (terminal T1), which also contains the central recorder. The camera in the baby room is connected to that same home gateway through a wireless link. The screen in the home office is a medium size 50 Hz screen (SD resolution) connected to terminal T2. Finally, the screen in the living room is a 100 Hz large screen (HD resolution), which is connected to terminal T3. The window sizes for the baby room supervision require QCIF format on the home office screen and SD format on the living room screen.

Movies are stored at the original movie picture rate of 25 HZ (adjusted to European standard display frequency). The 50-100Hz conversion, as well as the SD-HD conversion, can take place at the source (T1) or at the sink (T3).

4. QoS-based Resource Management for AmI

In an AmI system, coordinated effort is needed to make all resources (networks, power and terminal resources) cooperate to reach the common goal of meeting end-to-end application requirements and maximizing overall user satisfaction. In this section, we explore the possibilities for extending our framework towards AmI systems.

First, QoS network management techniques must be integrated with terminal QoS resource management. Our hierarchical approach is very well suited for a relatively closed environment such as the home, but not for fully open environments such as the Internet. Secondly, as described in [2], power management becomes a challenge for AmI devices, and has to be addressed at system level. Finally, terminals not only rely on processing resources (CPU, dedicated processors), but also on memory resources and transportation resources (bus, network on chip). The different resource types are very closely interrelated: for instance, a memory access by a processor involves all three resource types. Therefore, resource management should address all terminal resources in an integrated manner.

In the remainder of this section, we will discuss the above-mentioned issues in the given order.

4.1   Integrating In-home Network Management

4.1.1   Closed Networks

The successful addition of a stream to a loaded network depends on the capacity of the network, the average load, the load fluctuations, the bandwidth needs of the stream, and the medium access control. For the Internet, the packets associated with a given stream choose different paths depending on the rapidly changing load of the network. Mechanisms (e.g. Differentiated services and Integrated service) exist to reserve a given quality of a connection with respect to other connections, but most guarantees are probabilistic, unless the used communication lines are dedicated to a number of known applications (companies). Best effort techniques are used to adapt bandwidth needs to bandwidth availability. When the load is sufficiently low, the techniques assure that their use is close to optimal.

To guarantee a specific absolute quality of service, the load on the network needs to be known. The concept of closed network denotes those networks that allow the monitoring and control messaging of load changes to all interested devices within a bounded delay. Under those circumstances load changes can be measured and consistent network allocation decisions can be made.

Home networks are choice examples of closed networks. The networks size in number of network links ($<$ 4) and connected active devices ($<$ 100) is low. The low number of people in the home limits the number of applications. Applications are started and stopped with a rate at the human scale. Most applications are started within the home, and a limited number of applications access services outside the home. Therefore, we can assume that load changes can be communicated to all active devices at a rate that allows sending the corresponding control information with a marginal added network load.

4.1.2 Streams, Modes and Quality Levels

The execution model of an AmI application does not differ from the model described in Section 2.1. However, in the case of a distributed application, the application graph is partitioned into sub-graphs that reside in different terminals. The inter-subgraph connections represent inter-terminal transfer over the network. The resources involved in doing the transfer (network bandwidth) are to a large extent determined by the stream being transferred. We will use the term *stream transfer* for the inter-terminal transfer of a stream.

The *application mode* of an AmI application determines the terminal sub-graphs and the *stream modes* of the inter-terminal streams as well as the application graph and the RCE modes introduced in Section 2.2. For example, if a central recorder on a remote terminal would replace the disk in Figure 1, the decision to store the video image as QCIF or SD, in MPEG1 or MPEG2 format, would also affect the stream mode of the video stream between the two terminals.

Within a stream mode, scalable streams (see next section) can have different *quality levels*, allowing run-time tradeoffs between stream quality and network bandwidth. For the network bandwidth required by a stream in a particular mode and at a particular quality level, we define a *resource estimate*. Resource estimates can be computed at run time, or predetermined in the laboratory.

4.1.3 Scalable Video Streams

Just like tasks, streams can be scalable or non-scalable. Audio can also be scalable, but in the presence of video, scaling audio does not make sense. In this section, we will focus on scalable video only.

A *scalable video stream* consists of number of substreams incorporating a base layer (BL) and one or more enhancement layers (EL_i, $i = 1, \ldots, n$). The base layer contains enough information to make a low quality video visible to the user, just enough to understand the story told by this video. The base layer plus the first enhancement layer, $BL + EL_1$, allow the same video to be displayed with higher quality. Ever increasing quality is obtained by adding EL_i to $BL + \ldots + EL_{i-1}$ [18]. Note that a scalable video stream is always encoded. Raw video (which can be directly displayed) is not scalable.

4.1.4 (Scalable) Stream Transfer

The network is composed of a set of interconnected terminals. One or more streams can originate at a given terminal. Each stream is sent to at least one destination terminal. To transfer a stream, the source terminal packs the stream into packets that are sent over the communication network. We assume that a bandwidth allocation mechanism exists (e.g. as provided by IEEE 802.11e and IEEE 1394), which allocates a time slice to each terminal that serves as source, on a usually repetitive base [20]. Within the time slice the source terminal sends its packets. The time slice of the terminal is further partitioned over the streams this terminal has to send. The time slicing mechanism schedules, accounts and enforces the allocated time slots.

In a wired network, time slice allocation corresponds to guaranteed bandwidth reservation. In a wireless network, the number of bytes that can be transported during a given time window can be less than intended by the allocation. This temporary bandwidth reduction of the communication link is due to external causes, and cannot be prevented. As a consequence, the bandwidth reservations over wireless links are not guaranteed.

Scalable video streams allow scalable stream transfer. Each substream (= layer) is packed into packets separately. The *scalable transfer* is organized as follows. For a given picture in the video stream, the base layer packets are sent first, followed by the enhancement layer packets in order of decreasing importance (increasing index i). Packets that do not fit in the allocated time window are discarded.

4.1.5 Multi-layer Network Management

Network management is done at two control rates: slow and fast. The slow control partitions the *theoretically* available bandwidth over the different streams. The fast control compensates for fast network capacity changes that are easily provoked in wireless networks. It ensures that the *operationally* available bandwidth is allocated to the more important sub-streams when the operational bandwidth is lower than the theoretically available bandwidth, by using a simple priority algorithm when sending packets over the link.

To reduce the complexity of the network management problem, the complete network is divided into subnets. A subnet connects a number of terminals over a homogeneous network, and is separated from other subnets by bridges or routers. The available bandwidth in each subnet has to be partitioned over the streams, as described in the previous section. For this paper we restrict ourselves to the treatment of a single subnet.

Network controllers are distributed over the terminals in the subnet. The distributed controllers are replicated on each terminal. The replicated versions of each controller communicate with each other to maintain a common view on the decisions to be taken.

4.1.6 Integrated Solution

Figure 6 depicts the integrated framework including network management. The Application Manager, the Mode Manager, the Subnet Quality Manager (Subnet QM), and the Subnet Resource Manager (Subnet RM) are replicated in every terminal. The scope of the Application Manager and Mode Manager

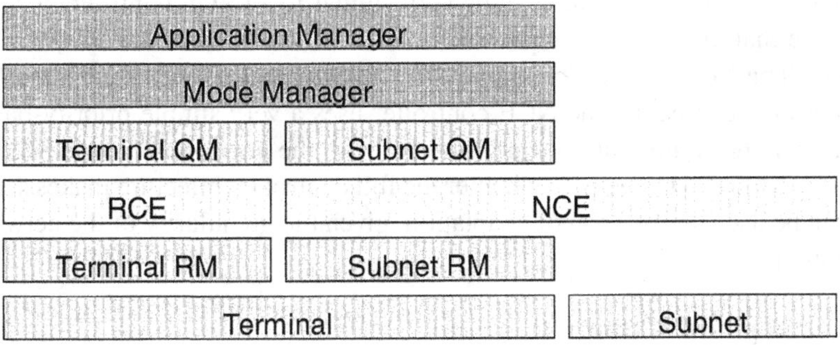

Figure 6.   Integrated framework: Layered view.

is the entire network (all terminals, all subnets); the scope of the Network QM and the Network RM is the subnet. Note that the Network Consuming Entity (NCE), the counterpart of the RCE in the subnet, is not an active component. It is a stream transfer over the network. However, every scalable NCE has a controller that resides on the source terminal of the stream transfer. Before going into more detailed descriptions of the network controllers, it is important to remark that our approach is only possible when all devices cooperate. If the behavior of a controller does not comply with the agreed control decisions, havoc may result. Moreover, as mentioned before, our hierarchical approach is only possible in a closed environment.

The distributed Subnet RM is comparable to the single-terminal terminal Resource Manager (Terminal RM). It performs the network bandwidth allocation algorithm described in Section 4.1.4. The main difference with the Terminal RM is the absence of the monitor interface.

The distributed Application Manager is exactly the same as the Application Manager described for the single terminal, except for the replication and the communication between the replicas. Not all replicas of the Application Manager will interact with the user(s).

The distributed Mode Manager is also a replication of the single-terminal Mode Manager. There is an additional difference in the application mode descriptions, which now contain the stream modes, and the mapping of tasks to terminals and stream transfers to subnets in addition to the application graph and the task modes.

The distributed Subnet QM is very similar to the single-terminal Quality Manager (Terminal QM). It determines the quality levels and the associated network reservations for the NCEs in the subnet, and orchestrates the transitions in case of quality-level and budget changes. The Subnet QM interfaces with the distributed Mode Manager, the distributed Subnet Resource Manager, and the NCE controllers. The scope of the Subnet QM is the subnet.

Finally, an NCE controller can be very similar to an RCE controller, except of course that the operational part is not a set of tasks, but a (scalable) stream transfer. The *NCE description* simply states the number of enhancement layers of the scalable stream. The NCE controller uses a very simple priority-based adaptation algorithm that does not interact with the environment (see Section 4.1.4). We have to investigate whether the full control architecture makes sense and can be used at this level of adaptation, given the granularity of the network adaptation.

4.1.7   Simplified Scenario

We use a simplified version of the scenario described in Section 3 to show some aspects of the framework's behaviour. The scenario focuses on a user-induced mode change at system level. We assume only two applications, A2, *John is watching the news*, and A3, *Mary is supervising the baby room*. The set-up is also different: the large living room screen is connected directly to the home gateway (T1), and the camera is connected to T2 through a wireless link.

The initial situation is depicted in Figure 7. We assume the simple set-up, where the living room screen is directly connected to terminal T1, and the camera is connected to terminal T2. The two terminals are connected through subnet S1.

Each terminal has a replica of the Application Manager, the Mode Manager, the Network Quality Manager, and the Network Resource Manager. Moreover, each terminal has its own Quality Manager and Resource Manager. The decoder and scaler tasks are scalable.

When Mary leaves the home office, application A3 goes on hold. When Mary enters the living room, the Application Manager reactivates A3, allowing Mary to continue supervising the baby room on the living room screen. The Mode Manager has to change the application mode; the picture should now be shown at a different location. Two modes are available: mode 1, upscaling the picture in T2, and mode 2, upscaling the picture in T1. Mode 2 consumes less network resources but more T1 resources. The following steps lead to the result in Figure 8:

1. T1 Mode Manager checks feasibility with T1 QM and S1 QM for mode 1.

2. T1 QM and S1 QM check availability of resources with T1 RM and S1 RM, respectively.

3. S1 RM informs the S1 QM that there is insufficient bandwidth to transmit the upscaled video.

4. T1 Mode Manager checks feasibility for mode 2.

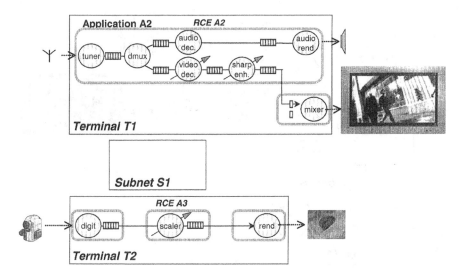

Figure 7. AmI scenario applications A2 and A3.

5. T1 QM and S1 QM check availability of resources.

6. T1 RM responds that there are not enough resources available to start the new RCE.

7. T1 QM reduces the quality level of A2 and tries again.

8. This time the feasibility check succeeds and A3 is reactivated in mode 2.

4.2 Multiple Resources in the Terminal

In a single terminal, there are different resources that are shared among AmI applications. These resources can be both on-chip and off-chip. Typical resources are for example, memory (on- and off- chip), processing resources (co-processors, DSPs, etc...) and transportation resources (bus, network on chip). The application utilization of any of these resources depends on the utilization of the other resources. Because of these interdependencies, efficient resource usage is only possible when the resource management is done at a global level (across resources), instead of managing each resource separately. Examples of the interdependencies between the resources are:

- Reserving memory space does not make sense if there is no intention to access it.

- Sharing a processor implicitly presumes the presence of memory space to store the states of the waiting tasks.

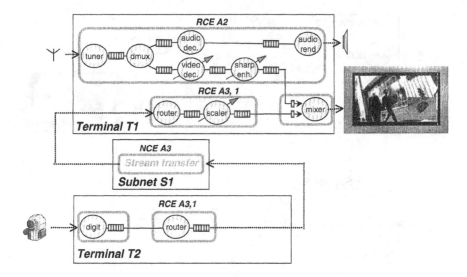

Figure 8. AmI scenario applications A1 and A3 after a mode change.

- Waiting for a memory access to complete creates a difference between guaranteed processing time and effective processing cycles available to a task.

- Sharing a processor that has a local cache creates interdependencies between the memory access times of the different tasks.

Memory space resources are non-volatile; they can be used, or occupied, but they have to be released in due time. Typical issues here are ageing, and fragmentation. The other resources are volatile; they have to be consumed when they are there; if not, they are lost. Therefore, bandwidth (bytes/sec, or cycles/sec) is an insufficient characterization for these volatile resources. Moreover, since every byte or cycle can only be provided once, competitors will have to wait, so latency should be part of the characterization as well.

The following subsections present our first considerations for resource management of multiple resources within one terminal.

4.2.1 Transportation Resources

Transportation resources are shared by all (co)processors in the system. From the transportation resource perspective, the (co)processors are the resource consumers. The transportation resource load strongly depends on the load allocation to the (co)processors. Conversely, the transportation resource load determines the effectiveness of resource usage at the level of the (co)processors, and influences this load allocation. The resource contract with transportation resources may be very different depending on the users of the resource and the applications nature. A special case of transportation resource is the memory access resource that is described below.

4.2.2 Memory

Memory can be used locally by a task, or by a buffer. In both cases, the use of the memory is closely related to the use of transportation resources. For an application involving tasks and buffers, it does not make sense to reserve memory space for the buffers without reserving transportation resources to access the allocated memory locations for the connected tasks.

4.2.3 Coprocessors

Coprocessors are hardware resources that deal with the processing of specific functions and can range from dedicated, non-shareable hardware blocks (e.g. MPEG decoder) to programmable components (e.g. DSP). When the coprocessor is non shareable, resource management is trivial. Resource management techniques come into play when resources can be shared. Pre-emption is an important characteristic. Without the ability to preempt a coprocessor, it is not obvious how to enforce budgets (and consequently guarantee budgets). We may have to look for budgets that are not enforced, but still provide a bounded approximation to a constant fraction of the coprocessor time. For this kind of budgets, non-preemptive scheduling algorithms are known to exist.

4.3 Power Management

Power saving allows prolonging battery life in mobile devices and reducing consumption in stationary devices. Power management is crucial for mobile devices to reduce power consumption while providing an acceptable computational capacity.

Modern processors and other hardware devices provide means for saving energy by reducing the clock frequency or by a dynamic variable voltage supply. Standards, such as the Advanced Configuration and Power Interface (ACPI) specification [3], provide a common way to access these capabilities from the operating system. Hence, advance power management techniques can be built based on considerations of the required hardware devices and computational capacity.

Power management must be integrated with the management of other hardware devices. In addition, its handling is a bit different, as changes in the system power state affect the behavior of the whole system. If the clock frequency of the CPU is reduced, the available processing capability will also be reduced. Hence, the resource budgets assigned to applications must be renegotiated.

Power management is an additional responsibility of the Quality Manager. When selecting quality levels and resource budgets, the Quality Manager should consider the power state: user energy saving preferences, the current battery state, etc. A change in the power state includes the following basic steps:

- Detection of a system state where a change in the power state is advisable, e.g. when system resources are underutilized, battery charge is low, etc.

- Determination of the next power state and selecting new quality levels and budgets.

- Recovery from a current power state.

- Setting the new power state and applications settings.

As can be easily deduced from this list of basic operations, a change in the power state may require a non-negligible overhead. Hence, it is necessary to evaluate in advance whether it is worth making the change.

Additional savings can also be made at the resource manager/operating system level. The basic mechanism is to detect situations were a device is idle, and to move it into a power saving mode. Whenever a request for using the device is detected, the previous state is restored.

5. Conclusion

Ambient Intelligence systems are heterogeneous, complex and resource constrained systems. The system requirements imposed by the user on the applications (responsiveness, low latency, quality) and on the platform (robust, predictable, stable), together with those imposed by the manufacturers (low power, cost-effective), are highly conflicting. When adding the dynamic fluctuations of resource needs of applications, the design of such systems becomes very challenging and requires an architecture that explicitly addresses these issues. Our QoS-based resource management approach aims at doing exactly this, for a limited class of AmI systems such as in extended-home environments.

We have applied this approach successfully to single processor systems. We believe our approach can be applied to distributed AmI systems. In this paper, we present our basic approach, and the initial research effort towards extending our framework to AmI. The extension of the framework is done at the level of the terminal, by including multiple resources, as well as at system level, by providing an integrated approach to the management of both subnets and terminals, and by addressing power management. The results of this preliminary study are encouraging, and the next step is to build a prototype where our ideas can be validated.

Acknowledgements

The authors would like to thank Wim van der Linden for his valuable comments on a previous version of the paper and Alina Albu for spotting one last mistake. This work was supported in part by the IST-2000-30026 project, Ozone.

References

[1] E. Aarts, R. Harwig, and M. Schuurmans, "Ambient Intelligence," in *The invisible future: The seamless integration of technology into everyday life*, P.J. Denning, Ed., pp. 235-250. McGraw-Hill, Inc. New York, NY, USA, 2001.

[2] E. Aarts and R. Rovers, "Embedded System Design Issues in Ambient Intelligence", this volume, 2003.

[3] Advanced Configuration and Power Interface Specification, Revision 2.0b, Compaq, Intel, Microsoft, Phoenix, Toshiba, 2002.

[4] R.J. Bril, C. Hentschel, E.F.M. Steffens, M. Gabrani, G.C. van Loo, and J.H.A. Gelissen, "Multimedia QoS in consumer terminals", in *IEEE Workshop on Signal Processing Systems (SIPS), Antwerp, Belgium, Proceedings*, Sept. 2001.

[5] R.J. Bril, E.F.M. Steffens, G.C. van Loo, M. Grabani, and C. Hentschel, "Dynamic behavior of consumer multimedia terminals: System aspects", in *ICME, IEEE International Conference on Multimedia and Expo, Tokyo, Japan, Proceedings*, pp. 597-600, Aug 2001.

[6] Y. Cui, D. Xu, and K. Nahrstedt, "SMART: A Scalable Middleware solution for Ubiquitous Multimedia Service Delivery", in *ICME, IEEE International Conference on Multimedia and Expo, NY, USA, Proceedings*, Aug. 2000.

[7] L.A. DaSilva, "QoS mapping along the protocol stack: Discussion and preliminary results", in *ICC, IEEE International Conference on Communications, Proceedings*, pp. 713-717, June 2000.

[8] P. Ferguson and G. Huston, *Quality of Service: Delivering QoS on the Internet and in Corporate Networks*, John Wiley and Sons, Jan. 1998.

[9] I. Foster, A. Roy, V. Sander, "A Quality of Service Architecture that combines resource reservation and application adaptation", in *International Workshop on Quality of Service*, 2000.

[10] M. García Valls, A. Alonso, J.F. Ruiz, and A. Groba, "An Architecture of a Quality of Service Resource Manager for Flexible Multimedia Embedded Systems", in *3rd International Workshop on Software Engineering and Middleware, Orlando (Florida), US, Proceedings*, vol. 2596 of Lecture Notes in Computer Science, Springer Verlag, May 2002.

[11] M. Grabani, C. Hentschel, E.F.M. Steffens, and R.J. Bril, "Dynamic behavior of consumer multimedia terminals: Video processing aspects", in *ICME, IEEE International Conference on Multimedia and Expo, Tokyo, Japan, Proceedings*, pp. 1220-1223, Aug. 2001.

[12] C. Hentschel, R. Braspenning, and M. Gabrani, "Scalable algorithms for Media Processing", in *ICIP, IEEE International Conference on Image Processing, Proceedings*, pp. 342-345, Oct 2001.

[13] C. Hentschel, R.J. Bril, Y. Chen, R. Braspenning, T.-H. Lan, "Video Quality-of-Service for Consumer Terminals - A Novel System for Programmable Components". *International Conference on Consumer Electronics, Digest of Technical Papers*, Los Angeles (USA), pp. 28-29, June 2002..

[14] G. Lafruit, L. Nachtergaele, K. Denolf, and J. Bormans, "3D Computational Graceful Degradation", in *ISCAS, IEEE International Symposium on Circuits and Systems, Proceedings*, vol. 3 of pp. 547-550, April 2000.

[15] C. Lee, J. Lehoczky, R. Rajkumar, and D. Siewiorek, "A scalable solution to the multi-resource QoS problem", in *RTSS, 20th IEEE Real-Time Systems Symposium, Proceedings*, pp. 315-326, Dec 1999.

[16] C.W. Mercer, S. Savage, and H. Tokuda, "Processor Capability Reserves: Operating System Support for Multimedia Applications", in *ICMCS, International Conference on Multimedia Computing and Systems, Proceedings*, pp. 90-99, April 1994.

[17] C.M. Otero Pérez and I. Nitescu, "Quality of Service Resource Management for Consumer Terminals: Demonstrating the Concepts", in *RTSS, 14th Euromicro Conference on Real-Time Systems, Vienna, Austria, Proceedings* Work in Progress Session, M. González Harbour, Ed., vol. 36/2002, pp. 29-32, June 2002.

[18] H. Radha, M. van der Schaar, Y. Chen, "The MPEG-4 Fine-grained scalable Video coding method for Multimedia streaming over IP". *IEEE Trans. On Multimedia*, vol. 3, no. 1, March 2001

[19] R. Rajkumar, K. Juwa, A. Molano, and S. Oikawa, "Resource kernels: A resource-centric approach to real-time and multimedia system", in *SPIE/ACM Conference on Multimedia Computing and Networking, Proceedings*, Jan. 1998.

[20] A.S. Tanenbaum, *Computer Networks*, 4th ed., Prentice Hall, 2003

[21] W. Van Raemdonck, G. Lafruit, E.F.M. Steffens, C.M. Otero Pérez, R.J. Bril, "Scalable 3D Graphics Programming in Consumer Terminals", in *International Conference on Multimedia and Expo*, Sept. 2002.

[22] C. Wüst, E.F.M. Steffens, W. Verhaegh, "Adaptive QoS Control for Real-Time Video Processing", *Euromicro Conference on Real time Systems*, Work in Progress, 2003.

Terminal QoS: Advanced Resource Management for Cost-Effective Multimedia Appliances in Dynamic Contexts

Jan Bormans

IMEC, Leuven, Belgium

jan.bormans@imec.be

Nam Pham Ngoc and Geert Deconinck

Katholieke Universiteit Leuven, ESAT/ELECTA, Leuven-Heverlee, Belgium

{ nam.phamngoc, geert.deconinck } @esat.kuleuven.ac.be

Gauthier Lafruit

IMEC, Leuven, Belgium

gauthier.lafruit@imec.be

Abstract Advanced multimedia applications such as those being developed within Ambient Intelligence typically share common characteristics, viz. the need to be able to access a wide variety of multimedia content using a heterogeneous communication and consumption infrastructure, in combination with low cost and low power requirements. The fact that a large variety of heterogeneous multimedia content has potentially to be dealt with (depending on user preferences and interaction) can lead to a cost inefficient overdimensioning of network and terminal resources. To tackle this issue, advanced resource management techniques are needed that make trade-offs on the fly to match the content bandwidth, the media coding and rendering complexity to the available network and terminal resources, while maximizing the overall perceived quality. This process is often referred to as Quality of Service (QoS) management. This paper illustrates the need to perform aspects of the overall QoS management on the terminal (Terminal QoS). The Terminal QoS for 3D graphics rendering on a software terminal (TriMedia set-top box) and reconfigurable platform is described in detail.

Keywords terminal resource management, quality of service, exploiting media scalability

1. Introduction

Advanced multimedia applications such as those being developed within Ambient Intelligence typically share common characteristics, viz. the need to be

T. Basten et al. (eds.), Ambient Intelligence: Impact on Embedded System Design, 183-201.
© 2003 *Kluwer Academic Publishers. Printed in the Netherlands.*

able to access a wide variety of multimedia content using a heterogeneous communication and consumption infrastructure, in combination with low cost and low power requirements.

The fact that a large variety of heterogeneous multimedia content has potentially to be dealt with (depending on user preferences and interaction) can lead to a cost inefficient over-dimensioning of network and terminal resources. To tackle this issue, advanced resource management techniques are needed that make trade-offs on the fly to match the content bandwidth, the media coding and rendering complexity to the available network and terminal resources, while maximizing the overall perceived quality. This process is often referred to as Quality of Service (QoS) management.

The most well-known aspect of QoS management is Network QoS, where the transmitted information is adapted to the network characteristics (bandwidth, latency, ...). However, when dealing with rich media in the Ambient Intelligence context, not all can be solved with Network QoS. For instance, simultaneously executing a multitude of media processes such as audio, video and 3D graphics, puts high computational demands that often cannot be met by high volume electronics consumer terminals, which - for cost reasons - are resource constrained to the average working conditions. Moreover, these computational requirements highly depend on the content richness and viewing conditions of the media (e.g. video window size, degree of motion, photo-realism of 3D graphics...), often leading to variations in rendering time of up to one order of magnitude, especially in 3D graphics [2, 6, 9, 14, 17]. Fluent, real-time 3D visualizations are therefore not always guaranteed, which seriously undermines the perceived output quality. To alleviate this problem, the 3D decoding and rendering is dynamically monitored and adapted (scaled down), in order to reduce the excess workload. For a more pleasant user experience, the adaptation process seeks optimal system operating points, by matching the computational requirements to the available processing power at minimal perceived quality degradation. It is clear that involving the network (or the server) will - because of the latency of the status information fed back from the client to the server - not enable to instantaneously react to the high load variations incurred by such dynamic applications. Hence at least part of the QoS process is to be performed at the terminal. This terminal-based adaptation management is referred to as Terminal QoS.

This paper describes the issues involved in Terminal QoS, focusing mainly on 3D graphics applications. Section 2 introduces the Terminal QoS framework, combining mainly (but not restrictively) video and 3D graphics applications. Section 3 gradually introduces the adaptation actions involved in 3D graphics. As an illustration, the application of 3D software rendering on a TriMedia set-top box is described. Subsequently, Section 4 shows how the Terminal QoS management can be provided with more axes of freedom by considering reconfigurable platforms. Finally, conclusions are presented in Section 5.

2. Terminal QoS Framework

The goal of Terminal QoS is to distribute the available resources (CPU time, memory, etc.) optimally among the different tasks (each task possibly contains several subtasks, also in demand for resources), possibly reducing the quality of the provided services in scarce resource conditions. The proposed Terminal QoS framework controls this task management in a layered way as shown in Figure 1. An in-depth description of this framework can be found in [18]. The rest of this section is merely devoted to a very brief summary of this framework.

The resource Manager (RM) provides guaranteed and enforced resource budgets, based on an admission test, which is similar to the functionality of a resource kernel. The resource budgets are provided to so-called Resource Consuming Entities (RCEs), which represent an indivisible collection of functional tasks (e.g. a 3D rendering engine or a video decoder) that will be controlled/adapted individually, consequently preventing direct interference between RCEs. RCE Controllers are added to make sure that the RCE performs acceptably well within the limitations of its budget. This is achieved by constantly adjusting the decoding parameters, possibly performing a decoding at lower (but still acceptable) quality for satisfying the budget constraint.

Obviously, the efficiency of the adaptation within an RCE can be greatly improved if the media to be rendered are scalable, since configuration choices are then no longer constrained to "ON/OFF" decisions only, i.e. the required resources and the resulting functionality can be controlled and adapted by a number of parameters. This is the case when coding for example 3D scenes with JPEG2000's [12] or MPEG-4's [10] wavelet-based still image coder for textures and MPEG-4's Animation eXtension Framework (AFX) tools [11] for 3D object shapes (called meshes).

The QoS managers determine the preferred quality settings and the budgets for the RCEs. The Global QoS Manager (GQM) is independent of domain semantics (e.g., 3D, Video), and takes decisions that encompass multiple domains. To avoid continuous (at small time constants) budget allocation decisions (and accompanying overhead), budgets are only renegotiated when a structural overload of one of the RCEs occurs, or when the services of the application layer are drastically changed. The GQM cooperates with one or more Domain QoS Managers (DQMs) that have domain-specific knowledge to take proper budget decisions.

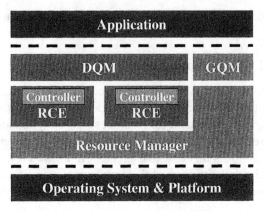

Figure 1.   The Terminal QoS framework between the Application and the Resource Manager.

For a given system, there may be several DQMs, one per semantic domain. For example, a system with 3D and video will require a 3D QoS Manager (3D-QM) and a Video QoS Manager (VQM).

This paper is focusing mainly on the DQM for 3D scene decoding, using MPEG-4 coding tools. The validity of the approach has been verified on a TriMedia set-top box (Section 3) and a reconfigurable platform (Section 4).

For the VQM, the reader is referred to [18] and [3].

3.   Terminal QoS for 3D Applications on TriMedia Set-Top Box

This section describes an example problem setting for Terminal QoS concepts in real life applications. The example described here was developed in the context of the ITEA EUROPA project [7] where one of the goals was to extend DVB-MHP (Digital Video Broadcasting - Multimedia Home Platform) with advanced 3D content rendering capabilities. DVB-MHP is middleware that allows broadcasters to offer interactive and digital television applications on consumer terminals [4]. This middleware is running on a set-top box connected to a television set. A set-top box is typically built around low cost media processors (e.g., TriMedia) that offer high quality video decoding and 2D graphics rendering capabilities. In our case, a set-top box with a TriMedia 1300 running at 166 MHz was used.

Extending the set-top box with 3D rendering capabilities could be envisaged by either considering a more powerful processor of by adding enough hardware to be sure that worst case processing requirements can be met. For the class of applications that had to run on the set top box, a variation of the processing requirements with a factor 6 was measured (depending on the complexity of the scene and the user interaction). This means that adding this amount of additional processing power would seriously increase the cost of the set top

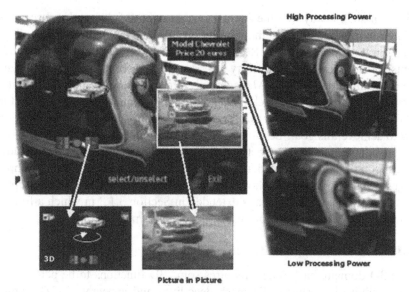

Figure 2.   Screenshot of the Terminal QoS enabled set top box application.

box, without a proportional increase in revenue for whoever bears the cost of the set-top box deployment.

Our approach consisted in using scalable 3D content whose processing complexity depends on a number of quality parameters. Terminal QoS principles were then applied to steer the adaptation of the 3D objects while keeping the overall perceived quality as high as possible, as will be explained in the next sections.

Figure 2 illustrates the final results of the QoS management implemented on the TriMedia set-top box platform, combining both video and 3D media. A screenshot of the resulting application is shown in the upper left corner. It is possible to pop up Picture in Picture functionality (middle bottom). When this is done and no sufficient resources are available, the Terminal QoS management switches the background video from high quality (upper right corner) to a lower quality setting (lower right corner). When the 3D objects are shown (lower left corner), the Terminal QoS management automatically adapts their characteristics to match the available processing resources.

The next sections are devoted to a more in-depth analysis of the 3D QoS Management, keeping in mind that concurrent video applications also consume a portion of the available processing power.

3.1   3D QoS Manager

The function of the 3D QoS Manager is to keep the rendering time of 3D objects (i.e. 3D RCE) within an assigned budget by adapting the quality of 3D objects in such a way that the overall visual quality of the rendered scene is as high as possible, despite the limitation on the available processing. This budget is assigned to the 3D RCE by the GQM based on the user's requirements and the resource availability. In order to perform the quality to frame rate adaptation, the 3D QoS Manager needs to know for each object and at a certain quality level, how much processing time is required and how good the quality perceived by the user is. The next subsections provide some ideas on how to achieve this goal. Section 3.1.1 introduces all necessary models to enable the adaptation control. Section 3.1.2 explains with a simple example how these models can be exploited appropriately for Terminal QoS. This example suggests that all 3D objects should be considered together during the adaptation decision taking. This idea is extended in Section 3.1.3 in which a formal definition of the adaptation optimisation problem is provided.

3.1.1 The Workload and Benefit Models

In the 3D domain, each object requires some 3D functions to be performed: the decoding of its shape (3D mesh decoding) and appearance (2D texture decoding) and the projective transformation of this 3D information onto the 2D display (3D rendering): see Figure 3(b).

In the 3D semantic domain, the natural entity for determining resource requirements, or *workload*, is a 3D functional task t_{ij}, which is a 3D function f_i applied to an object O_j, e.g. 3D rendering applied to the foreground race car O_1 in Figure 3(a). The workload of a given functional task t_{ij} is modelled through a Task Workload Model (TWM), providing a high-level relation between the CPU load and the parameters having an impact on the workload variation. For instance, for the software 3D renderer running on the TriMedia processor used in our experiments, Figure 4 shows how the rendering time of object O_j is influenced by its number ζ_j of 3D mesh triangles (polygons) and its screen *coverage* σ_j (in pixels). In [13] it is shown that ζ_j and σ_j uniquely determine the workload of the 3D rendering functional tasks on a wide variety of platforms. Note that for low and moderate-end platforms without hardware-accelerated cards, parameter σ_j has a large impact on the total rendering time. However, for powerful 3D graphics hardware accelerated platforms, the number σ_j of projected pixels for object O_j has little impact on the execution time, since the bottleneck is located around the 3D mesh processing, rather than in the texture processing of the 3D graphics pipeline. These sensitivities can be precisely modelled in TWMs by off-line black-box model calibration methods [1].

However, the prediction of σ_j is itself a hard problem to solve: in principle σ_j is only known after rendering, leading to a chicken-and-egg problem in estimating the 3D rendering workload. For 3D objects, this problem can be solved by adding a limited amount of meta-data to each object describing

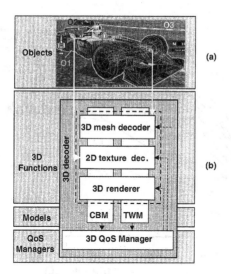

Figure 3. 3D semantics.

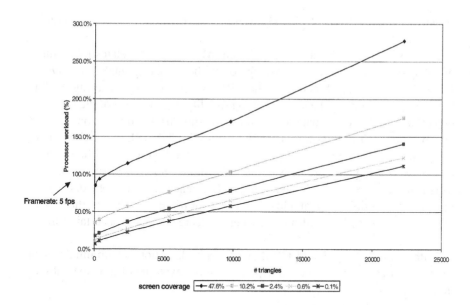

Figure 4. Workload variation with number of triangles and screen coverage, in percentage.

the relationship between its shape, the user's viewpoint and its parameter σ_j. This solution has also been adopted in MPEG-4 Synthetic and Natural Hybrid Coding [10].

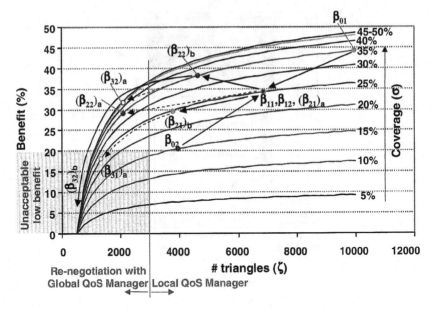

Figure 5.   Content Benefit Model, and potential benefits during the 3D graphics animation.

Besides of the TWM, a Content Benefit Model (CBM) relates the number of triangles ζ_j and the screen coverage σ_j to the rendering quality, or *benefit*, β_j of object O_j. In [20] and [21], the CBM is obtained by measuring off-line the overall error due to the simplification (quality degradation) of the rendered object in a variety of viewing conditions. To simplify the discussion, we will however consider Funkhouser's [5] and Gobetti's CBM [8], in which the benefit is an increasing function of ζ_j and σ_j, similar to Figure 5. For the same ζ_j, objects with a large σ_j are considered to be more important than smaller objects and therefore have a higher benefit. The benefit of small objects rapidly reaches a saturation level, at a low number ζ_j of triangles, whereas the benefit of larger objects suddenly drops off when ζ_j decreases below a certain threshold.

With both the TWM and CBM, workload adaptations at minimal visual degradation can be achieved, as explained in the example of the next subsection.

3.1.2 Adaptation Example

In this subsection, it will be shown with a simple scenario how, thanks to the TWM and CBM, the 3D QoS Manager fulfils its goal of staying within budget when possible, and (re)negotiating a budget when needed. An interactive 3D game with three objects: the 3D foreground race car O_1, the background race car O_2, and the landscape O_3, each coded in separate MPEG-4 streams, is

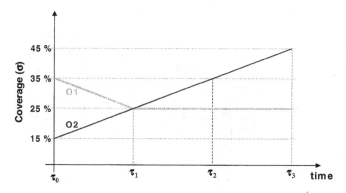

Figure 6.   Variation of the screen coverage in time for the objects O_1 and O_2.

taken as an example (see Figure 3(a)). In this scenario, it is assumed that the objects' screen coverage changes according to Figure 6.

It can also be assumed that all user interactions occur in the 3D game window, in other words, the 3D graphics window has user focus. For simplicity, it can be assumed that, despite its large coverage, the background object O_3 is sufficiently simple (very limited number of triangles) to be rendered with 12% of the processing power. Also, for simplicity, all 3D function workloads are combined into the simple linear model of Figure 7, which is in accordance with Figure 4. Instead of three functional tasks per object, as shown in Figure 3(b), only one is considered.

Initial 3D budget negotiation.   Since the 3D window has user focus, the video service has lower relative importance, and will get a smaller piece of the resources. In a negotiation with the 3D-QM and VQM, at time instance τ_0 (see Figure 6), the GQM allocates 70% of the processing power to the 3D service and only 30% to the video service. Objects O_1 and O_2 have operation points ζ_{0_1} and ζ_{0_2}, screen coverage 35% and 15%, workload 45% and 13% (see Figure 7), and benefit 44% and 21% (β_{0_1} and β_{0_2}, see Figure 5), respectively.

3D Budget re-distribution.   At each time instance τ_i, the 3D QoS Manager adjusts the number ζ_{ij} of triangles for each object O_j ($j = 1, 2$), in order to keep the total 3D workload constant, which leads to benefit β_{ij}. Any significant workload change Δ_{i1} resulting from the screen coverage variations of object O_1 between time instances τ_{i-1} and τ_i, must be compensated by an opposite workload change Δ_{i2} for object O_2, and vice versa.

At time instance τ_1 a new balance is achieved by selecting the operating points $\zeta_1 1$ and $\zeta_1 2$, yielding the same benefit for both objects: β_{11} and β_{12}. Both objects now have a workload of 29%. From that moment on, the size of O_1 remains constant, while that of O_2 increases (see Figure 6). At time

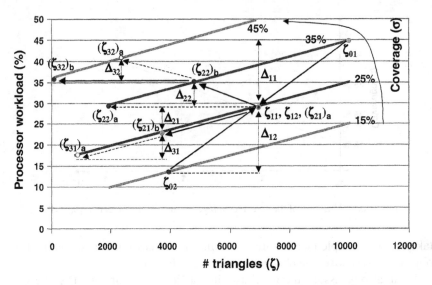

Figure 7.   Workload allocation during the 3D graphics animation.

instance τ_2, to avoid overload, i.e. the sum of the workloads exceeds the budget, it may seem reasonable to decrease the number of triangles of O_2 to $(\zeta_{22})_a$. Unfortunately, the resulting benefit $(\beta_{22})_a$ would then be lower than $(\beta_{21})_a$, the benefit of O_1. This is an unacceptable situation, because O_1 is now the smaller, and thus the less important object. Therefore, the number of triangles is modified for *both* objects, so that the more important object has the highest benefit: O_1 reduces to $(\zeta_{21})_b$, with workload 23% and benefit $(\beta_{21})_b$, while O_2 goes to $(\zeta_{22})_b$, with workload 35% and benefit $(\beta_{22})_b$. The workload changes Δ_{21} and Δ_{22} clearly compensate each other.

This simple example shows that the workload adaptations should be jointly optimised over all 3D objects together in order to achieve an acceptable over-all quality. The next subsection generalizes these concepts for more realistic conditions, i.e. real benefit functions, applied on a possibly large number of 3D objects populating the 3D scene.

3.1.3   Formalizing the Adaptation Framework

Let $T_{ij}(V)$ be the time needed to render object i at quality level j and at view-point V. In general the following relation holds:

$$T_{ij}(V) = f(\zeta_i, \sigma_i)$$

where function f is specific for each platform. This function is used by the 3D QoS Manager to estimate the rendering time of 3D objects at different quality levels and viewpoints.

To measure how good the quality of a 3D object is, instead of using the conceptual CBM of Figure 5, a more realistic quality metric is used: the Peak Signal-to-Noise Ration (PSNR), a well-known quality measure in the 2D image processing domain. The PSNR approximates the quality degradation between the object's image, rendered at full mesh resolution and the object's 2D image obtained after rendering with degraded mesh resolution. By using PSNR, the dependency of quality degradation on texture quality can also be modelled. To represent the viewpoint dependency of quality degradation, we propose that at each mesh resolution, PSNRs are measured in advance for a set of viewpoints around the object, from which the PSNR for any other viewpoints can be estimated by barycentric interpolation, in very much the same way as σ_i is estimated through meta-data in Section 3.1.1.

Denote $\beta_{ij}(L_i, V)$ to be the benefit of object i, rendered at quality level j and viewpoint V. The benefit value represents the amount of perception contribution of the object to the overall scene. It depends on the $PSNR_{ij}(V)$, the size of the object σ_i (in % screen coverage), the semantic meaning B_i of the object in the scene (e.g. a foreground object is more important than a background object) and the viewing angle α_i from the user viewpoint to the object. The view-dependent benefit function can thus be represented as follows:

$$\beta_{ij}(V) = B_i \cdot \sigma_i \cdot \cos(\alpha_i) \cdot PSNR_{ij}(V)$$

The objective of the quality to frame rate adaptation of the 3D QoS Manager is to select a combination of quality levels of all visible 3D objects in such a way that the overall benefit of the scene is maximized while the total rendering time does not exceed the frame time budget. Mathematically, the optimisation problem thus consists in finding quality levels j for all objects, such that:

$$\sum_{j=1}^{N} \beta_{ij}(V)$$

is maximized, subject to:

$$\sum_{i=1}^{N} T_{ij}(V) \leq T_{max}$$

where N is the number of visible objects at the viewpoint V under consideration and T_{max} is the execution time budget allocated to rendering one single frame.

This optimisation problem is an NP-hard optimisation problem for which an optimal solution cannot be found in real time. Heuristics are thus needed to find approximated solutions as fast as possible within an acceptable accuracy. The execution time of such an approximation algorithm should be much smaller, compared to frame time budget, in order to be used at every frame. In [19] an approximation algorithm that satisfies this constraint is presented.

In the next section, this framework is extended by including additional flexibility provided by the reconfiguration capabilities of the platform. The optimisation problem complexity will become large, therefore requiring even more heuristic approaches for proper decision taking.

4. Terminal QoS on Reconfigurable Platforms

Recent work on dynamic platform reconfiguration, such as reported in [15], extends the optimisation space available for Terminal QoS management. Indeed, the Terminal QoS described above in fact reconfigures the (rendering of) content to match the available platform (resources), but with reconfigurable platforms, it also becomes possible to match the platform to the rendering task to be performed. The Terminal QoS manager for a hybrid processor(s)/FPGA platform could e.g., decide to configure the FPGA as a MPEG-2 decoder when the main task is video decoding or to configure the FPGA as a 3D pipeline when mostly 3D objects have to be shown (remaining tasks are then run on the processor(s)). The next subsections explain how applications can be mapped on a reconfigurable platform, which consists of a instruction set processor (ISP) and run-time reconfigurable hardware, achieving a maximized QoS. Section 4.1 presents a task level model for applications. Section 4.2 describes a QoS middleware interfacing the applications and the software(SW)/hardware(HW) multitasking operating system running on the reconfigurable platform. Section 4.3 demonstrates the middleware framework with a case study of a 3D game.

4.1   Extended Adaptation Framework

Applications are represented by a scalable directed acyclic task graph $G(V, E)$ as shown in Figure 8. The vertices of the graph are tasks, which can be implemented in hardware or software. The edges represent data dependencies between tasks. Tasks may be scalable to provide different quality levels at different processing requirements. The tasks in dotted lines are optional: they provide additional functionalities to improve the quality of the application and hence may be dropped when needed. This kind of scalability is called layered scalability, where the quality of the application is enhanced by adding one or more enhancement layers and their accompanying tasks. For a different kind of scalability, called input-data driven scalability, the quality level of a task depends on the input data: the execution time and other resource requirements of the task are also a function of the input data. Usually, for this kind of scalability, the code sequence executed by the task is fixed, but the amount of time spent in the code depends on the richness of the incoming data. For example, in

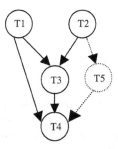

Figure 8. Task graph of applications.

3D rendering, the rendering algorithm is fixed and the rendering time depends on the mesh resolution, corresponding to the quality of the 3D object.

Two types of tasks are considered: SW only and HW/SW tasks. SW only tasks are implemented only in SW, while HW/SW tasks have been implemented in both HW and SW versions at the service provider side and one of these versions is selected at the user side.

Consider one application which has N different objects $O_1, O_2, \ldots, O_N$ sharing M tasks $T_1, T_2, \ldots, T_M$. Each quality level j of object i corresponds to a benefit value β_{ij} which represents the degree of user satisfaction when receiving this quality level. Let us further define:

- Q_i is the number of quality levels of object i.

- For each task T_k, a variable γ_k, $\gamma_k = 1$ if T_k is implemented in HW, $\gamma_k = 0$ if T_k is implemented in SW.

- $t_{T_{kij}}(\gamma_k)$ is the execution time of task T_k when processing object i at quality j with implementation γ_k. This time is the sum of the actual execution time and the communication time to send the data from this task to other tasks or to its environment.

- $A_{T_{kij}}(\gamma_k)$ is the hardware area needed by task T_k when processing object i at quality j with implementation γ_k. Therefore $A_{T_{kij}}(0) = 0$.

With these definitions, the mathematical framework of Section 3.1.3 can now be extended to include SW/HW partitioning. For applications whose tasks are executed sequentially one obtains the following optimisation problem: Maximize

$$\sum_{i=1}^{N} \sum_{j=1}^{Q_i} \beta_{ij}$$

subject to:

$$\sum_{i=1}^{N}\sum_{k=1}^{M}\sum_{j=1}^{Q_i} t_{T_{kij}}(V, \gamma_i) < T_{max} \text{ and } \forall i \sum_{k=1}^{M} A_{T_{kij}} < A_{max}$$

Although only time and hardware area constraints are considered here, other constraints like memory and power consumption can easily be added.

Observe that this constrained optimization problem is far more complex than the one encountered in Section 3.1.3, which was already recognized as NP-hard. Explaining the heuristics for solving such problems would lead us far beyond the scope of the paper. The interested reader is referred to [19, 15] for more details. The next sections will therefore restrict themselves to giving a feeling on how decisions can be taken and enforced by the associated middleware, based on a simple, pragmatic adaptation example.

4.2   QoS Middleware for Reconfigurable Platforms

Figure 9 shows the QoS middleware, interfacing with the application and the SW/HW multitasking operating system. This middleware consists of: (i) a QoS based SW/HW partitioner (QHSP); (ii) a resource estimator (RE); and (iii) a SW/HW translator (HST). The QoS based SW/HW partitioner basically performs the functionalities of the QoS Managers presented in Section 2. In addition it decides which tasks should be put in SW, which tasks should be put in HW and at which quality level, in such a way that the user receives the highest possible quality level of the application given the resource constraints (e.g., hardware area). The resource estimator is in charge of estimating the resource requirements for each task based on high-level information from the application (e.g., frame rate, resolution, metadata etc.) and information specific to the platform (which typically can be obtained by profiling). These resource requirements are the input for the QoS based HW/SW partitioner to make the partitioning decision. The HW/SW translator translates the platform-independent SW and HW code to platform-specific code. The translated SW code will run on the processor while the translated HW code will be used to configure the FPGA.

4.3   3D Game on a Prototype Reconfigurable Platform: A Case Study

A prototype platform has been developed which consists of a Compaq iPaq[TM] PDA, running RT-Linux on its Strong-Arm processor SA-1110 (206 MHz) and controlling a Xilinx Virtex[TM] XC2V6000 FPGA (Figure 10). The FPGA is the reconfigurable hardware, which is divided into logical tiles of coarse granularity with fine-grain reconfiguration. For allowing run-time reconfiguration,

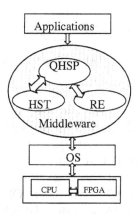

Figure 9. QoS middleware on reconfigurable platforms.

Figure 10. Prototype reconfigurable platform.

these tiles are connected by a packet-switched interconnection network. The iPaq and the FPGA board are connected together via the Expansion Bus of the iPaq. The communication between the processor and the network on the FPGA is performed by buffering messages in DPRAM. The glue logic on the FPGA board allows the processor to perform a partial reconfiguration of the FPGA. On top of RT-Linux is the middleware layer, which performs HW/SW partitioning and QoS management.

A Quake-alike 3D shooting game was developed as driver application for this platform. The 3D scene is composed of a number of walls. The aim of the game is to shoot the targets on the wall. The user can move left and right to go around the scene and can move the gun up and down to aim the targets. Water-ringing effects were added on the (wet) floor of the 3D scene. Figure 11 shows the task graph of the game. Tasks T1, T2, T3 do the transformation of the 3D scene from the object space to screen space. The texture mapping task

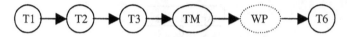

Figure 11. Task graph of the 3D game.

(TM) does the mapping of textures on the walls. The water processing task (WP) does repetitive 2D filtering operations to create the water-ringing effects. WP is an optional task. Finally, task T6 displays the 3D scene on the screen.

Furthermore, the window size of the game can be adjusted (scalable) from LARGE (320 × 240) to MEDIUM (210 × 160) or SMALL (160 × 120). Therefore, there are six different quality levels in total for the game, these are QL1(LARGE, water), QL2(LARGE, no water), QL3(MEDIUM, water), QL4(MEDIUM, no water), QL5(SMALL, water), QL6(SMALL, no water).

Tasks T1, T2, T3 and T6 are software only tasks and are implemented in C, while TM and WP are HW/SW tasks which are implemented in both C and VHDL. The VHDL code of the TM and WP tasks have been generated by OCAPI-XL [16], a C/C++ system-level hardware design environment. The synthesis has been performed with Synplicity^{TM} Synplify on the Virtex XC2V6000.

Analysis of the HW and SW code reveals that the execution time of the TM and WP tasks in HW and SW can be estimated based on the window size (X×Y), as follows:

$$t_{TM_{HW}} = (44 + (25 \cdot Y + 28) \cdot X) \cdot 0.02 \; \mu s$$
$$t_{TM_{SW}} = 1.95 \cdot X \cdot Y \; \mu s$$
$$t_{WP_{HW}} = 0.125 \cdot X \cdot Y \; \mu s$$
$$t_{WP_{SW}} = X \cdot Y \; \mu s$$

In our implementation, all tasks of the 3D game are executed sequentially and thus the time to execute one frame is the sum of the execution times of all tasks. The total execution time for tasks T1, T2, T3 and T6 is approximately 10% of the frame time. Table 1 shows some possible HW/SW partitioning solutions based on the model presented in Section 4.1 (here we have $N = 1$, $M = 6$, $Q_1 = 6$).

A number of simulations have been done by varying the user requested framerate and the number of available configurable logic blocks (CLB) on the FPGA. For example, for a framerate of 15 fps and the number of available CLB equal to 1300, the game will be played with LARGE window, without the water effects, while the TM task is running in hardware. However, if the user also wants to have the water effects, the game is played with MEDIUM window, while the TM task is running in hardware and the WP task is running in software (see Table 1). It is clear that the partitioning solution strongly depends on the quality of service requirements of the user.

Table 1. Different SW/HW partitioning solutions.

Quality level	Mapping	T_{TM} (ms)	T_{WP} (ms)	A_{TM} (CLB)	A_{WP} (CLB)	fps
QL1	$\alpha_{TM} = 0, \alpha_{WP} = 0$	150	77	0	0	4
	$\alpha_{TM} = 1, \alpha_{WP} = 0$	39	77	1200	0	7.7
	$\alpha_{TM} = 0, \alpha_{WP} = 1$	150	9.6	0	586	7.7
	$\alpha_{TM} = 1, \alpha_{WP} = 1$	39	9.6	1200	586	18
QL2	$\alpha_{TM} = 0$	150		0		6
	$\alpha_{TM} = 1$	39		1200		23
QL3	$\alpha_{TM} = 0, \alpha_{WP} = 0$	67	34	0	0	9
	$\alpha_{TM} = 1, \alpha_{WP} = 0$	17	34	1200	0	17
	$\alpha_{TM} = 0, \alpha_{WP} = 1$	67	4.3	0	586	12
	$\alpha_{TM} = 1, \alpha_{WP} = 1$	17	4.3	1200	586	42
QL4	$\alpha_{TM} = 0$	67		0		13
	$\alpha_{TM} = 1$	17		1200		52
...	...	...	...	...	...	...

One can thus consider the new solutions given by platform reconfiguration as new degrees of freedom that can be taken into account by the Terminal QoS management, leading to following possible actions for 3D Terminal QoS:

- Selecting/deselecting objects depending on their processing requirements and their importance in the scene (often resulting in a poor user perception);

- Using media scalability to attribute to objects a quality that is in line with the user perception and the processing requirements (as in the example of Section 3.2);

- Selecting between different implementations of the same functionality, without considering platform reconfiguration (e.g., selecting between a SW implementation requiring a lot of processor cycles and few memory transfers for a SW implementation requiring few processor cycles and a lot of memory transfers); and

- Selecting between different implementations of the same functionality, taking into account the possibility of platform reconfiguration.

5. Conclusions

In Ambient Intelligence, the need exists to be able to access a wide variety of multimedia content using a heterogeneous communication and consumption infrastructure, in combination with low cost and low power requirements. In this context, the network and terminal resources can be overdimensioned to be able to deal with worst case conditions (leading to cost inefficient solutions), or advanced resource management can be applied that aims to match the content resource requirements to the available resources, while maximizing the overall perceived quality.

The paper discusses how media scalability and platform reconfiguration extend the possible actions that can be taken by the Terminal QoS management: (i) selecting/deselecting objects depending on their processing requirements and their importance in the scene; (ii) using media scalability to attribute to objects a quality that is in line with the user perception and the processing requirements; (iii) selecting between different implementations of the same functionality, without considering platform reconfiguration; and (iv) selecting between different implementations of the same functionality, taking into account the possibility of platform reconfiguration.

Two designs were presented illustrating the feasibility of Terminal QoS in real life applications. Extending this work to be able to cope with any media and any platform, in concert with network QoS techniques still presents some though challenges and is hence the subject of ongoing research.

Acknowledgements

This work was supported in part by the IST-2000-30026 project, Ozone, and by the ITEA/EUROPA project.

References

[1] G. Bontempi and G. Lafruit, "Enabling multimedia QoS control with black-box modeling", Soft-Ware 2002: EUNITE Workshop Computational Intelligence in Telecommunications and Multimedia, In D. Bustard, W. Liu, R. Sterritt (Eds.) *Soft-Ware 2002: Computing in an Imperfect World*, Lecture Notes in Computer Science, LNCS2311, pp. 46-59, 2002.

[2] S. Brandt, G. Nutt, T. Berk and M. Humphrey, "Soft Real-Time Application Execution with Dynamic Quality of Service Assurance," *6th International Workshop on Quality of Service (IWQoS 98)*, pp. 154-163, 1998.

[3] R.J. Bril, C. Hentschel, E.F.M. Steffens, M. Gabrani, G.C. van Loo and J.H.A. Gelissen, "Multimedia QoS in Consumer Terminals," Invited paper *IEEE Workshop on Signal Processing Systems (SIPS)*, Antwerp, Belgium, September 26-28, 2001.

[4] http://www.dvb.org/ last visited on April 2, 2003.

[5] T.A. Funkhouser, *Database and Display Algorithms for Interactive Visualization of Architectural Models,* Ph.D. thesis, University of California, Berkeley, 1993.

[6] T.A. Funkhouser and C.H. Séquin, "Adaptive Display algorithms for Interactive Frame Rates during Visualization of Complex Virtual Environments," *Proc. of the 20th Annual Conference on Computer Graphics*, pp. 247 - 254, 1993.

[7] J.H.A. Gelissen, "The ITEA project EUROPA, A software platform for digital CE appliances," *Proc. Int. Conf. Consumer Electronics*, Los Angeles, CA, June 2002, pp. 22-23.

[8] E. Gobbetti and E. Bouvier, "Time-Critical Multiresolution Scene Rendering", *Proc. IEEE Visualization 1999*.

[9] H. Hoppe, "View-Dependent Refinement of Progressive Meshes," *Proc. of SIGGRAPH '97*, pp. 189-198, 1997.

[10] ISO/IEC 14496-2:2001: *Coding of audio-visual objects - Part 2: Visual*.

[11] ISO/IEC 14496-16: *Coding of audio-visual objects - Part 16: Animation Framework eXtension (AFX)*.

[12] ISO/IEC 15444-1: *JPEG 2000 image coding system, Part 1: Core coding system*.

[13] G. Lafruit, L. Nachtergaele, K. Denolf and J. Bormans, "3D Computational Graceful Degradation," *Proceedings ISCAS2000 Workshop and Exhibition on MPEG-4*, pp. III-547 - III-550, 2000.

[14] K.L. Ma and S. Parker, "Massive parallel software rendering for visualizing large-scale data sets", *IEEE Computer Graphics and Applications*, Vol. 21, No. 4, pp. 72-83, July/August 2001.

[15] J-Y. Mignolet, S. Vernalde, D. Verkest and R. Lauwereins, "Enabling hardware-software multitasking on a reconfigurable computing platform for networked portable multimedia appliances," *Proc. of the International Conference on Engineering of Reconfigurable Systems and Algorithms (ERSA)*, pp. 116-122, Las Vegas, June 2002.

[16] http://www.imec.be/ocapi.

[17] *The OpenGL Performance Characterization (SPECopcSM) organization*, http://www.specbench.org/gpc/opc.static/.

[18] C.M. Otero Pérez, E.F.M. Steffens, P. van der Stok, G.C. van Loo, A. Alonso, J.F. Ruíz, R.J. Bril and M. García Valls, "QoS-based Resource Management for Ambient Intelligence," this volume, 2003

[19] N. Pham Ngoc, G. Lafruit, G. Deconinck and R. Lauwereins, "A Fast QoS Adaptation Algorithm for MPEG-4 Multimedia Applications," *Proc. of Joint International Workshop on Interactive Distributed Multimedia Systems-Protocols for Multimedia Systems (idms-proms 2002)*, pp. 92-105, 2002 .

[20] N. Pham Ngoc, W. Van Raemdonck, G. Lafruit, G. Deconinck, and R. Lauwereins, "A QoS Framework for Interactive 3D Applications," *Proc. 10th International Conference on Computer Graphics and Visualization'2002, WSCG'2002* , February 4 - 8, 2002 .

[21] P. Yuan, M. Green and R.W.H. Lau, "A Framework for Performance Evaluation of Real-time Rendering Algorithms in Virtual Reality," *Proc. of the ACM symposium on Virtual Reality Software and Technology*, pp. 51-58, 1997.

Scalability and Error Protection - Means for Error-Resilient, Adaptable Image Transmission in Heterogeneous Environments

Adrian Chirila-Rus, Gauthier Lafruit and Bart Masschelein

IMEC, Leuven, Belgium

{ adrian.chirila-rus, gauthier.lafruit, bart.masschelein } @imec.be

Abstract   Ambient Intelligence aims to provide the means to transparently adapt the environment to the user needs. This paper focuses on the aspect of transmitting images over heterogeneous, unreliable networks to a diversity of terminals. The transparent adaptability to the heterogeneous environment is obtained through scalability concepts, providing support to the "Encode Once, Decode Everywhere" paradigm. The scalability is implemented using the intrinsic properties of the wavelet based coding, and data protection is based on turbo codes. Optimized hardware implementations of both technologies are described and intelligently exploited to provide support to the aforementioned Ambient Intelligence goals. All proposed solutions devote special attention to compliance with standards: MPEG-4 for compression, MPEG-21 for scalability and adaptation approaches and UMTS/IEEE 802.11 for error protection.

Keywords   scalability, source coding, wavelet coding, channel coding, turbo coding, unequal error protection

1.   Introduction

There is no doubt that within the coming decade computers will permanently assist human beings, eventually leading to the very interactive man-computer environment depicted in the visionary Star Trek series. Nowadays however, our high-technological environment can only be apprehended by very skilled technicians, putting us very far from the final goal of Ambient Intelligence, where our Ambient environment Intelligently interacts to the user's requests. Ambient Intelligence should thus provide means to let the environment transparently adapt the user's needs, through terminals that are preferably hidden (embedded) in the surrounding world. In real life the environments can be very complex, including inputs from many different data sources, different ways of encoding this data, different types of interconnections, and above all an immeasurable amount of users with very diverse profiles, capabilities and requests. Intelligently dealing with this immersive heterogeneity points toward

T. Basten et al. (eds.), Ambient Intelligence: Impact on Embedded System Design, 203-227.

the solution and aim of Ambient Intelligence. So far, the problem described is philosophical and as philosophical as it is, it offers very little practicality.

In the future, still images and video terminals will be certainly a part of such heterogeneous environment envisaged by Ambient Intelligence. The adaptation of the terminal to the user's requests and other changes will have to be done transparently, with minimal user interaction. For example, if the user changes priorities (e.g. switch to a 3D game and let the video play 'passively' in the background), the decoding of the incoming video images will have been assigned reduced resources. Nevertheless, the user should be able to preview or view the images at reduced quality that fit the new reduced resource allocation. Another example, of seamless adaptation is where a user changes the environment (e.g. leaving the room): he/she should be able to watch the incoming images in the new and unknown environment, e.g. PDA (Personal Digital Assistant) over an unreliable wireless network. Scalability of the transmitted data is a key technology in implementing the transparent adaptation required by the Ambient Intelligence. This paper therefore focuses mainly on 'intelligent' image/video transmission, for which a practical solution will be presented.

Every user has one or more terminals that allow him/her to interact with the environment, which is composed from data of various audio-visual sources, on-line or off-line, providing means to a sustained interaction with many other users. Inter-user connections are heterogeneous, as are the user terminals themselves, having very diverse local resources (display size, storage and processing capabilities, interconnection bandwidth,...). Moreover, since such environments evolve to higher mobility scenarios, the power consumption of the associated portable and handheld devices becomes increasingly important. It therefore does not suffice to provide means to adaptation to a variety of environmental conditions; it becomes also mandatory to propose low-complexity solutions, easily matched to miniaturized, low-cost devices. Recently, a very elegant solution to this problem has emerged in literature: the concept of "Encode Once, Decode Everywhere", through scalable data representation. The data stream is packetized according to the significance of the data, allowing fast transmission of low-resolution, though critical information, amended by progressively transmitting additional detail information, through additional data packets. This scalability mechanism provides means to optimally exploit the data transmission for recovering the audio-visual information at its highest quality under the imposed system resource constraints. Moreover, the coexistence of different scalability axes (Spatial resolution and Quality level per image frame, Temporal resolution in a video sequence) supports a variety of trade-offs between image quality and multi-variable resource costs. The control management between the complementary/competing system configuration options is an integral part of the Ambient Intelligence decision taking. The paper addresses this issue through simple application scenario examples.

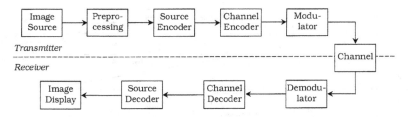

Figure 1. Classical architecture of an image transmission terminal.

2. End-to-end Transmission Architecture

A simplified architecture of the end-to-end delivery chain is depicted in Figure 1. Images gathered from an image source, preprocessed to remove unwanted noise and aliasing, are processed through (i) a Source Encoder for reducing the information redundancy prior to transmission, (ii) a Channel Encoder for protecting critical data segments against channel errors and (iii) a Modulator to fit the electrical data characteristics to the underlying physical transmission medium. The receiver performs all operations in reverse order to recover the original image (or video sequence) at its highest possible fidelity. A well-designed/controlled system should provide the highest possible immunity to non-idealities of the overall end-to-end delivery chain. In particular, spurious channel errors should minimally degrade the perceived quality of the received image, and resource constraints should not cause any failure of the system, i.e. the system performances should gracefully degrade under increasingly constrained resources (lower network bandwidth, decreased processing capabilities,...). Figure 2 shows an instantiation of the architectural view of Figure 1, with specific technologies that serve our purpose of Ambient Intelligence in Image transmission. Eventually, all technology presented in the paper provides an overall solution for the transmission chain in Figure 2. In particular, special attention is devoted to compliance with standardized technologies, e.g. MPEG-4[11] / JPEG2000[12] for the image coding, UMTS[21] / IEEE 802.11[10] for the channel coder and OFDM technology for the channel modulation. A wireless communication example is followed for its challenging issues w.r.t. error protection.

Each terminal is therefore additionally equipped with an RF-front-end. The image source is a flexible CMOS sensor, providing its raw data to a wavelet-based encoder further described in Section 3. The channel encoder is a turbo coder, for which more details are given in Section 4. The channel modulator is an OFDM modem, which sends data over a wireless link. The receiver part is the reversed processing chain of the encoder using the appropriate decoding technologies. A global controller performs the dynamic changes of the encoder/decoder parameters to best fit the communication system conditions,

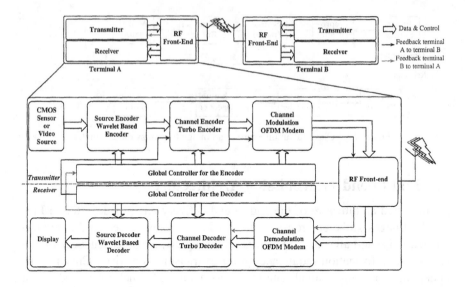

Figure 2. Wireless communication chain.

aiming at an adaptive Ambient Intelligence image/video transmission system. The overall system management relies on network (e.g. channel statistics, error rates, latency, ...) and terminal statistics (e.g. processing load, memory/storage topology, ...), satisfying the optimized implementation characteristics of the hardware or embedded system modules.

To better understand the control intelligence of Figure 2, the reader is invited to first assimilate the algorithmic issues and implementation specifics of the source and channel coders (Sections 3 and 4), before analyzing the overall control algorithm that maximizes the image quality for the given resources (Section 5), relying on implementation characteristics.

3.   Scalable Source Coding with Wavelets

Because of its good intrinsic features, the Discrete Wavelet Transform (DWT) took over the autocracy from the Discrete Cosine Transform (DCT) in the domain of image decorrelators. These features include both spatial and quality scalability. Transfer of large images over a network, whether it is a wireless link, the Internet or a downlink from a satellite, benefits from both features. With little downloaded data, a low quality/resolution image can already be recovered, allowing the user to decide whether or not the current image should be further downloaded.

Wavelets have recently also made their way into multimedia standards, e.g.

- MPEG-4 (Moving Picture Experts Group) in the area of Visual Texture Coding (VTC), mainly targeted to texture coding of 3D objects [11];

- JPEG2000, the successor of the Discrete Cosine Transform (DCT) based technique, better known as JPEG (Joint Photographic Experts Group) [12];

- CCSDS (Consultative Committee for Space Data Systems) is in the preparation of new space data system standards, using the wavelet transform for lossy image compression [24].

- The SVC (Scalable Video Compression) group of MPEG-4 envisages wavelet techniques to scalably encode video sequences [20].

This section is devoted to briefly explain these wavelet-based techniques, their advantages in the context of scalability for Ambient Intelligence (Section 3.1) and their associated implementation issues (Section 3.2).

3.1 Algorithmic Description

3.1.1 Why Wavelets?

In the current state of the art, compression techniques are typically targeted to compression performance, i.e. achieving the highest possible information reduction without perceived quality penalty. The well-known DCT-based coding techniques originate from this generation. The recently developed wavelet techniques do not dramatically outperform DCT in this field; only the image artifacts are slightly less annoying (slight edge smoothing and ringing effects in wavelets compared to sometimes very annoying blocking artifacts in DCT techniques).

However, recently, there is a higher demand for scalable coding techniques that support adaptation to varying environmental (network) conditions. Several solutions for scalable DCT coding have been proposed, but they are recognized as highly unsatisfactory. MPEG-4's Fine Grained Scalable (FGS) video coding for instance, provides only quality scalability, which incurs a bitrate overhead of 30 to 80%, compared to classical DCT coding [13]. Moreover, spatial scalability (scalability in the image size) is not supported by FGS.

In contrast, the Wavelet Transform is by its very nature a multi-resolution representation, having inherently the spatial scalability feature. The lowest resolution representation of the original image - the smallest image size - can be successively refined until the original image size is obtained without any cost in terms of additional bandwidth requirements. Also by using an efficient entropy encoder, which successively refines and encodes the wavelet image, the quality scalability can be obtained as well, at virtually no bitrate overhead.

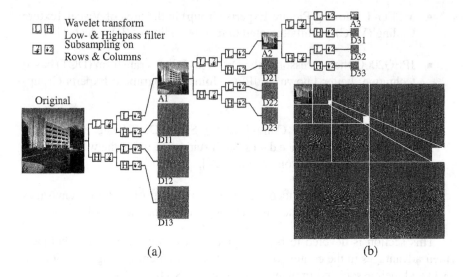

Figure 3.   (a) Three levels Wavelet Transform decomposition (b) The Wavelet Transform image and the parent - children tree structure.

Consequently, wavelet techniques are now widely adopted in scalable transmission applications.

3.1.2   The Wavelet Transform

The Wavelet Transform (WT) provides a multi-resolution/multi-level data representation of the original image, therefore supporting spatial scalability. Each level of the WT is obtained by recursively decomposing the previous (Lowpass-Lowpass) resolution level into four subimages: the Lowpass-Lowpass (LL), the Lowpass-Highpass (LH), the Highpass-Lowpass (HL) and the Highpass-Highpass (HH) subimages, through associated low- and Highpass filters in conjunction with downsampling. The LL-subimages are often referred to as the DC-images (or average images Ai in Figure 3 (a)), whereas the LH-, HL- and the HH-subimages are the AC or detail subimages (D11 - D33 for the 3-level WT of Figure 3 (a)). The lowest resolution DC image, grouped together with all AC subimages, creates the actual WT data structure (see Figure 3 (b)), which is as large as the original image. Starting from the DC image, successive higher resolutions of the image can be obtained by 'injecting' (through filtering and upsampling - not shown in the figure) the detail images into the DC image.

3.1.3 Embedded Zero Tree Coding

A common data entity used in the wavelet transform for image compression is the parent-children tree. The parent-children tree shown in Figure 3 (b) is a group of pixels with spatial correlation that is composed of one wavelet coefficient from the highest-level subimages, together with its four children in the lower level and all wavelet coefficients collected by recursively repeating this process until the lowest level is reached. Shapiro [18] has observed that whenever a zero-valued wavelet coefficient is encountered in the data structure of Figure 3 (b), there exists a high probability that also its children are zero. A high data reduction is obtained by exploiting this unique feature, gathering zero-valued wavelet coefficients into one single symbol: the Zero-Tree (ZT). This process may be regarded as an intelligent run-length coding.

The exceptions from the ZT are coded using as set of auxiliary, so-called dominant symbols: Isolated Zero (IZ) meaning that the coefficient is zero but at least one of its children has a non-zero value; Positive (POS) and Negative (NEG) indicates that the current coefficient has non zero value and it specifies its sign. Assuming a sign-magnitude representation of the coefficients and the usage of the POS and NEG symbols, the identification process of the dominant symbols is performed only on magnitude bits, including also the sign. Each wavelet coefficient tagged as dominant is separately and progressively transmitted through successive quantisation/approximation phases. In essence, the digits of the radix M representation of the wavelet coefficients are transmitted one by one, in so-called subordinate passes. Power-of-two quantisation corresponds to a binary radix representation. All digits of the same weight over different wavelet coefficients represent a so-called bitplane, as shown in Figure 4. Quality scalability is obtained by decoding the wavelet data structure of Figure 4 from the Most Significant Bitplane down to the (non necessarily) Least Significant Bitplane.

The global wavelet data structure of Figure 4 is thus finally decomposed in small information entities, called also contexts, which are coded through interleaved dominant and subordinate passes, creating for each subimage as many dominant and subordinate substreams as there are bitplanes and subimages. Considering the DC image, the dominant and the subordinate data as separate streams, the total number of possible contexts is given by $2 \times NoBitplanes \times NoSubimages + 1$. A 5-level wavelet transform, using a 16-bit sign-magnitude representation, has therefore a total of 451 ($2 \times 15 \times 15 + 1$) substreams that are coded separately and that can be reordered at will. Spatial scalability (through levels) and Quality scalability (through bitplanes) information can thus be interleaved for obtaining the best overall, combined bitstream.

The fact that the encoded data is embedded ensures that the decoding can stop at any point. This property is useful when the transmission bandwidth

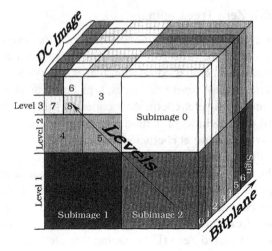

Figure 4.   Wavelet encoding contexts for a 3-level Wavelet Transformed image represented on 8 bits.

is strictly limited: the sender can simply stop transmitting when the target bandwidth has been reached.

The next subsection describes more refined approaches in exploiting the scalability options of the wavelet data structure of Figure 4.

3.1.4   Adaptation through Scalable Coding

A wavelet-based image encoder as described so far generates a set of sub-streams that can be clustered to build an embedded stream that targets not only a fixed compression ratio, but also a scalability target. The scalability can be spatial scalability (targeting users with different display sizes), image quality scalability (inherent to reduced processing power), or a combination of these two. For better understanding let us take some simple setup examples for different scenarios. In all cases, no constraints are imposed on the encoder, therefore being able to provide an embedded bitstream at highest possible quality. Starting from the same 451 substreams, described previously, all application scenarios can easily be supported.

In a first application scenario with varying and limited transmission bandwidth, one client connection is limited to 100Kb/s, another one is limited to 50Kb/s and the goal is to transmit to both of them 1 image per second. The classical approach would consist in creating totally different streams, each targeted to one specific user, yielding a total source bitrate of 150 Kb/s. According to the scalable stream coding paradigm, only one embedded bitstream of 100 Kb/s would be transmitted, the first user decoding the complete bitstream, while the second one only receives and decodes half of the bitstream.

Since image quality is the main user target, the substreams should be ordered in such a way that Peak Signal to Noise Ratio (PSNR) is maximized for the amount of data transmitted/decoded. In this application scenario, DC data is put first in the embedded bitstream, followed by the data of the higher significant bitplanes and gradually adding the lower significant bitplanes. This results in the quality scalability of Figure 5 (a) for the typical Lena image. The image PSNR is constantly growing with the number of transmitted bits, and the most important information is clearly concentrated at the start of the bitstream, suggested by the sharp, left-most rising of the curve. At 0.03 bits per pixel only the DC image is decoded. Starting with compression ratios around 0.4 bits per pixel the image quality is acceptable, while at 0.75 bits per pixel the differences compared with the original image becomes almost visually indistinguishable.

In a second application scenario with limited display size and decoding processing power, it is sufficient to create a bitstream with embedded spatial scalability. The user with the bigger display decodes the full bitstream, while the other one decodes only the portions corresponding to his/her display/decoding capabilities. This application scenario with spatial scalability is represented in Figure 5 (b): the image size increases successively with a factor 2, from the DC image size to the highest spatial image resolution.

As a third application let us consider a broadcasting scenario, with a wide range of devices that is potentially decoding the transmitted data. A bitstream with balanced embedded spatial and quality scalability can be created to serve all users regardless of their resources e.g. bandwidth, display size, processing power. For such combination of the cases illustrated in Figure 5 (a) and (b) a bitstream is built by adding first DC data, followed by interleaved contexts that gradually increases the image size, as well the image quality.

These simple examples clearly point out that several important conditions should be satisfied for appropriately using a scalable bitstream:

- It should be possible to cut the stream at any point (embedded).

- The most sensitive image information should be positioned at the beginning of the stream.

- It should have a degree of granularity in order to allow selective decoding.

3.2 Optimized Hardware Implementation

3.2.1 Wavelets versus DCT

Section 3.1.1 already pointed out that DCT techniques are not really appropriate for scalable transmission scenarios, from the compression point of view. But also from the implementation point of view, DCT exhibit severe drawbacks compared to wavelets.

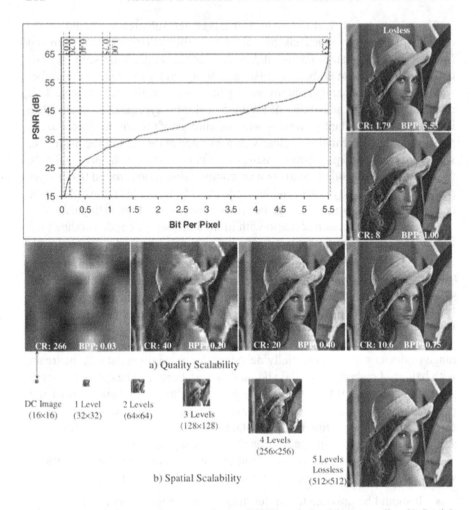

Figure 5.   a) SNR scalability - quantitative (PSNR) and visual image quality; b) Spatial scalability of a 512 × 512 image.

To provide spatial scalability one can always use a multi-resolution/multi-level Laplace Pyramid representation [5], constructed out of the original image. Unfortunately, this data representation incurs a storage overhead of 33% by collecting the all lower-resolution images in separate memory locations. In wavelets, thanks to some peculiar, advantageous mathematical reasons (critical downsampling capabilities), the overhead can be avoided, a condition which cannot be satisfied with DCT. Extending current DCT codecs with spatial scalability inevitably leads to a frame buffer storage increase of 33% compared to the wavelet-based techniques. This difference is reflected in the Wavelet ver-

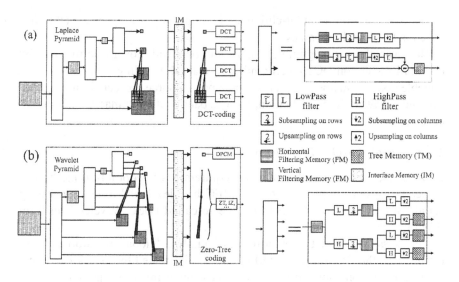

Figure 6. Scalable codecs: (a) scalable DCT encoder based on the Laplace Pyramid, (b) scalable wavelet encoder.

sus Laplace pyramidal structure of Figure 6, which shows the algorithmic flow graph of the multi-level Wavelet and DCT coding [14].

Despite of their differences, both multi-level coding schemes use essentially the same approach: a first stage transforms the image into a multi-resolution representation by successive filtering operations, the second stage performs the actual coding: parent-children coding for the WT, 8×8 block-oriented transform (DCT)-coding for the DCT. Note that the WT coding requires parent-children tree information throughout the levels of the multi-resolution representation, while the DCT technique codes the blocks in each level separately. This suggests that at first sight the DCT coding has less severe data dependencies for the decoding process, which would reflect itself in a lower implementation complexity. However, a more in-depth analysis reveals that the DCT decoding also requires a parent-children tree approach for the decoder memory optimization. All the DCT blocks that after decoding correspond to one particular 8×8 block in the decompressed image should indeed be processed simultaneously and should therefore be transmitted to the decoder as one cluster. Thus, even if the DCT encoding does not require the parent-children trees, a memory optimized DCT decoding process is inevitably bound to the data-dependencies of a parent-children tree.

As a consequence, the data processing in the DWT and DCT coders are essentially similar from the memory optimization point of view: a successive filtering stage for obtaining the multi-resolution representation is followed by a coding stage with a parent-children data-dependency graph, inevitably imposed in the decoding process.

However, the non-negligible storage and bandwidth overhead imposed by the multi-resolution DCT clearly indicates that DCT coding is not optimal for scalable coding: memory remains a precious resource, especially for embedded and hardware implementation, as does bandwidth in a massive number of users network.

3.2.2   Local Wavelet Transform

In the quest of reducing the memory requirements for optimized hardware/-embedded system implementations, IMEC has developed a new processing paradigm for the wavelet transform: the so-called Local Wavelet Transform (LWT).

The classical wavelet transform is essentially a 'global' transform, where for instance the DC image A3 in Figure 3 (a) is created by successive global filtering operations applied on the complete input image. This introduces a huge latency - and hence high memory requirements - between the availability of the input image and the extraction of the first parent-children tree, for further use in a compression chain.

One way to reduce this latency consists in dividing the image into blocks, called tiles. Tiling approaches, as the one proposed in JPEG2000 [12], reduce the effect of the global transform by independently processing the smaller rectangular regions of the original image. Each tile is then symmetrically extended along its four borders for providing sufficient data for proper successive filtering operations within the tile. The wavelet-transformed images that are then created are glued together as shown in Figure 7. Because adjacent tiles don't take information from their surrounding tiles, the so-obtained inverse wavelet transform typically results in visible blocking artifacts at the borders of the tiles, especially at high compression ratios: see Figure 8(a) for a fragment of San Diego image compressed at 0.2 bits per pixel. Figure 8(b) shows the results obtained with IMEC's FlexWave-II compression core (see section 3.2.3), which includes the LWT: blocking artifacts have completely disappeared and details are better distinguishable; the only remaining artifacts are inherent to the high compression ratio of 0.2 bits per pixel.

The Local Wavelet Transform (LWT) is a good compromise between the global transform and the tiling approach: its end result is the global wavelet transform (with the original image as one, indivisible entity), but the internal processing of the LWT mimics the tiling approach. In the LWT architecture, intermediate data between tiles used during the filtering process, are stored in an intermediate memory. The intermediate data is marked in Figure 7 as gray shaded area.

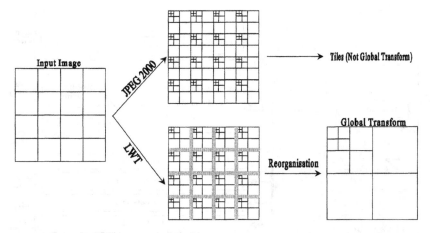

Figure 7. Differences between LWT and JPEG2000 Wavelet Transform.

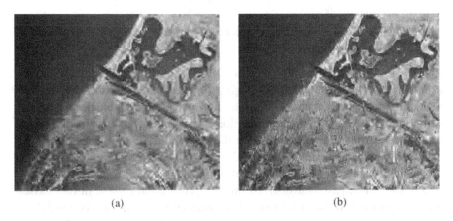

(a)   (b)

Figure 8. Compression artifacts using (a) JPEG2000 and (b) FlexWave-II.

The difference between the LWT and JPEG2000's tiling approach is that instead of performing symmetrical tile extensions, each tile in the LWT is extended with (overlapping) data from adjacent tiles, which has previously been stored in intermediate memory for the horizontal (row-wise) and the vertical (column-wise) filtering operations respectively. Consequently, tiles are seamlessly glued together for creating the same net result as if the filtering operations of the global wavelet transform were applied.

3.2.3 The FlexWave-II

FlexWave II [22] is IMEC's dedicated hardware for wavelet based scalable encoding of images using the Local Wavelet Transform and Embedded Zero Tree (EZT) algorithm, described in sections 3.2.2 and 3.1.3 respectively. The

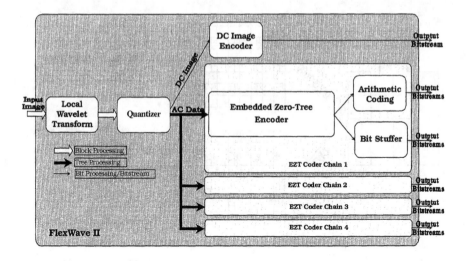

Figure 9.   FlexWave II architecture - A hardware Embedded Zero Tree encoder chain.

FlexWave II architecture, shown in Figure 9 takes as input the image divided in blocks of typically 32x32. The high programmability of the LWT module makes it possible to match the production of the output blocks to the consumption of the encoding chain, therefore minimizing the intermediate memory. The quantiser is able to perform the thresholding and the scaling of the each subband independently. The DC image is coded by simply packing the bits since it represents only a small amount of the total encoded information. The quantised wavelet transformed image is coded and refined in two passes producing so-called dominant and subordinate information (see section 3.1.3). Each dominant substream, is encoded using a multi-context 4-symbol arithmetic encoder [16], while the subordinate substreams are built by collecting and packing the bits, directly provided by the bitplane representation of the wavelet coefficients. Because of the high encoder throughput of 10 Mbit/sec at a relatively low clock speed of 40 MHz, the EZT encoder has been parallelised and instantiated four times, as shown in Figure 9.

FlexWave II has been implemented as an IP block, using 14.571 slices (43%) of a Xilinx Virtex II - XC2V6000 FPGA at 40 MHz. An estimation of the complexity for the 0.18 μm UMC technology yields preliminary synthesis results of 175K equivalent NAND2 gates. More implementation results are presented in Table 1.

Table 2 gives the lossless (i.e. at full bitstream decoding) compression performances for some representative images. All images have been encoded using a 5-3 taps filter, performing a 5 levels wavelet transform. The resulting typical lossless compression performance is around a factor of 2, depending on the

Table 1. Implementation details of FlexWave II.

Block	Technology		Memory
	FPGA XC2V6000 (Slices) Max 33,792	*0.18 μm UMC (Equivalent Gates)*	*(Words 16 bit)*
LWT	3,626+30 Multipliers	51,070	118,027
Quantiser	762	6,733	0
EZT Coder Chain	2355	26,760	3,684
DC Encoder	624	7,508	0
FlexWave	**14,571**	**174,776**	**132,361**
Operating Frequency	40 MHz	83 MHz	

Table 2. Lossless compression performances for various images.

Image	Uncompressed Size	Compressed Size (bytes)	DC Image Size (5 levels)	Compression Ratio
Akiyo	$352 \times 288 \times 8$ bit	49241	$11 \times 9 \times 10$ bit	2.0587
Lena	$512 \times 512 \times 8$ bit	145862	$16 \times 16 \times 10$ bit	1.7972
Sun Spot	$512 \times 512 \times 12$ bit	190216	$16 \times 16 \times 12$ bit	1.3781
Solar	$1024 \times 1024 \times 12$ bit	811528	$32 \times 32 \times 14$ bit	1.2920
North Atlantic	$1024 \times 1024 \times 12$ bit	638743	$32 \times 32 \times 12$ bit	1.6416

image content. Better compression ratios are obtained for natural images with smooth areas, while for satellite images with many details the compression performance is slightly smaller. Of course, higher compression performances can always be obtained by only partially decoding the embedded bitstreams. Compression ratios of 10 to 20 are readily achieved with acceptable image quality.

4. Scalable Channel Coding with Turbo Codes

Channel Coding is concerned with making the transmitted information immune to channel errors, which is a very challenging issue in wireless links. Well-established approaches like TCP/IP propose to retransmit any corrupted or non-received information, which is a perfectly satisfactory solution for non-real-time file transfers. Unfortunately, the resulting round-trip delays make such solutions unviable in the context of low-latency, real-time applications. Forward Error Correction (FEC) methods remedy to this situation by adding code redundancy, which are sufficiently matched to the channel characteristics for providing full error correction.

In this section, a special instantiation of FEC will be discussed: the turbo codes, which incrementally reduce the probability of information corruption through an iterative processing. Turbo codes are also referred to as parallel-concatenated convolutional codes (PCCC). The reader should keep in mind that this technique does not guarantee the correct decoding of the information, but that the probability of correctly recovering the information is maximized, close to the theoretical limit.

Section 4.1 gives a brief algorithmic overview of turbo encoders and decoders respectively, while Section 4.2 provides some specifics w.r.t. the hardware implementation that is plugged into the end-to-end transmission chain of Figure 2.

4.1 Algorithmic Description

Turbo codes are a method of intelligently adding redundant information to the transmission that yields to a maximum correct decoding probability at the receiving side. The decoding process, called also iterative turbo decoding, uses different pieces of the received information to recursively reconstruct, verify and improve (in case of errors) the transmitted information. The principle of iterative decoding is, simplistically, similar to the crossword solving, where one first tries to find the horizontal words, obtaining additional information to find vertical words, which themselves help in finding horizontal words, etc, until the crossword puzzle is completely solved. This idea is analyzed in more detail in the next subsections.

4.1.1 Turbo Encoder

A turbo encoder is formed by parallel concatenation of a copy of the input data, called systematic code, together with two Recursive Convolutional encoders, [23, 3, 7] separated by a random interleaver as shown in Figure 10. The component Recursive Convolutional encoder is based on a delay line convolution with feedback and feed forward. The structure is called parallel because the two encoders operate on the same set of input bits, rather than one encoding the output of the other. The multiplexing of the resulting sequences generates the turbo code: the original information sequence and two encoded versions of it, which can be seen as a set of parity-check sequences. The first encoder operates directly on the input sequence, while the second encodes a scrambled version of the original. The additional amount of redundant information is given by the encoder's code rate. For example, the code rate 1/3 indicates that for every input bit, 3 output bits are produced. The block diagram of a code rate 1/3 turbo encoder is shown in Figure 10, which has been standardized by "Universal Mobile Telecommunications System" (UMTS) committee [21].

The interleaver of the turbo coder is a pseudo-random block scrambler. The first role of the interleaver is to generate a new data block starting from the original input data. Secondly, it decorrelates the input to the two encoders so that

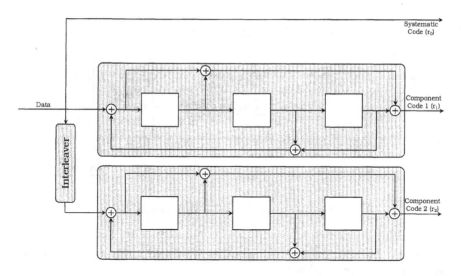

Figure 10. A typical 1/3 turbo encoder.

an iterative sub-optimal decoding algorithm based on information exchange between two component decoders can be applied. If the input sequences to the two component decoders are decorelated, a high probability exists that the correction of some of the errors in one decoder makes it possible to initiate the correction of the remaining errors in a second decoder, in the same line of thoughts as the horizontal and vertical passes of the cross-word puzzle mentioned before. The pseudo-random interleaving pattern must be available at the decoder as well.

Since any error correcting code increases bitrate over the network, it may be more appropriate to vary the code rate while keeping its designed structure intact. Typical examples are unequal error protection schemes that protect sensitive/critical data more intensely. This topic will be revisited in Section 5.2.

4.1.2 Iterative Turbo Decoder

A distinguishing feature of the turbo decoders [1] is that they rely on feedback information for minimizing the symbol or bit error probability, in the same manner as a turbo engine improves its efficiency through feedback.

Each iteration, the decoder generates optimum reliability estimates, in terms of probabilities that will be used in the next stage as starting point for a new iteration. The estimates are also known as a posteriori symbol probabilities and the algorithm used is known as the Maximum A posteriori Probability (MAP) decoding. It uses one encoded code and the a posteriori probability

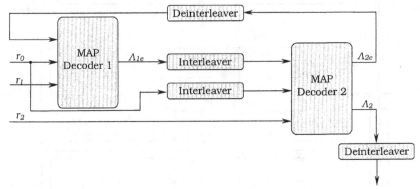

Figure 11. Typical iterative turbo decoder.

to improve the correctness likelihood of the systematic code, and generate the new a posteriori probabilities.

The iterative decoding consists of two components decoders serially concatenated via an interleaver, identical to the one in the encoder, as shown in Figure 11. The first MAP decoder takes received information sequence r_0 and the received parity sequence generated by the encoder r_1 of Figure 10 as input. The decoder then produces an estimate output Λ_{1e}, which is interleaved and used to produce an improved a posteriori probability of the information sequence for the second decoder.

The other two inputs to the second MAP decoder are the interleaved received information sequence and the received parity sequence produced by the second encoder r_2 of Figure 10. The second MAP decoder also produces a soft output Λ_{2e}, which is used to improve the estimate of the a priori probabilities for the information sequence at the input of the first MAP decoder. The decoder performance can be improved by this iterative operation relative to a single decoder. After a number of iterations, the soft outputs of both MAP decoders stop to produce further improvements, which triggers a hard decision Λ_2 on the recovered symbol through the final deinterleaving.

Since turbo decoders are iterative, they may also be considered as scalable in the sense that it is possible to perform any trade-off between quality (probability of correct decoding) and processing cycles. This approach is complementary to the unequal code rate distribution mentioned in previous subsection and further described in Section 5.2.

4.2 Optimized Hardware Implementation

The T@MPO core [9] is IMEC's efficient implementation of a low power, low latency and full duplex turbo coder/decoder based on parallel-concatenated convolutional codes described in Sections 4.1.1 and 4.1.2.

Table 3. T@MPO implementation characteristics.

Characteristic	Value
Max. Throughput	80.7 Mbit/sec
Latency	$< 10\,\mu s$
Max Clock Frequency	170.0 MHz
Power Consumption	< 10 nJ/bit
Number of equivalent gates (0.18 μm UMC)	373 K gates
Total Area	14.7 mm^2
Interleaver Block Sizes	432, 288, 256, 192, 144, 128, 96, 72, 64, 48, 32
Code Rates Available	1/3, 1/2, 2/3, 3/4, 7/8
Memory Size	2376 K Bit

Classical implementations of the turbo codes hardly allow a throughput of 2 Mbit/sec, which is acceptable for the UMTS recommendations [21], but upcoming generations of broadband applications might require up to 100 Mbit/sec, as suggested by the IEEE 802.11 standard [10].

The decoding operation can be massively parallellized, exploiting the block-based operations (interleaving and deinterleaving) and the double recursion inherent to the turbo decoder, resulting in increased throughput and reduced latency. T@MPO's innovative parallel decoder architecture [15] achieves a throughput of up to 80 Mbit/s in 0.18μm UMC technology, at a latency below 10μs and an operating power of less than 10 nJ/bit. The latter is obtained through a power conscious control of the internal activity using a dedicated clock strategy, and avoids unnecessary iterations thanks to an early stop criterion. The T@MPO implementation characteristics are presented in Table 3.

The programmability of the T@MPO core makes it possible to be used in various contexts, including unequal error protection schemes that will be tackled in more details in Section 5.

5. Transmission over Unreliable Channels

Previous sections have been devoted to analyzing the characteristics and specifics of each hardware/embedded module in the end-to-end system of Figure 2. This knowledge is about to be exploited to establish the best approach in reliably transmitting visual information over a wireless link. Section 5.1 is devoted to analyzing the impact of errors on the transmitted data, while Section 5.2 sketches a methodology for optimal error protection, achieving the best balance between recovered image quality, bitrate overhead and processing cycles.

5.1   Error Impact on the Encoded Bitstream

Different error types - with possible very diverse impacts - may appear in an unreliable network: single bit toggles, burst errors over consecutives bits, or even full packet errors (equivalent to long burst errors). Moreover, if the coder contains a variable length coding procedure, based on Huffman or arithmetic coding, it is impossible to a priori relate a bitstream position to a spatial location in the original image; this information can only be recovered by correctly decoding all previous bits of the bitstreams. Without special synchronization markers, any information beyond an error in a variable length coded bitstream is irremediably lost. This phenomenon will be referred to as a non-synchronization/cut error. This drawback does not occur in fixed length coding schemes, where each pixel corresponds to a fixed number of bits in the bitstream e.g. in the DC subimage and the subordinate substreams of the FlexWave-II output (see Section 3.2.3).

The aforementioned error types have also to be correlated against the characteristics of the embedded bitstreams provided by the FlexWave-II core: their substreams contain information portions at different levels of importance and inter-data dependency.

The DC image substream, for instance, represents only a small percentage of the total bitstream, but its content is very sensitive to errors, since a single error spreads over large portions of the decoded image. Despite the fact that non-synchronization/cut errors cannot occur in the DC image, its information should nevertheless be highly protected.

On the very opposite, subordinate substreams are not of critical importance, since (i) they only represent refinements on single wavelet coefficient values and (ii) their fixed length coding cannot produce non-synchronization/cut errors. Subordinate pass information can therefore probably be loosely protected. Be aware that data dependencies might cause errors in dominant substreams to ripple through in subordinate substreams, but this fact does not change the non-criticality of subordinate substreams.

Dominant pass information always uses a variable length-coding scheme, as a result of grouping data in specialized, single symbols (e.g. the Zero-Tree coding of a parent-children tree of zero values), and/or through the arithmetic coding process (see Section 3.1.3). Non-synchronization/cut errors may therefore occur, invalidating the remaining portion of the substream. Additionally, a high data-dependency exists between different dominant substreams and subordinate substreams, as suggested by Figure 12: by construction of the FlexWave-II core, correctly decoding dominant data for bitplane n and subimage s ($D_{n,s}$) is mandatory in order to correctly decode the dominant streams of the lower subband $s - 3$ ($D_{n,s-3}$), the dominant stream of the lower bitplane $n - 1$ ($D_{n-1,s}$) and the subordinate stream of that context ($S_{n,s}$). Any error in a dominant

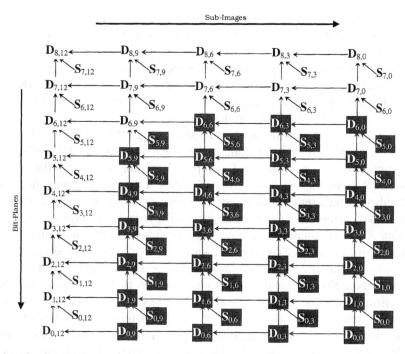

Figure 12. Substreams correlation in the FlexWave II encoded bitstream. Arrows point from current substream to the ones on which its correct decoding depends.

substream (e.g. $D_{6,9}$ indicated in shaded gray in Figure 12) has a devastating, cascading effect, further invalidating other substreams (indicated in black in Figure 12), that depends on it through the data dependency arrows of Figure 12. Consequently, dominant pass information should definitely be highly protected. In summary, it is clear that the FlexWave-II embedded output bitstream contains portions with diverse levels of importance. The DC image is critical, followed in importance by dominant streams, to which a single error can produce a cascading effect. The subordinate stream is not of such a critical importance because there are no other dependent streams (e.g. $S_{7,9}$ indicated in shaded gray in Figure 12). Consequently, for achieving the best trade-off between error vulnerability and bitrate overhead, an unequal error protection mechanism is preferred. This is analyzed in more details in the next section.

5.2   Unequal Error Protection Methodology and Results

Previous section has analyzed how errors can propagate and cause devastating effects in image decoding and reconstruction, as exemplified by Table 4 for the unprotected transmission (second image from left) of the Akyio test image in an error-prone wireless link. All figures of Table 4 are average results obtained through a large number of simulations, using a random Additive White Gaussian Noise model, corrupting on average 1 out of 60 transmitted bits (i.e. the energy per bit is 1.7 dB higher than the noise energy). To quantify the obtained image quality, the objective PSNR metric is used: it measures the energy of the difference between original and decoded image. As a rule of thumb, PSNR values below 30 dB relate to unsatisfactory image reconstructions, PSNR values above 40 dB yield high-quality reconstructions.

Error-free image transmission and decoding (high PSNR) can be obtained through a high, uniform protection yielding e.g. a data rate 1/3 (rightmost image of Table 4): the Turbo coder outputs 3 bits for each single data bit to protect. This perfect image transmission comes at the cost of a huge increase in bandwidth requirements: a factor 3! However, knowing that not all data transmission errors have the same impact on the recovered image quality (see Section 5.1), a non-uniform data protection can achieve the same qualitative image reconstruction at a considerable lower bandwidth increase, as suggested by the two rightmost images of Table 4. The methodology for distributing the protection bits into the data bitstream is the key to the success of the Unequal Error Protection scheme. The approach relies on the embedded bitstream scalability features, described in Section 3: all substreams are ordered in such a way that - for the specific application under test (see Section 3.1.4) - the quality of the reconstructed image increases monotonically with the number of transmitted bits, as generically represented in Figure 13(a). The expected behavior of this curve is that the quality starts increasing significantly at the beginning when the most important data is decoded and slowly reaches the maximum when all refinement bits are decoded (e.g. Figure 13(a)). W.r.t. the user's perceived quality criterion, the instantaneous speed of increase in image quality - i.e. its derivative (see Figure 13(b)) - is a measure of the information importance. Conceptually, this means that Figure 13(b) is also a measure for the protection level per transmitted bit. Unfortunately, only a discrete set of protection levels can be attributed to each bit. Consequently, as shown in Figure 13(c), each bit in the x first bits of the bitstream should be allocated the highest number of protection bits (data rate 1/3), the y next bits get a protection level yielding a data rate 1/2, etc. with x, y, etc chosen such that the total surface under the staircase curve of Figure 13(c) remains below a pre-established bandwidth budget. The Lloyd-Max [8] technique provides a solution to find the optimal combination protection (code rate) - protected bits (x, y, z) by minimizing the difference between the ideal protection distribution and the stair-step approximation. This technique has been applied for obtaining the unequal error protection results of Table 4.

In summary, exploiting the data error sensitivity of scalable bitstreams, an Unequal Error Protection scheme yields very immune image transmission capabilities with acceptable bandwidth increase.

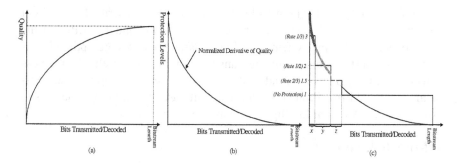

Figure 13. Generic quality measure (a) and its corresponding ideal protection bits distribution - infinite protection levels (b). Distributing limited protection bits over a limited number of protection levels by optimizing x, y, and z sizes (c).

Table 4. Unequal Error Protection effect.

Protection	Original	Unprotected	Unequal	Unequal	Uniform
Size (bits)	391,680	391,680	415,180	470,016	1,175,040
Code Rates	-	-	1/3, 1/2, 2/3	1/3, 1/2, 2/3	1/3
Increase	0%	0%	6%	20%	200%
Mean Error	0	3276	446	12	5
PSNR (dB)	Not Relevant	8.58	30.23	49.80	51.60

6. Conclusions

The ultimate goal of the Ambient Intelligence is to create an environment that automatically adapts to the user, which is a challenging problem because of the network/terminal heterogeneity and network unreliability. Although wavelets hardly outperform well-established DCT techniques for compression, their outstanding scalability features provide high flexibility to match the environmental conditions. In particular, wavelet compression techniques provide embedded bitstreams that by simple packet truncation and prioritization adapt to the available resources, such as bandwidth and processing power. Forward Error Correction has emerged as a solution to protect data over noisy channels for real-time and low-latency applications. Turbo coding is a highly efficient Forward Error Correction technique that additionally has an iterative processing approach, allowing also adaptation through scalability. Optimized hardware implementations for both technologies have been described throughout the paper: FlexWave-II for wavelet based image coding and T@MPO for the turbo codec. The flexibility features of both hardware cores are optimally exploited to support (i) immune image transmission through unreliable, wireless networks, and (ii) transparent adaptation to heterogeneous networks and terminals. Unequal Error Protection and scalability are the core functionalities to provide unique Ambient Intelligence adaptation capabilities.

Acknowledgements

The implementation of the FlexWave-II would not have been possible without the support of (and discussions with) other colleagues in the DESICS division at IMEC. The authors would also like to express their special thanks to Ph. Armbruster of ESA/ESTEC and W. Wijmans (former employee with ESA/ESTEC) for their valuable feedback. Part of this work was funded by the STARS WO3 project (ESTEC contract 13716/99/NL/FM).

References

[1] L.R. Bahl, J. Cocke, F. Jelenick, J. Raviv, "Optimum Decoding of Linear Codes for Minimising Symbol Error Rate", *IEEE Transactions on Information Theory*, vol. IT-13, pp. 284-287, 1974.

[2] S. Benedeto, G. Montorsi, "Design of Parallel Concatenated Convolutional Codes", *IEEE Transactions on Communications*, vol. 44, no 5, May 1996.

[3] S. Benedeto, G. Montorsi, "Unveiling Turbo Codes: Some Results on Parallel Concatenated Coding Schemes", *IEEE Transactions on Information Theory*, vol. 42, no. 2, March 1996.

[4] C. Berou, A. Glavieux, P. Thitimajshima, "Near Shannon Limit Error-Correcting Coding and Decoding: Turbo Codes", *Proceedings ICC'93*, Geneva, Switzerland, pp. 1064-1070, May 1993.

[5] P.J. Burt and E.H. Adelson, "The Laplacian pyramid as a compact image code," *IEEE Trans. Commun.*, vol. 31, pp. 532-540, April 1983.

[6] I. Daubechies, *Ten Lectures on Wavelets*, CBMS-NSF Regional Conference Series in Applied Mathematics, 1992.

[7] P. Elias, "Error-free Coding", *IRE Transactions on Information Theory*, vol. IT-4, pp. 29-37, September 1954.

[8] M.R. Garey, D.S. Johnson, and H.S Witsenhausen "The complexity of the generalized Lloyd-Max problem", *IEEE Trans. Inform. Theory*, 28(2), pp. 255-256, 1982.

[9] A. Giulietti, B. Bougard, V. Derudder, S. Dupont, J.-W. Weijers, L. Van der Perre "A 80 Mb/s Low-power Scalable Turbo Codec Core," *Custom Integrated Circuits Conference (CICC)*, Orlando, FL, May 12-15, 2002.

[10] IEEE Std. 802.11a - 1999, part 11, *Wireless LAN Medium Access Control (MAC) and Physical Layer (PHY) Specifications*.

[11] ISO/IEC 14496-5/FPDAM1, *The MPEG-4 Audio-Visual Compression Standard* ISO/IEC JTC1/SC29/WG11/MPEG99/N3309, March 2000.

[12] ISO/IEC 15444-1, *JPEG 2000 Image Coding system; Part 1 - Core Coding System*, JTC1/SC29, 07-2002.

[13] G. Lafruit, J. Bormans, *Complexity comparison between scalable wavelet codec and scalable DCT codec*, ISO/IEC JTC1/SC29/WG11/MPEG97/m2654, Fribourg, October 1997.

[14] G. Lafruit, L. Nachtergaele, J. Bormans, M. Engels, I. Bolsens, "Optimal Memory Organization for scalable texture codecs in MPEG-4", *IEEE Transactions on Circuits and Systems for Video Technology*, Vol. 9, No. 2, pp. 218-243, March 1999.

[15] F. Maessen, A . Giulietti, B. Bougard, L. Van der Perre, F. Catthoor, M. Engels, "Memory power reduction for the high-speed implementation of turbo codes," *IEEE Workshop on Signal Processing Systems (SIPS) Design and Implementation*, Antwerp, pp. 16-24, September 2001.

[16] R.R. Osorio, B. Vanhoof, "200 MBit/s 4-symbol Arithmetic Encoder Architecture for Embedded Zero Tree-based compression," *IEEE workshop on Signal Processing Systems. SIPS'01*, Antwerp, pp. 397-405, September 2001.

[17] P. Robertson, "Illuminating the structure of parallel concatenated recursive systematic (TURBO) codes", *Proc. GLOBECOM'94*, San Francisco, CA, pp. 1298 - 1303, November 1994.

[18] J.M. Saphiro, "Embedded image coding using zerotrees of wavelet coefficients," *IEEE Transactions on Signal Processing*, Vol. 41, No. 12, pp. 3445 - 3462, December 1993.

[19] D.S. Taubman, M.W. Marcellin, *JPEG2000 image compression fundamentals, standards and practise*, Kluwer Academic Publishers, 2002.

[20] S. Tsai, H. Hang, and T. Chiang, *Exploration Experiments on the Temporal Scalability of Interframe Wavelet Coding* ISO/IEC JTC1/SC29/WG11 MPEG2002/M8959, Shanghai, China, October 2002.

[21] Universal Mobile Telecommunications Systems (UMTS), *Multiplexing and Channel Coding (FDD)*, 3G TS 25.212 version 3.6.0 Release 1999.

[22] B. Vanhoof, B. Masschelein, A, Chirila-Rus, R. Osorio, "The FlexWave-II: a Wavelet-based Compression Engine," *ESCCON-2002*, pp. 301-308, Toulouse, September 24-27 2002.

[23] B. Vucetic, J. Yuan, *Turbo Codes. Principle and Applications,* Kluwer Academic Publishers, 2000.

[24] P.S. Yeh, G.A. Moury, P. Armbruster, "CCSDS data compression recommendation: development and status", *Applications of Digital Image Processing XXV, Proc.* SPIE 4790, Seattle, 2002.

Metaprogramming Techniques for Designing Embedded Components for Ambient Intelligence

Vytautas Štuikys and Robertas Damaševičius

Kaunas University of Technology, Software Engineering Department, Kaunas, Lithuania
vystu@if.ktu.lt, damarobe@soften.ktu.lt

Abstract   Design for Ambient Intelligence (AmI) requires development and adoption of novel domain analysis methods and design methodologies. Our approach is based on domain analysis methods adopted from software engineering, Generic Embedded Component Model (GECM) and metaprogramming (MPG). A novelty of our approach is that we apply MPG systematically in order to deal with a vast quantity, diversity and heterogeneity of embedded components, manage variability and raise the level of abstraction in embedded system design, as well as achieve higher flexibility, reusability and customizability for AmI-oriented design. We discuss applicability of the MPG techniques for designing embedded components (ECs) for AmI and provide three case studies.

Keywords   ambient intelligence, embedded systems, domain analysis, metaprogramming, generic embedded component, software generation

1.   Introduction

Ambient Intelligence (AmI) is usually defined as a combination of ubiquitous computing, ubiquitous communications and intelligent interfaces [10]. Ubiquitous computing anticipates a great quantity, diversity and heterogeneity of collaborating embedded systems (ES) in a distributed environment. Ubiquitous communications anticipate a communication infrastructure that is flexible, reliable and omnipresent. Intelligent interfaces require self-adaptability and mobility of advanced user interfaces that enable interaction between *nomadic users* and environment in a natural and personalized way. Design for AmI involves convergence and integration of software (SW) and hardware (HW) design methodologies, embedded and distributed design, as well as consumer and microelectronics technology. AmI may become the next paradigm for ES design.

Currently, ESs are applied as subsystems in a wide variety of the so-called smart products: mobile phones, MP3 and DVD players, TV sets, kitchen appliances, etc. These ESs implement a large diversity of functions, however, they are composed of a limited number of common SW/HW components such

229

T. Basten et al. (eds.), Ambient Intelligence: Impact on Embedded System Design, 229-250.

as DSP, MPEG, codecs, etc. These basic blocks of an ES can be called *embedded components* (ECs). We use this term as a generic name for IP (*Intellectual Property*) components (IPs), (embedded) SW components, HW components (soft IPs), and SW/HW subsystems.

The AmI paradigm raises the stakes in ES design. Smart products, additionally to their basic functionality, must be able to communicate and co-operate with other smart products within a common AmI-oriented environment. Therefore, every ES must be designed as an EC for a larger context of application. ES designers have to reconsider the existing ES design methodologies or develop the new ones.

Already, ECs must be developed on an industrial scale in order to match huge customer demand and varying requirements, as well as maintain quality-of-service and shorten time-to-market. One of the solutions that can be used for achieving these aims is a concept of *product lines* [9]. The concept considers SW systems in the same way as traditional industrial products, and aims at achieving greater configurability, variability and adaptability to particular user and application requirements. Adoption of this concept for AmI-oriented ES design can be particularly useful.

ES Design Roadmap [12] outlines major opportunities for action in ES design as follows. (1) Promotion and facilitation of IP reuse on a broad scale in order to enable increase in design productivity. (2) Development of methods and tools that capture ideas in high-level models and bridge the gap between requirements and specification. (3) Design space exploration with the purpose of obtaining high-quality solutions.

Our contribution is based on (1) analysis of requirements and constraints for designing embedded components (ECs) for AmI-oriented ES and the usage of modern domain analysis methods, (2) Generic Embedded Component Model (GECM) for designing generic ECs that encapsulate families of EC instances, and (3) Metaprogramming (MPG) paradigm for managing variability across the domain and describing generic ECs.

We propose to use MPG for (1) dealing with diversity and heterogeneity of ECs, (2) raising the level of abstraction in ES design, and (3) achieving higher flexibility, reusability and customizability for AmI. This paper continues our previous research on soft IP design using MPG techniques [33].

The structure of this paper is as follows. In Section 2, we review the related works. In Section 3, we analyze requirements and constraints for designing ECs. In Section 4, we describe the domain analysis methods that can be particularly useful for EC design. In Section 5, we describe the EC design framework based on MPG and Generic Embedded Component Model (GECM). In Section 6, we deliver three case studies that demonstrate applicability of the MPG techniques for designing ECs for HW, embedded SW and SW domains.

In Section 7, we evaluate the proposed design methodology. Finally, in Section 8, we conclude.

2. Related Works

Embedded system (ES) design is a very wide topic. We are mostly interested in research that covers generalization, customization and specialization of ES and their components. For example, Edwards *et al.* [11] argue that higher-level models are necessary to specify ES architectures and implement the underlying models of computation. Lee [23] underscores importance of *frameworks* (abstract models of computation) for embedded SW design and examines the problem of adapting SW design techniques to meet the challenges of the physical world. Paulin *et al.* [29] examine the following trends in embedded SW and HW design: (1) extreme diversity of requirements, architectures and applications, (2) rapid growth of complexity of design, (3) emergence of new standards for ES subsystems. Goossens *et al.* [16] underline importance of system-level algorithmic optimizations and high-quality code generation in embedded SW design. Ernst [13] surveys typical ES architectures and discusses the ES specialization problem that deals with adaptation of ES architectures to a given domain application. Rau and Schlansker [30] discuss specialization and customization of ES that involves tailoring the reusable but not efficient general-purpose ES to the application domain, and propose to automate this process using the parameterized component libraries and design frameworks.

Several authors suggest designing product lines of ES. Diaz-Herrera and Madisetti [8] describe a product line methodology for designing ES that focuses on controlling variability within a product line and designing commonality across product lines. Männistö *et al.* [26] investigate embedded SW as a configurable product family that may include millions of variants from that individual products are configured to meet particular customer needs. Variability in SW families can be implemented using customization, pre-processing and modularization.

One of the major challenges for designing a product line is to identify, implement and manage variability. To tackle this problem, several approaches are proposed. For example, Czarnecki and Eisenecker [7] propose Generative Programming approach that uses the built-in MPG features of programming languages to implement generic components and configuration generators for developing embedded SW. Beuche *et al.* [4] describe the techniques based on Aspect-Oriented Programming (AOP) for customization of embedded SW. The AOP methodology is based on the separation of aspects and their integration using the generative techniques. Cross and Schmidt [6] use the MPG techniques to provide the distributed real-time ES with open interfaces through which they can be configured in order to optimize their performance and re-source consumption.

AmI only shortly has emerged as a vision for consumer electronics, thus there is still little research in the area. Aarts and Roovers [1] describe basic AmI functions, generic classes of devices for AmI and survey HW design challenges for AmI. Riva *et al.* [31] present an extensive overview of AmI paradigm and formulate the technological requirements, challenges and capabilities that an AmI environment should provide. Lindwer *et al.* [25] examine the challenges to AmI vision, acknowledge the existence of a gap between AmI systems and components that are needed to implement those systems, and call for development of design methodologies and tools based on new paradigms that take the AmI vision into account. Basten *et al.* [2] explore scalability of various AmI enabling technologies. The authors acknowledge that one of the most challenging and least understood issues about AmI systems is how to design them, and advocate for the development of new programming models and the usage of the generative techniques.

The related works can be summarized as follows. The authors emphasize the role of generalization, customization and specialization in ES design, and acknowledge the need for higher-level design abstractions (frameworks, models) to deal with complexity and diversity of ECs. The emerging AmI paradigm further raises requirements for ES design and underscores importance of new domain analysis methods, design methodologies and techniques such as product lines and metaprogramming. These approaches usually focus on separation of design concerns, management of domain variability, and usage of the generative techniques. Analysis of domain constraints and requirements is vital. We consider these in the next section.

3. Analysis of Constraints and Requirements for Embedded Component Design

Currently, ES design faces the following technological and market challenges. (1) Increasing diversity and complexity of products, services, and applications. (2) Increasing number of non-functional constraints. (3) Increasing degree of integration and networking. (4) Growing importance of flexibility and embedded SW. (5) Shrinking time-to-market.

Any ES architecture must balance the following technological, product, and market constraints. (1) *Technological constraints*: Moore's law and *design productivity gap* [17], the divide between available gates on a chip and design tool capability. (2) *Product constraints*: smart products demand various combinations of high performance, low cost, and low power. (3) *Market constraints*: time-to-market and the need to support a sharp increase in number and complexity of ES designs.

Requirements for embedded component (EC) design from the designer's viewpoint can be summarized as follows. (1) Easier reuse of heterogeneous HW/SW modules. (2) Design space exploration considering non-functional constraints (e. g., memory size, power usage). (3) Transformation techniques for obtaining minimum power or maximum speed from available specifications. We expect the following requirements to predominate for future AmI-oriented EC design.

(1) *Higher complexity*. AmI will require billions of tiny devices packed into millions of systems handling any number of tasks. Such systems will have to serve many masters at the same time, straining the existing design methodologies. Increasing heterogeneity and networking of ES further complicates EC design.

(2) *Specialization*. Developers design general-purpose ES for reuse in numerous smart products. Since their specific applications are unknown these designs must incorporate both generality and completeness. Specialization involves departures from both characteristics with respect to structure and functionality. The more ECs a system has, the larger design space for optimization and specialization is. Therefore, there is a need for tailoring every aspect of ES to meet specific demands.

(3) *Diversity of ECs*. Specialization of ECs leads to a vast diversity of ECs available for ES design. Already, there is a great diversity of ECs with the same functionality but different characteristics (high performance, low power, etc.). In the AmI environment, diversity of ECs will only grow. AmI will require subsystems that span 9 orders of magnitude in bandwidth or computational load, and 6 orders of magnitude in terms of power dissipation [14]. The realm of embedded processors already ranges from simple 4-bit controllers to 64-bit desktop processors.

(4) *Heterogeneity*. Assembly of ES from ECs that were designed using the *orthogonal* requirements leads to heterogeneity of the designed systems. Furthermore, ECs are often designed using different design styles, methodologies and interface standards, and those are very hard to integrate.

(5) *More complicated component selection and design space exploration.* A prerequisite for successful ES design is selection of ECs that match the requirements of ES. Component selection requires successful design space exploration. The growing design complexity requires performing design space exploration in a structured and (semi-) automatic way.

(6) *Further raising the level of abstraction.* In HW design, it has proved necessary to introduce a new level of abstraction every 8 years in order to allow designers to cope with growing complexity [12]. As the AmI paradigm emerges, it is a high time to introduce a new level of abstraction that must be suitable for HW as well as embedded SW design in order to allow using it before partitioning of HW and SW.

(7) *Increased quality, productivity and reuse content.* Complexity of modern ES design requires designers to reuse ECs. Reuse of existing ECs is expected to increase from current 20% to 80% in 2011 [12]. To make ECs more reusable there is a need of introducing the high-level component models and standardizing EC properties such as area and power dissipation. Automatic EC customization, wrapping and interface synthesis should increase productivity and reusability in ES design.

(8) *Standardization.* There is an urgent need of new standards for EC design, verification and integration into ES.

(9) *Time-to-market.* Principal requirement to ES design is to shorten time-to-market in order to stay ahead of competition. Major hurdles are EC qualification, exchange, and seamless integration into the designed ES. In order to solve these problems more research is needed on formal verification methods, collaboration-based design, program transformation and optimization techniques, and design automation.

(10) *Fault-tolerance.* ECs designed for AmI should be more fault-tolerant. Various architectures and redundancy schemes should be explored.

(11) *Intelligence.* AmI requires designing more intelligent ES that are personalized, pro-active and user-oriented. ES designers have to focus on designing more and higher-quality ECs for speech, face and gesture recognition for user-friendly communication with smart products.

We focus on mapping of the most part of stated requirements onto the design framework described by the domain and application-specific analysis methods, the metaprogramming techniques, and the Generic Embedded Component Model (GECM). We detail these in the following sections.

4. Domain Analysis Methods for Embedded Component Design

Although a variety of domain analysis methods exists, we focus on two kinds of the methods: the general and application-specific ones. We apply them for well-understood domains where well-proven models exist (e.g., triple redundancy model for fault-tolerant design). We consider *Multi-Dimensional Separation of Concerns* (MDSOC) [28] as a general domain analysis method. The related domain analysis methods are *Feature-Oriented Domain Analysis* (FODA) [19] that deals with analysis and documentation of product *features*, *Feature-Oriented Reuse Method* (FORM) [20] that is an extension of FODA for *product lines*, and *Family-oriented Abstraction, Specification and Translation* (FAST) [34] for *SW families*.

MDSOC [28] is based on the idea that *design concerns* should be represented independently. A designer develops domain programs by composing the separated concerns according to systematic rules. Concerns can be understood in terms of an *n*-dimensional design space, called a *hyperspace*. Each dimension is associated with a set of similar concerns, such as a set of component instances; different values along a dimension are different instances. A *hyperslice* is a set of instances that pertain to a specific concern. A *hyper-module* is a set of hyperslices and integration relationships that dictate how the units of hyperslices are integrated into a program. MDSOC is especially useful where a great variety of requirements exist at different levels of abstraction such as in ES design. Suitability of the explicit separation of concerns for platform-based ES design is demonstrated in [22].

FODA [19] focuses at the identification of distinctive *features* of SW systems. These features are user-visible domain characteristics that define both common aspects of the domain as well as differences between related domain systems. The underlying concepts of FODA are as follows:

(1) *Aggregation and decomposition*. Aggregation means composition of separated concerns into a generic component. Decomposition means abstracting and isolating domain commonalities and variations.

(2) *Parameterization*. Parameterization means that components are adapted by substituting the values of component parameters that are defined to uniquely specify the application-specific concern.

(3) *Generalization and specialization*. Generalization means capturing the domain commonalities, while expressing the variations at a higher level of abstraction. Specialization means tailoring a generic component into a specific component that incorporates the application-specific details.

Applicability of FODA for ES design is demonstrated by [3] for developing the family of embedded operating systems.

FORM [20] starts with feature modeling to discover, understand, and capture commonalities and variability of a *product line* in terms of *product features*, and organizes them into a model that serves for deriving reusable and adaptable artifacts (e.g. components). FORM suggests the following guidelines for developing components for a product line:

(1) *Separation of common features from product specific features*. A product line includes components implementing the common features of the product line and those that implement the product-specific features.

(2) *Encapsulation of product specific features*. Features specific to each product are separated from the components implementing the core features and encapsulated in separate components with appropriate interfaces.

(3) *Separation of additive features*. The product line components must be reusable and composite, and accommodate variability among products through different composition configurations.

FORM has been applied to several industrial application domains, including phone and elevator control systems [20].

FAST [34] focuses on grouping similar entities or concepts into *families*. A family is a collection of system parts that are treated together because they are more alike than different. Grouping is based on the *commonalities* that fall along many dimensions (structure, algorithm, etc.). Commonality defines the shared context that is invariant across abstractions of the application. The individual family members (instances) are distinguished by their differences, called *variability*. Variability captures the properties that distinguish abstractions in a domain. The aim of FAST is to uncover variability and provide means for its implementation.

Some actions performed during domain analysis can be automated. We consider *parsing* as an application-specific domain analysis method that is concerned with automatic analysis of abstract domain representations - source code. Parsing have been used for a long time in SW compilers. It allows better understanding of the domain and extracting the application-specific information for EC customization of and/or domain code generation.

We assume that the analyzed domain has *variant* and *invariant* parts. These parts represent variability and commonality in the domain. The primary goal of using domain analysis methods is to recognize the parts, and express them in a suitable form for further refinement. At a higher level, we use domain concerns along with domain-oriented abstractions to describe the variant and invariant parts more precisely. Finally, we formulate the issues of domain analysis either explicitly or implicitly. The *explicit* artifacts are taxonomy of domain objects and their features (requirements, parameters), methods, processes and models for ES design. The *implicit* artifacts are domain knowledge in the form of a conceptual model used for further refinement of the obtained domain artifacts and models.

5. Embedded Component Design Based on Metaprogramming

5.1 Outline

In this section, we explain principles of heterogeneous metaprogramming (MPG). We consider the MPG paradigm as a bridge for combining the domain analysis methods, Generic Embedded Component Model (GECM), and generative techniques. First, we describe a framework of the MPG paradigm that is aimed at developing metaspecifications, in Subsection 5.2. Then, we describe GECM and its implementation using MPG in Subsection 5.3. We illustrate applicability of the MPG techniques by presenting a simple example in Subsection 5.4. Finally, we summarize our findings in Subsection 5.5.

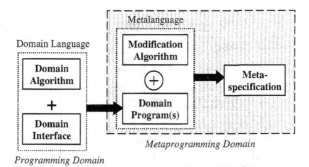

Figure 1. Framework of the heterogeneous MPG paradigm.

5.2 Framework of Metaprogramming (MPG) Paradigm

(a) *Definition and objectives.* MPG is a higher-level programming technique that provides a means for manipulating (generating, customizing, etc.) domain programs. The main aim of MPG is to develop a *metaspecification* - a program generator for a narrow domain of application. Such metaspecification describes commonalities and variations in a particular application domain. Commonalities reflect the shared context that is invariant across a set of similar components, whereas the variations that capture the distinguishing properties of the components are specified at a higher level of abstraction and represented via generic parameters.

(b) *Principles.* Here we consider the principles of heterogeneous MPG only (see Figure 1): (1) *explicit separation of concerns,* (2) *usage of multi-language paradigm,* (3) *external parameterization* (see Subsections 5.3 and 5.4).

(c) *Role of languages.* The role of a *domain language* is to express basic domain functionality (domain algorithms). The role of a *metalanguage* is to express generalization, to describe modifications of domain code and to generate the modified domain component instances. The modification algorithms range from the basic *meta*-constructs such as *meta-if* (conditional generation) and *meta-for* (repetitive generation) to the sophisticated application-specific *meta*-patterns that are composed of the nested combinations of the simpler *meta*-constructs.

5.3 Generic Embedded Component Model (GECM)

(a) *Objectives.* GECM (Figure 2) serves for developing metaspecifications (representations of generic ECs). GECM is common for both HW and SW domains: it may represent either HW IPs (soft IPs) or SW IPs, respectively.

(b) *Structure.* GECM describes the structure of metaspecifications that represent the generic ECs. Each metaspecification has a metalanguage interface

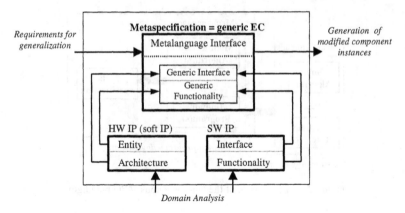

Figure 2.   Generic Embedded Component Model (GECM).

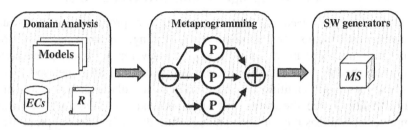

Figure 3.   Framework for development of a metaspecification.

for describing the generic parameters, and a body that contains generic interface and generic functionality of ECs. Metaspecifications serve for (1) concise
representation of the families of qualified EC instances that have the related
functionality, and (2) selection and generation of the particular domain component instances depending upon the values of the generic parameters introduced
through metalanguage interface.

(c) *External parameterization* (see Figure 3). (1) The domain (usually represented by one or more available component instances (ECs), models and the
requirements (R) for modification and generation) is analyzed, and the modification concerns are identified and separated ($\ominus$). These concerns represent
the *variable* aspects in a domain that *depend* upon generic parameters. (2) The
separated concerns are expressed through generic parameters, implemented using the parameterization-based MPG techniques (P), and then integrated ($\oplus$)
back with the *fixed* aspects of a domain that are *orthogonal* with respect to the
generic parameters. The result is a metaspecification (MS) that encapsulates a
family of the related EC instances and implements a generic EC.

(d) *Generation.* A metaspecification that describes a generic EC is used
as a set of instructions for a metalanguage processor to generate the domain

language code (modified EC instances) depending upon the values of generic parameters specified by a designer or other program (metaspecification).

5.4 Application of Metaprogramming: An Example

Below we present an example of a generic gate (see Figure 4) described using Open PROMOL [32] as a metalanguage, and a VHDL as a domain language.

(a) *PROMOL functions*. Open PROMOL is a functional metalanguage. In a PROMOL metaspecification, all modifications of a domain program are represented as a specific composition of the PROMOL functions with the domain language code to be modified. Each PROMOL function has a list of arguments and returns a string of the domain language code. The arguments are either a constant, a parameter implicitly declared in the *PROMOL interface* of a metaspecification, an expression, or a composition of PROMOL functions and domain language code. The functional programming paradigm allows us to achieve a great deal of flexibility, since the domain language code may be gradually modified by several nested functions, until it reaches its final form.

(b) *PROMOL interface*. Interface of a PROMOL metaspecification describes generic parameters that represent different aspects of design. In this example (see Figure 4), these are a number of gate inputs (num) and a logical function (f). Each generic parameter has an explanation in a natural language, a set of feasible values specified, and a default value assigned.

(c) *Modifications*. Necessary modifications of the domain language code are implemented using a set of PROMOL functions. In this example, modification of a domain component (gate) concerns the name of a VHDL *entity*, a *port* statement and a *signal assignment* statement. We used the **@sub** function to insert the value of the generic parameter into a domain program, and the **@for** function to implement a *meta-for* loop.

5.5 Summary

(a) The example shown in Figure 4 represents only the simplest case of the parameterization model. Tasks from the real world may require more complex parameterization models such as conditional and hierarchic ones that we do not consider here.

(b) A conventional programming language (C++, Java, etc.) can be used in the role of a metalanguage, as we will demonstrate in our case studies in Section 6. Therefore, contrary to AOP and MDSOC, our approach generally does not require any new languages or tools to support the MPG paradigm.

(c) A metalanguage does not bring new domain content, but rather describes modification of the existing domain content gained through domain analysis at a higher level of abstraction.

```
@- PROMOL interface of a generic gate
$
"A number of inputs:"  {2..16}              num := 3;
"A logical function:"  {AND, OR, XOR}       f := AND;
$
@- Generic interface of a VHDL component
-- Interface of a VHDL component
ENTITY  GATE_@sub[f]_@sub[num]  IS
   PORT (X1 @for[i, 2, num, {, X@sub[i] }]: IN  BIT;
         Y : OUT  BIT);
END  GATE_@sub[f]_@sub[num];

@- Generic functionality of a VHDL component
-- Functionality of a VHDL component
ARCHITECTURE  BEH  OF  GATE_@sub[f]_@sub[num]  IS
   BEGIN
      Y <= X1 @for[i, 2, num, { @sub[f] X@sub[i] }];
END  BEH;                                                (1)
```

```
-- Interface of a VHDL component
ENTITY  GATE_AND_3  IS
   PORT (X1, X2, X3: IN  BIT;
         Y : OUT  BIT);
END  GATE_AND_3;

-- Functionality of a VHDL component
ARCHITECTURE  BEH  OF  GATE_AND_3  IS
   BEGIN
      Y <= X1  AND  X2  AND  X3;
END  BEH;                                                (2)
```

Figure 4.   An example of MPG: (1) a generic gate metaspecification (in Open PROMOL), and (2) a generated instance (in VHDL) for default generic parameter values (num = 3; f = AND).

(d) The MPG paradigm that is considered here as a specific composition of a metalanguage with a domain language brings generative technology and bridges domain analysis with SW generators.

6.   Case Studies

To demonstrate applicability of the MPG techniques for EC design, we deliver three case studies, i.e. the generation of (1) fault-tolerant (F-T) soft IPs; (2) specialized Discrete Cosine Transforms for embedded SW; and (3) intelligent interfaces between PC and Pocket PC. Our intention is to show suitability of the MPG techniques for different application domains (HW, embedded SW and SW), as well as contribution of MPG for essential elements of AmI: ubiquitous computing and intelligent user interfaces.

6.1   Generation of Fault-Tolerant (F-T) ES

(a) *Motivation.* For critical applications such as airbag or brake control, reliability of ES is essential. The embedded HW should be able to handle every possible system state under normal and exceptional conditions. The demand

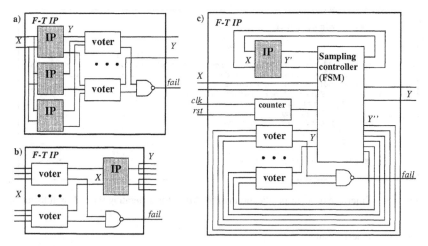

Figure 5. Simplified architectures for the triple-redundant F-T: a) space, b) data, and c) time.

for such systems is quickly growing. In this case study, we demonstrate how to generate the F-T soft IPs using the MPG techniques.

(b) *Analysis.* The F-T applications usually rely on *redundancy* [18], i.e., the addition of resources, time, or information beyond that is needed for normal system operations as follows (see Figure 5). (1) *Space* redundancy - the addition of extra HW, e.g., in Triple Modular Redundancy, we have three instances of the same soft IP and their outputs are voted using a majority voter. (2) *Data* redundancy - the addition of redundant data (e.g., auxiliary data path, error-correcting codes) to ensure reliability during a transfer of data via system interconnections. (3) *Time* redundancy - the usage of additional time to perform system functions, thus achieving soft error tolerance.

As we see in Figure 5, there is a great deal of similarity in the domain of F-T architectures. We treat the soft IP, whose reliability we increase, as a black-box component. This soft IP is wrapped with a trivial circuitry that performs majority voting of redundant signals in search of a possible error.

(c) *Implementation.* (1) We analyze soft IP source code described in VHDL to extract the values of the parameters for wrapping. These parameters represent the soft IP interface signals. (2) Extracted values are used to generate the F-T architecture models from the metaspecifications implemented using heterogeneous MPG (Open PROMOL [32] as a metalanguage, and VHDL as a domain language). Each metaspecification encapsulates a generic domain object (counter, voter, FSM, etc.). (3) The generation process is controlled via the top-level PROMOL metaspecification, whose interface is presented in Figure 6. This interface describes the set of generic parameters that are used to select the soft IP and specify the concrete implementation of the F-T architecture.

```
$
        "Select a component:
                1 - dragonfly core
                2 - i8051 microcontroller
                3 - Free6502 core"                              {1,2,3}         c := 1;
[c = 1] "Enter address bus width for program memory"            {8,16}          prgm := 16;
[c = 1] "Enter address bus width for data memory"               {4,8}           dm := 8;
[c = 1] "Enter address bus width for i/o"                       {4,8}           io := 8;
[c = 3] "Generate Free6502 core version for:
                1 - synthesis
                2 - debugging"                                  {1,2}           g := 1;
        "Select the type of redundancy:
                1 - space
                2 - data
                3 - time"                                       {1,2,3}         type := 3;
        "Enter the order of redundancy"                         {3,5}           order := 3;
$
```

Figure 6.   Interface of metaspecification in PROMOL for generating the F-T architectures.

Table 1.   Synthesis results.

IP	Area, cells (IP)	Area, cells (space F-T IP)	Area, cells (data F-T IP)	Area, cells (time F-T IP)	Power, μW (IP)	Power, μW (space F-T IP)	Power, μW (data F-T IP)	Power, μW (time F-T IP)
Free-6502	4670	14688	5006	6696	8.2693	25.8447	11.1202	12.2117
Dragonfly	5883	18347	6907	9856	19.9421	58.2476	23.6390	21.7217
i8051	24258	72774	24345	28578	50.5518	100.2620	63.9123	56.5174

(d) *Results.* For F-T applications, we have customized the freely available third-party soft (HW) IPs. (1) Free-6502 core [21] is a CPU core compatible with 8-bit 6502 microprocessor. (2) DRAGONFLY core [24] is a controller that can be used for serial communication management, FLASH and SDRAM control, etc. (3) i8051 micro-controller [15] is compatible with 8-bit microprocessor designed by Intel. Synthesis results of the original and generated cores (Synopsys; CMOS 0.35 μm) are presented in Table 1.

The synthesis results show that an average increase of chip area is about 209% for space redundancy, 8% for data redundancy and 43% for time redundancy. An average increase of power consumption is about 201% for space redundancy, 26% for data redundancy and 23% in time redundancy for the generated F-T components with respect to the original soft IPs. Note that space redundancy means that there are 3 instances of the original soft IP.

6.2   Generation of the Specialized DCT Algorithms

(a) *Motivation.* One can not overestimate the importance of DSP algorithms such as *Discrete Cosine Transform* (DCT) in ES design. Many smart products use DCT as an EC for various applications such as image and speech processing, face recognition, etc. The overall performance of ES design depends

```
#include <iostream.h>                                              (a)

template <class T_RET, class T_PARAM>
void gen_table(T_RET func_ptr(T_PARAM), T_PARAM start, T_PARAM finish,
               int num, char *name) {
  cout << typeof(start) << " " << name << "[" << num << "] = {\n";
  for (int i=0; i<num; i++) {
    cout << "\t" << (T_RET)(*func_ptr)((T_PARAM)(start+(finish-start)*i/num));
    if (i!=num-1) cout << ",\n";
      else cout << "\n};\n\n";
  }
}
```
```
  gen_table(&cos, (double)0, (double)2*PI, 16, "cos_table");       (b)
```

Figure 7. (a) Metaspecification in C++ for generating look-up tables, and (b) an example of the generation of the 16-value cosine look-up table.

upon embedded SW quality. Different applications usually require different performance/memory trade-offs. For this reason, optimization of embedded SW is one of the most important issues in ES design. In this case study, we demonstrate how to significantly speedup the DSP applications at the cost of the larger data memory size.

(b) *Analysis.* The most performance-costly part of DCT is calculation of the *cosine* function. According to Amdahl's law, the most effective way to improve performance of a program is to speed-up the most time-consuming part of it. To achieve better performance, we apply the *data specialization* [5] approach: beforehand known cosine values are stored in a generated look-up table. The trade-off here is that accuracy of the result may depend upon the size of the table. However, in many applications such as JPEG, the results of DCT are rounded-off to the integer values anyway.

(c) *Implementation.* We perform specialization of the given DSP algorithms as follows: (1) We analyze application source code to identify references to the trigonometric library-based C functions (cos, in this case). (2) We generate a look-up table for the cosine function using heterogeneous MPG: we use C++ as a metalanguage to generate C code (Figure 7). Homogeneous MPG (C++ templates) is used to generalize a C++ metaspecification for generating look-up tables of any 1-argument C function. (3) Then, all references to the cosine function are replaced by the reference to the cosine look-up table.

(d) *Results.* We have performed the experiments with the one-dimensional forward DCT (DCT) and inverse DCT (iDCT) algorithms. The source code was compiled for 32-bit 300 MHz *Infineon TriCore*[TM] MCU-DSP using TASK-ING EDE for TriCore and simulated using TASKING CrossView Pro - TriCore instruction accurate simulator. We have measured the cumulative code execution time, code memory size, and required data memory size for the specialized algorithm. The simulation results for both original (generic) and specialized implementations are presented in Table 2.

Table 2. DCT simulation results.

SW algorithm	No. of points	Execution time, cycles (generic)	Execution time, cycles (specialized)	Code size, B (generic)	Code (+data) size, B (specialized)
DCT	4	315,823	17,962	352	228+64
iDCT	4	308,051	17,001	358	264+64
DCT	8	1,078,458	57,890	362	228+128
iDCT	8	1,077,309	55,497	346	264+128
DCT	16	3,989,454	203,026	364	228+256
iDCT	16	4,007,161	198,129	350	264+256
DCT	32	15,383,480	754,418	372	238+512
iDCT	32	15,450,803	743,769	358	274+512

The simulation results show the drastic speedup of about 19 times, the decrease of 134 B in code size that is attributed to elimination of cos function call, and increase of $16n$ B (n - number of points) in the required data memory size. (Note that we used *float* type instead of *double* to save data memory, as it had no impact on the accuracy of the DCT).

6.3   Generation of Intelligent Interfaces

(a) *Motivation.* Researchers envision an AmI environment with a platform of communicating smart devices that will have to be controlled by remote control. User interfaces should be simple enough to be accommodated by an average mobile consumer (sometimes called *nomad*), and intelligent enough to be able to control the complex functions those appliances perform. A recent trend is that handheld devices such as PALMtop or PocketPC provide an intermediary interface with that the user interacts as a remote control for any appliance [27]. We use the MPG techniques to generate the intelligent interfaces between PC and NomadPC (PocketPC) to control the database application.

(b) *Analysis.* The architecture of the system is presented in Figure 8.

Below, we explain the architecture in detail: *RDB* - a remote database of the nomadic user. *MDB* - a meta-database, where the context-sensitive meta-data (i.e., information about hierarchies of nomadic user's databases, attributes of tables and fields, etc.) is stored. *SQL* - SQL queries that are used to extract content from a database. *ASP* - Active Server Page that includes one or more scripts that are processed by a Script processor before the page is sent to the user. *HTML* - the user is working with an Internet browser, where all data and management elements are displayed using HTML elements. *GUI* - the user interface visible on the screen of Nomad PC. *Http* - a browser sends data back to Local PC using HTTP (HyperText Transfer Protocol). *XML* - XML file that is generated for every user's database. XML file includes meta-data from the MDB and data from the user's RDB. *XSL* - a style sheet file that describes how

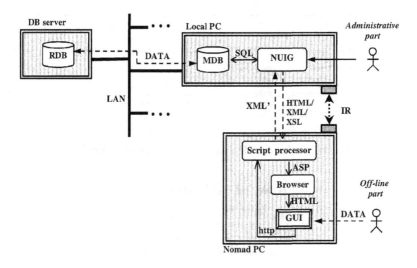

Figure 8. Architecture of the remote control application.

data that is sent over using XML, is to be presented to the nomadic user. XSL file is generated automatically for every user's database separately. *XML'* - user's off-line edited data is saved in the XML file using ASP and updated on the RDB. *IR* - XML/XSL/XML' files are transmitted between Local PC and Nomad PC using Infrared wave gateway. *DATA* - database records that are read from the RDB, edited by the nomadic user, and written back to the RDB. *NUIG* - Nomadic User Interface Generator that generates the web-based user interfaces for a Nomad PC.

(c) *Implementation.* The NUIG was implemented using C# programming language. The implementation consists of several parts as follows:

The *Administrative* part is used to: (1) query the MDB using generated SQL queries to extract context information (hierarchy, content, etc.), (2) create and edit attributes of tables and fields in order to customize interfaces to the particular preferences of a nomadic user; (3) export data from the nomadic user's RDB into the XML file; (4) generate the XSL file that is used to render XML data on a web browser; (5) import data from edited XML file into the nomadic user's RDB.

The *Off-line* part is used to (1) visualize generated user interfaces using locally stored XML and XSL files; (2) edit database content.

We have used MPG (C# as a metalanguage, and HTML, XML, XSL as domain languages) to generate graphical web-based user interfaces. The metaspecification (see Figure 9) describes a generic HTML INPUT tag.

(d) *Results.* In Figure 10, we present a simplified example of an interface that was generated for the database application using the MPG techniques.

```
enum _INPUT { BUTTON, CHECKBOX, HIDDEN, IMAGE, PASSWORD, RADIO, SUBMIT, TEXT};
char* get_type_value (_INPUT  t);

void generate_html_input(_INPUT t, char *value, char *name, char *function)
{
        xWr.WriteStartElement("input");
        xWr.WriteAttributeString("type", get_type_value(t));
        if (value) xWr.WriteAttributeString("value", value);
        if (name) xWr.WriteAttributeString("name", name);
        if (function) xWr.WriteAttributeString("onClick", function);
        xWr.WriteEndElement();
}
```

Figure 9.   An example (fragment) of code generation in C#.

Rows of table Region

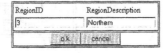

Figure 10.   Database interface generated for Pocket PC.

7.   Evaluation of the Proposed Methodology

Design for AmI requires a great deal of common ECs with slightly different characteristics. MPG allows (1) describing the generic ECs at a higher level of abstraction, (2) parameterizing them with respect to the variety of user- and application-specific requirements, (3) generating the customized ready-to-use EC instances for well-understood domains, and (4) representing the generic architectures and generating the specialized architectures for ES design. Thus, higher quality and productivity in ES design can be achieved.

The areas of application of the MPG techniques for ES design can be stated as follows:

(a) *Specialization* of EC functionality for a particular context of application: (1) data path specialization such as adaptation of word length, e.g., 8 bit and 16 bit-wide ALU for micro-controllers, 16 bit ALU for DSP, 32 bit ALU for processor cores. (2) Instruction set specialization, e.g., different ALU instruction sets for different applications. (3) Memory specialization such as selection of a memory size, word length and access mode.

(b) *Wrapping* of ECs with additional functionality: (1) wrapping for fault-tolerance using various redundancy schemes; (2) communication interface synthesis using various data protocols, e.g., handshake, FIFO.

(c) *Generalization* of ECs with similar functionality: (1) packaging of different ECs for convenient component selection from a reuse library; (2) generalization of the specialized EC instances such as DSP algorithms into a single generic EC.

Table 3. Summary of our case studies.

Case study	Aim is to increase...	MPG used for...	Contributes in...	Domain	Meta-language	Domain language
(1)	reliability	generation, wrapping	ubiquitous computing	HW	Open PROMOL	VHDL
(2)	performance	generation, specialization	ubiquitous computing	embedded SW	C++	C
(3)	interaction	generation, generalization	intelligent interfaces	SW	C#	HTML, XML, XSL, SQL

(d) *Generation* of domain programs, e.g., data look-up tables, instances of parameterized domain objects such as adders, multipliers, DSPs, etc.

Finally, we evaluate the usage of MPG for ES design in the context of AmI. A metaspecification, the result of applying MPG paradigm, contains the variant and invariant parts woven together. The variant part (types of domain entities, their varying characteristics obtained from domain analysis, etc.) is expressed at a higher level through metaparameters described in a generic interface of a metaspecification (see, e.g., Figure 5). In general, a metalanguage is used to express all possible variability in a domain. While a domain language is used to express the invariant part or commonality in a domain (e.g., in case study (1), it would be the triple-redundancy model (see Figure 6) in VHDL). As metaparameters obtain values from a restricted set relevant to a domain, a metaspecification describes the family of the related component instances in a domain. Thus, a metaspecification that expresses in total commonality and variability of a domain is a *domain generator*.

Summary of our findings and case studies (see Table 3) is as follows. (1) MPG paradigm is a methodology for bridging domain analysis and SW generation. (2) Our methodology is invariant of the application domains. It can be applied for SW, embedded SW and HW domains in ES design no matter what kind of language (meta- and domain) is used. (3) Metaspecification is a concise representation that allows reducing the heterogeneous design space of ECs, thus simplifying design space exploration. (4) MPG can be implemented using the conventional programming environments and tools only. (5) MPG techniques should be applied for the well-proven applications in the first place.

8. Conclusions

The emerging *Ambient Intelligence* paradigm requires new domain analysis methods and methodologies for successful embedded system design. The designers face a need to design a large number of embedded components with common functionality, but extreme diversity of non-functional requirements. General-purpose components and architectures often must be specialized for a specific context of application. Metaprogramming paradigm combined with modern domain analysis methods adopted from software engineering and *Generic Embedded Component Model* allows us (1) to design parameterized generic embedded components, and (2) to generate the specialized component instances tailored to the user- and application-specific requirements. This enables achieving higher flexibility, reusability and productivity in embedded system design for Ambient Intelligence.

Acknowledgements

We would like to thank to G. Rapacioli, A.M. Caponi and M. Magistrali from H&S qualità nel software (Italy) for providing necessary resources for our MSc student J. Valančius, who implemented the system described in case study (3). We also thank the anonymous reviewers for their valuable comments that allowed us to improve the paper.

References

[1] E. Aarts and R. Roovers, "Embedded System Design Issues in Ambient Intelligence," this volume, 2003.

[2] T. Basten, L. Benini, A. Chandrakasan, M. Lindwer, J. Liu, R. Min, and F. Zhao, "Scaling into Ambient Intelligence", in *Proc. of DATE 03*, N. Wehn and D. Verkest, Eds., München, Germany, 3-7 March 2003, pp. 76-81.

[3] D. Beuche, "Feature Based Composition of an Embedded Operating System Family", in *Feature Interaction in Composed System*, E. Pulvermuller, A. Speck, J.O. Coplien, M. D'Hondt, and W. DeMeuter, Eds., pp. 55-60. Universität Karlsruhe, Karlsruhe, Germany.

[4] D. Beuche, O. Spinczyk, and W. Schröder-Preikschat, "Finegrained Application Specific Customization of Embedded Software", in *Design and Analysis of Distributed Embedded Systems*, B. Kleinjohann, K.H. (Kane) Kim, L. Kleinjohann, and A. Rettberg, Eds., pp. 141-151. Kluwer Academic Publishers, Boston, USA, 2002.

[5] S. Chirokoff, C. Consel, and R. Marlet, "Combining Program and Data Specialization", *Higher Order and Symbolic Computation*, 12(4), pp. 309-335, 1999.

[6] J.K. Cross and D.C. Schmidt, "Meta-programming Techniques for Distributed Real-time and Embedded Systems", in *Proc. of 7th IEEE Int. Workshop on Object-Oriented Real-Time Dependable Systems, January 7-9, San Diego, CA, USA*, pp. 3-10, 2002.

[7] K. Czarnecki and U.W. Eisenecker, "Separating the Configuration Aspect to Support Architecture Evolution", in Proc. 14th European Conference on Object-Oriented Programming (ECOOP'2000), Workshop on Aspects and Dimensions of Concerns, Cannes, France, June 11-12, 2000.

[8] J.L. Diaz-Herrera and V.K. Madisetti, "Embedded Systems Product Lines", in *Proc. 22nd Int. Conference on Software Engineering (ICSE), Workshop on Software Product Lines: Economics, Architectures, and Implications, Limerick, Ireland*, pp. 90-97, June 2000.

[9] P. Donohoe, Ed., *Software Product Lines: Experience and Research Directions*, Kluwer Academic Publishers, Boston, USA, 2000.

[10] K. Ducatel, M. Bogdanowicz, F. Scapolo, J. Leijten, and J.C. Burgelman. *Scenarios for Ambient Intelligence in 2010*, IST Advisory Group Report, IPTS, Seville, Spain, 2001.

[11] S. Edwards, L. Lavagno, E.A. Lee, and A. Sangiovanni-Vincentelli, "Design of Embedded Systems: Formal Models, Validation, and Synthesis", in *Proc. of the IEEE*, 85(3), pp. 366-390, March 1997.

[12] E.D.J. Eggermont, Ed., *Embedded Systems Roadmap 2002*, STW Technology Foundation/PROGRESS, Utrecht, The Netherlands, March 2002.

[13] R. Ernst, "Embedded System Architecture", in *System-Level Synthesis*, A.A. Jerraya and J.P. Mermet, Eds., pp. 1-43. Kluwer Academic Publishers, Dordrecht, Holland, 1999.

[14] B. Fuller, "Ambient Intelligence gains traction", *EE Times*, February 6, 2002.

[15] T. Givargis. *Intel 8051 micro-controller.* http://www.cs.ucr.edu/~dalton/i8051.

[16] G. Goossens, J. Van Praet, D. Lanneer, W. Geurts, A. Kifli, C. Liem, and P.G. Paulin, "Embedded Software in Real-Time Signal Processesing Systems: Design Technologies", in *Proc. of the IEEE*, 85(3), pp. 436-454, March 1997.

[17] *International Technology Roadmap for Semiconductors.* International Sematech, 2001.

[18] B.W. Johnson. *Design and Analysis of Fault-Tolerant Digital Systems.* Addison-Wesley, Reading, MA, USA, 1989.

[19] K. Kang, S. Cohen, J. Hess, W. Nowak, and S. Peterson, "Feature-Oriented Domain Analysis (FODA) Feasibility Study", Technical Report CMU/SEI-90-TR-21, Software Engineering Institute, Carnegie Mellon University, Pittsburgh, Pennsylvania, Nov. 1990.

[20] K.C. Kang, K. Lee, J. Lee, and S. Kim, "Feature-Oriented Product Line Software Engineering: Principles and Guidelines", in *Domain Oriented Systems Development - Practices and Perspectives*, K. Itoh, S. Kumagai, T. Hirota, Eds., Taylor & Francis, 2003.

[21] D. Kessner. *Free-6502 core.* http://www.free-ip.com/6502/.

[22] K. Keutzer, A. Newton, J. Rabaey, and A. Sangiovanni-Vincentelli, "System-level design: orthogonalization of concerns and platform-based design", *IEEE Trans. on Computer-Aided Design of Integrated Circuits and Systems*, 19(12), pp. 1523-1543, 2000.

[23] E.A. Lee, "What's Ahead for Embedded Software?" *IEEE Computer Magazine* 33(9), pp. 18-26, September 2000.

[24] LEOX Team. *DRAGONFLY micro-core.* http://www.leox.org.

[25] M. Lindwer, D. Marculescu, T. Basten, R. Zimmermann, R. Marculescu, S. Jung, and E. Cantatore, "Ambient Intelligence Visions and Achievements; Linking abstract ideas to real-world concepts", in *Proc. of DATE 03*, N. Wehn and D. Verkest, Eds., München, Germany, 3-7 March 2003, pp. 10-15.

[26] T. Männistö, T. Soininen, and R. Sulonen, "Product Configuration View to Software Product Families", in *Proc. 23rd Int. Conference on Software Engineering (ICSE), Workshop on Software Configuration Management, Toronto, Canada, May 14-15, 2001.*

[27] J. Nichols, B.A. Myers, M. Higgins, J. Hughes, T.K. Harris, R. Rosenfeld, and M. Pignol, "Generating Remote Control Interfaces for Complex Appliances", in *Proc. of ACM Symposium on User Interface Software and Technology (UIST 2002)*, pp. 161-170, 2002.

[28] H. Ossher and P. Tarr, "Multi-Dimensional Separation of Concerns and The Hyperspace Approach", in *Software Architectures and Component Technology: The State of the Art in Software Development*, M. Aksit, Ed., Kluwer Academic Publishers, Boston, 2001.

[29] P.G. Paulin, C. Liem, M. Cornero, F. Nacabal, and G. Goossens, "Embedded Software in Real-Time Signal Processesing Systems: Application and Architecture Trends", in *Proc. of the IEEE*, 85(3), pp. 419-435, March 1997.

[30] B.R. Rau and M.S. Schlansker, "Embedded Computer Architecture and Automation", *IEEE Computer*, pp. 75-83, April 2001.

[31] G. Riva, P. Loreti, M. Lunghi, F. Vatalaro, and F. Davide, "Presence 2010: The Emergence of Ambient Intelligence", in *Being There: Concepts, effects and measurement of user presence in synthetic environments*, G. Riva, F. Davise, and W.A. IJsselsteijn, Eds., pp. 59-82. IOS Press, Amsterdam, The Netherlands, 2003.

[32] V. Štuikys, R. Damaševicius, and G. Ziberkas, "Open PROMOL: An Experimental Language for Domain Program Modification", in *System on Chip Design Languages*, A. Mignotte, E. Villar, and L. Horobin, Eds., pp. 235-246. Kluwer Academic Publishers, Boston, USA, 2002.

[33] V. Štuikys, R. Damaševičius, G. Ziberkas, and G. Majauskas, "Soft IP Design Framework Using Metaprogramming Techniques", in *Design and Analysis of Distributed Embedded Systems*, B. Kleinjohann, K.H. (Kane) Kim, L. Kleinjohann, and A. Rettberg, Eds., pp. 257-266. Kluwer Academic Publishers, Boston, USA, 2002.

[34] D.M. Weiss and C.T.R. Lai. *Software Product-Line Engineering: A Family-Based Software Development Approach*. Addison-Wesley, Reading, MA, USA, 1999.

Application-Domain-Driven System Design for Pervasive Video Processing

Zbigniew Chamski and Marc Duranton
Philips Research Laboratories Eindhoven, Eindhoven, The Netherlands
{ zbigniew.chamski, marc.duranton } @philips.com

Albert Cohen and Christine Eisenbeis
INRIA Rocquencourt, Le Chesnay, France
{ albert.cohen, christine.eisenbeis } @inria.fr

Paul Feautrier
Ecole Normale Supérieure de Lyon, Lyon, France
paul.feautrier@ens-lyon.fr

Daniela Genius
Université Paris 6, Paris, France
daniela.genius@lip6.fr

Abstract   Pervasive video processing in future Ambient Intelligence environments sets new challenges in embedded system design. In particular, very high performance requirements have to be combined with the constraints of deeply embedded systems, frequently changing operating modes, and low-cost, high-volume production. By leveraging upon the key properties of the application domain, we devised a computation model, a hardware template, and a programming approach which provide a natural mapping from application requirements to a complete system solution. Our approach enables the direct exploitation of concurrency and regularity in achieving the combined challenge of adaptability, performance, and efficiency.

Keywords   SANDRA, video processing, timed process networks, hierarchical architecture, piecewise static control

1.   Introduction

Vision plays a dominant role in human perception, placing pervasive visualization and video processing at the heart of the Ambient Intelligence concept. The underlying properties of adaptability, anticipation, and ubiquity make the

T. Basten et al. (eds.), Ambient Intelligence: Impact on Embedded System Design, 251-270.
© 2003 *Kluwer Academic Publishers. Printed in the Netherlands.*

video processing sub-systems operate in a changing environment, and require run-time flexibility. While video processing is essentially regular, user interactions and communications with other devices introduce variability in operation modes, system loads, and quality-of-service requirements.

In media streaming applications of un-encoded data (before encoding/after decoding), most events can be predicted and anticipated. Whenever the execution latency of individual tasks can also be predicted (or imposed), asynchronous (interrupt-triggered) control can be entirely eliminated, leading to a fully predictable, real-time system. This in turn enables a tighter dimensioning of the system, reducing the difference between average and peak performance, and thus, directly increasing its efficiency.

At the same time, upcoming display technologies and ever improving compression standards enable a dramatic increase in content and display resolutions. The corresponding performance requirements are beyond the reach of general-purpose processor architectures, implying the use of domain-specific, multi-processing solutions and an increased system complexity. Yet to become commercially viable, these solutions must additionally satisfy the criteria of silicon efficiency (area and device utilization, power dissipation), affordable system design effort, and low manufacturing costs.

In the SANDRA (Stream Architecture eNgine Dedicated to Real-time Applications) project, we tackled this challenge using a global approach driven by the key characteristics of the application domains (video pre and post-processing): massive amounts of parallelism, piecewise regular processing of structured data, predictability of events, multiple processing rates, and explicit temporal requirements in applications. These characteristics were used to identify a suitable computation model, which in turn determined many aspects of system hardware and software. Ultimately, this led to the definition of a system template and a programming flow in which the requirements of the applications are driving the entire design process. The requirements of embedded systems are also taken into account: the SANDRA system is designed to be silicon efficient, satisfy hard real-time constraints and have the lowest possible power consumption and memory bandwidth.

The scalability of the architecture is also important to cope with various applications and instances of embedded systems: a base SANDRA sub-system can be easily extended with new blocks using an intra- or inter-chip network, while still using the same model and representation of applications: it is seen as a single entity, with higher performances. The "connected" nature of Ambient Intelligence systems enables to think of even more sophisticated systems: for example a SANDRA system inside a camera could use the resources of another SANDRA system in the TV set for increasing its computational power during, e.g., video segmentation or depth reconstruction (if a suitable communication channel is available).

To address the complexity of programming such inherently concurrent systems, we propose to move away from sequential application descriptions towards *timed process networks* [5], which are much more suitable to our target application domain. Process networks directly capture the concurrency available in the applications, and temporal annotations attached to processes (or groups thereof) provide a natural way of representing the timing requirements of the applications. It also helps in distributing tasks onto separate instances of the system, allowing networking at the SoC level and at the multi-chip/multi-device level. The introduction of the quantitative time representation, important for real-time guarantees, also helps in characterizing the communication links (bandwidth, latency, buffer requirements). The resulting system can process data "on-time" and not necessarily as fast as possible, allowing to determine the slowest possible clock required for performing each function, and thus reducing the global power consumption.

The hierarchical organization of the SANDRA architecture reflects the structure of both applications and data they manipulate. From the programmer's point of view, applications are seen as computations on different levels of data structures. From the system design point of view, the hierarchy of control and communications exploits temporal and spatial locality to enforce storage and bandwidth requirements. It also helps addressing the issue of on-chip signal propagation delays.

Another challenge of the project was to support the hard real-time requirements of "live" stream processing in combination with the dynamic reconfiguration inherent to most Ambient Intelligence applications. We propose to address this issue through piecewise static control: the scheduling and mapping of computations and communications in SANDRA is made statically for a range of "scenarios" defined by the different levels of maximum load guarantees and throughput/latency requirements. Within each scenario, the schedules and resource allocations guarantee the respect of performance and resource *requirements* while ensuring the best possible *efficiency*.

This paper is further organized as follows: section 2 introduces the application domain and the representation of applications used in SANDRA. Section 3 presents the overall structure of the SANDRA compilation chain. Key issues in code generation are presented in Section 4. Section 5 describes the key system architecture concepts of SANDRA.

2.   Representation of Applications

When designing a domain-oriented system suitable for a range of applications, the characterization of the application domain is a key success factor: it makes possible to exploit application properties in an efficient way.

The target domain of SANDRA is real-time media stream processing. We provide a tentative solution to the system design issues in presence of:

- massive amounts of parallelism;

- piecewise regular processing of structured data;

- predictability of events;

- multiple processing rates;

- explicit temporal requirements in applications.

The application model must capture the concurrency and real-time attributes of the applications and of the SANDRA hardware. In addition, the applications operate on structured data whose size has to be taken into account in the model. When combining a machine description of the target system with the information of an application's degree of concurrency, data size, clock rates and hierarchy, it is possible to determine the peak and average bandwidth values, end-to-end and partial latencies, intermediate buffer sizes, utilization rates of target system elements, etc.

2.1 Multi-Periodic Process Networks

To enable fast retrieval of time and resource properties at every stage of the design process, we developed a process-based application model called Multi-Periodic Process Networks (MPPN; a detailed presentation and discussion of the model can be found in [5]). The MPPN model is inspired by Kahn process networks [10], Petri nets variants such as event graphs [2], and by Control/Data-Flow Graphs (CDFG, [15]). It also shares some motivations with the COMPAAN project [11] within the PTOLEMY environment [4].

The MPPN provides four distinctive concepts: (1) explicit synchronizations between processes, (2) bounded-size communication channels, (3) a quantitative notation for delays, latencies and periods of processes, and (4) a hierarchical composition mechanism for building aggregate processes from elementary ones. In particular, the two latter features of MPPN are fundamental when distributing applications onto networks of SANDRA instances. The hierarchical composition helps in partitioning, and the quantitative modeling of delays and latencies allows the MPPN network analysis tools to check if the communication channels are suited to the proposed partitioning. Figure 1 shows a simple MPPN for a two-dimensional polyphase filter, applied to the downscaling of video frames from a high definition (1920×1080) to a low definition (720×480) screen; the filtering process is decomposed into a horizontal stage (sub-process P_5) and a vertical stage (sub-process P_6).

The HARRY verifier tool that we have built checks the coherency of the input data and timing constraints and computes the required buffer sizes, process

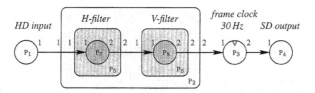

Figure 1.   MPPN representation for a downscaler.

latencies, bandwidth and resource usage. It uses an XML representation of a MPPN (cf. Figure 2). This representation is designed such that an MPPN can easily be abstracted from a SALLY program (see 3.1). At present, it handles cyclic networks and clocked processes but not the splitter/selector extensions.

We tested MPPN models for five typical applications: downscaler with a polyphase filter, horizontal split-screen display (splitter and selector), picture-in-picture (full example with deep hierarchy), advanced anti-aliasing filter (pipelined execution and complex acyclic graph), and noise correction (cyclic graph). Using HARRY, we solved the network's equations to check for soundness and compute the missing parameters. Finally, we deduced resource requirements (memory sizes, bandwidths, functional unit counts).

The MPPN model was designed to leverage upon the key properties of the domain of pixel-stream processing (predictable behavior, regular processing, extended stream semantics with a steady state, timing and bandwidth constraints, few data dependent control loops). Its main limitations are the fixed token size and the restricted splitter/selector semantics. Token sizes are bounded to allow the design of predictable systems, possibly leading to a worst case design (this is also the case with ASIC design.) The splitter/selector model can hardly be extended without losing static schedulability, but applications that have non-deterministic (or data dependent) splits and selects also have an upper bound in their activation frequency, allowing for an approximate MPPN model. Extensions of the model are possible, at the expense of a big complexity increase. We therefore preferred to use simplifications and perhaps less optimal solutions to model the few applications that escape from the common characteristics of the domain.

3.   Compilation Chain

The compilation process consists in mapping the timed process network representation of the application to the hierarchy of control, memory, and processing units which form a SANDRA instance. Both the mapping of an application to the SANDRA architecture and the validation of the resource constraints for this application rely on a model of system. The model provides quantitative target system models at multiple levels of refinement and precision. By combining

```
<!DOCTYPE MPPN SYSTEM "MPPN.dtd">
<MPPN>
    <!-- The first process -->
    <Process id="1"
             Type="Normal">
        <Name>HDInput</Name>
        <OutPort ChannelID="1"
                 Start="true"
                 End="false"
                 Q="2073600">
            <Bandwidth/>
            <Message/>
            <Access a="1"/>
            <Latency l="0"/>
            <Burstiness/>
        </OutPort>

        <!-- Process parameters -->
        <Period/>
        <Burstiness N="1"/>
        <Latency/>
        <PipelinedExecution/>
        <Activation/>
    </Process>

    <!-- ... -->
```

```
                    HPN XML Document
    ♀ ▣ Element: Process
        ▯ Text:
    ⊙▣ Element: Name
        ▯ Text:
    ⊙▣ Element: InPort
        ▯ Text:
    ♀ ▣ Element: OutPort
        ▯ Text:
        ▯ Element: Bandwidth
        ▯ Text:
        ▯ Element: Message
        ▯ Text:
    ⊙▣ Element: AccessLatencyRange
        ▯ Text:
    ⊙▣ Element: LatencyRange
        ▯ Text:
    ⊙▣ Element: BurstinessRange
        ▯ Text:
        ▯ Text:
        ▯ Comment: Process parameters
        ▯ Text:
        ▯ Element: Period
        ▯ Text:
```

Figure 2.   Sample XML representation and parse tree.

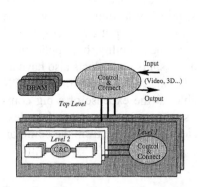

Figure 3.   Hierarchical control and storage.

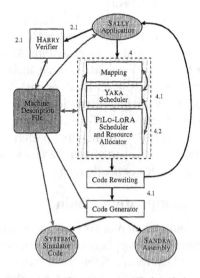

Figure 4.   SANDRA compilation chain.

this information with the timed process network representation of the *requirements* of the application, it is possible to carry out performance estimations in a systematic way, even at early design stages of either applications or the target system.

The purpose of the compilation chain of SANDRA is threefold:

- providing timed simulation models to be used in design-space exploration and debugging of functional and temporal behavior of SANDRA applications;

- generating code and parameters for general-purpose and dedicated units, for each controller, at each layer of the architecture; this must be fully automatic because the execution model is too complex to be handled at the application level;

- optimizing the code such that the time constraints are satisfied while minimizing memory, computation and communication resources; tedious optimizations and transformations are automatic, but the engineer can still drive the design space exploration using an abstract process-network model of the application.

In the high-level synthesis community, Control/Data-Flow Graphs (CDFG) have been a successful representation for data-intensive applications with timing and resource constraints [15]. Indeed, CDFGs can be simulated for design-space exploration, they serve as a basis for optimizing transformations, and of course, they enable code or circuit generation. Well-known research tools such as HYPER [15] or PTOLEMY [4] (with alternative data-flow graph models) have been developed in this area.

Several compiler techniques developed for resource-constrained scheduling of loop nests can reconstruct the control and data structures completely through algebraic loop-nest transformations [9, 16]. These techniques can distinguish between each iteration of a loop or each value of a stream/array, enabling much more aggressive transformations.

However, the better efficiency of such techniques comes at a price: some of the versatility of CDFGs and other flow-graph approaches is lost, like the ability to handle arbitrary control flow or the natural integration of timing and resource constraints. But since our applications do not rely on arbitrary control flow, and despite the lower versatility and the higher computational complexity, we believe that only aggressive techniques can efficiently harness the resources of the highly parallel architecture.

The structure of the SANDRA compilation chain is sketched in Figure 4, where numbers link transformation phases and code representations to the relevant sections. Compilation starts with an application description specified in the high-level language SALLY (section 3.1) and checks real-time properties with the HARRY verifier (section 2.1). During the design space exploration, the YAKA multidimensional affine scheduler (section 4.1) and the PILO-LORA software-pipelining tool (section 4.2) produce one or several schedules and re-source allocations of the concurrent program; the programmer may drive the exploration in suggesting a coarse-grain mapping of (some) processes to SAN-DRA controllers. The code rewriting phase (section 4.1) regenerates SALLY code from the abstract schedule, allowing for iterative refinement of the sched-ule. Finally, cycle-accurate simulation code and SANDRA assembler code are generated from the fully scheduled SALLY program. All communications be-tween software modules are done via XML files, while the tools use their own internal formats.

To accommodate the flexibility of the hardware template, software tools supports parameterization by a machine description file that contains the in-formation necessary for the code generation stage. It also seamlessly interacts with HARRY's evaluation of communication latencies, parallelism, buffer and bandwidth requirements, and the results may be fed back into machine descrip-tions of higher-level operations. Eventually, the machine description file feeds YAKA and PILO-LORA for resource allocation information, enables the auto-matic generation of simulation models, and provides a reference for regression testing of the actual hardware. This pervasive use of the machine description is a major governing principle in the SANDRA architecture and compilation chain.

3.1 The SALLY Language

In order to capture both functional and non-functional requirements of appli-cations at the program level, we designed a small, domain-oriented language called SALLY. Rather than extending a sequential language such as C, C++, or Java, we propose to use a clean set of concepts tailored to the characteris-tics of the application domain, and we provide constructs which are familiar to domain specialists. This in turn makes it possible to formalize application re-quirements, while still allowing the compiler to perform domain-specific anal-yses, verifications, and optimizations.

SALLY is a declarative language. At the core of SALLY is the synergy be-tween structured data types (arrays and records), iterators, and processes.

Variables in SALLY are *streams* of array or scalar values, indexed by iterator values. Each variable has an associated *index domain* (possibly unbounded), which identifies all index values for which this variable is defined [13]. SALLY variables correspond directly to channels in MPPN.

Iterators are a uniform concept for expressing loops (either parallel or se-quential) and event-based processing. Three types of iterators are available: *indices* correspond to unordered, potentially concurrent iterations; *counters* correspond to ordered, i.e., serial iterations; *clocks* are counters with quantitative time distribution, they are used to capture the real-time requirements of the application. Processes are sets of equations or networks of other processes that are evaluated in response to a change of value of a special input, called the trigger, mapped to an iterator defined outside the process.

The basic statements of SALLY are equations and process activations. Statements may be explicitly associated with an iterator. An *equation* defines the value of a variable as the result of evaluating an expression in the current context of iterator values. A process definition consists of an *interface definition* and a *body*: the body lists the local variables of the process followed by its equations and subprocesses, whereas the interface definition provides type signatures and names for the input/output ports, along with per-invocation parameters of the process. A process activation instantiates the process, binds the ports and parameters of the process with actual variables, and maps the *trigger* of the process to an actual iterator.

SALLY programs can express parallelism in three ways:

- by triggering multiple statements/processes on the same iterator (there is no explicit sequential ordering; instead, the dependencies are extracted and checked at compile time);

- through unordered iterators (for all i do ...);

- through array-wide operators.

The first method provides a natural expression of control parallelism, while the last one is specifically directed at data parallelism. The unordered iterators provide a means of trading control parallelism for data parallelism.

Example: Figure 5 shows a short excerpt from a SALLY implementation of a two-dimensional polyphase filter.

The first three lines define iterators (clocks) used by the main process: frameStart runs at 30Hz and is provided by the environment; output_line_clk is a clock running at 660 times faster than FrameStart and is reset to zero at every tick of FrameStart; visible_line_clk is a sub-sampling of output_line_clk and is only active when output_line_clk value is between 100 and 579 inclusive.

The process Vstage takes one input value (frame_after_HFL array) per activation, using VFL_coefs and VFL_offset as per-activation parameters.

When triggered, process Vstage will activate process VFL_stage at every tick of clock FrameStart, and will activate OUTPUT for every tick of visible_line_clk. Each activation of VFL_stage consumes the current value of frame_after_HFL, produces a new value of frame_after_VFL and uses the current value of VFL_coefs and VFL_off as per-invocation pa-

```
extern clock frameStart        30Hz
clock          output_line_clk  660 @ FrameStart (* visible + Vsync *)
clock          visible_line_clk output_line_clk[100 .. 579]

node Vstage(param float VFL_coefs[64][6], param int VFL_off[64][6],
            input  pixel frame_after_HFL[1080][720])
{
 decls
   (* after Vfilter: 480x720 pixels *)
   pixel frame_after_VFL[480][720]

 code
   (* frame-level V filter invocation *)
   frame_after_HFL -> VFL_stage(VFL_coefs, VFL_off) -> frame_after_VFL
      every frameStart

   (* output ctrl at line level *)
   frame_after_VFL[visible_line_clk - 100] -> OUTPUT -> VOID
      every visible_line_clk
}
```

Figure 5.   SALLY application example: filtering and output.

rameters. Each activation of OUTPUT selects an appropriate line from the latest value of frame_after_VFL, and acts as a sink node (output to VOID).   ■

SALLY programs form a concrete representation of MPPN, with the addition of complete information on process internals. This information is critical to the precise evaluation of MPPN parameters such as process and channel latencies, based on a machine description of the underlying SANDRA architecture. In this way, SALLY program analysis and transformation can leverage on all techniques developed for MPPN.

The actual computation of bandwidth, buffer, latency and resource usage properties is done by the HARRY verifier. The MPPN abstraction of a SALLY source code enables fast estimation of these properties, which is critical to the design-space exploration of application and the target system. HARRY output is also used to find the necessary sequential and timing constraints to be included in the SALLY program, and to identify over-constrained applications not amenable to parallelization. It gives also a quick go/no-go answer in case of distributed application onto an network of resources: the available bandwidth and latency of the communication channels are checked against the current application partitioning.

4.   Scheduling and Code Generation

The scheduling phase benefits of the domain-specific semantics of SALLY: array and iterator structures are constrained such that memory dependences (i.e., causality constraints) can easily be captured at the level of each iteration using classical *array dependence analysis* techniques [8]. Dependences are described by systems of *affine constraints* enforcing sufficient conditions to make a schedule valid. In addition, SALLY processes explicitly communicate through FIFO channels following the semantics of MPPNs. Extending array dependence analysis to communicating processes requires to match every send with its corresponding receive, i.e., to count the number of sends and receives; this may lead to polynomial expressions when communications are nested within multiple loops. To get back to a classical array dependence analysis problem, we convert each send/receive statement into a store/load reference into a cyclic buffer. This corresponds to a candidate implementation for the channel, assuming that the buffer is bounded and that the bounds are known at compile-time, which is easily checked on the MPPN model.

The resulting affine constraints can be handled by Feautrier's scheduling algorithm for "static-control" loop nests [9] (a class that includes most streaming algorithms), proven optimal in terms of asymptotic parallelism extraction [17]. This method uses an efficient constraint solver based on Parametric Integer Programming, PIP [7]. In theory, the result should be a multidimensional affine schedule for the whole program, telling when each iteration, operation or communication should occur. In practice, PIP may not scale to large systems generated from real-world streaming applications (its complexity is exponential in the worst case). Instead, we can benefit of the hierarchical decomposition of the SALLY program to cut down the scheduling problem to tractable pieces. This approach has already been studied and implemented in the context of the Alpha language [6].

4.1 The YAKA Scheduler

The YAKA scheduler is the first step in the construction of the target program. It is invoked as soon as a first sketch of the architecture (number and type of the operators, size of the memory) is available. This information may be given, e.g., by a preliminary analysis using MPPN or be generated by the designer. It has two input interfaces. First, a C-like programming language (sYAPI) with process, channel, port and send/receive extensions; its ease of use makes it suitable as a development tool. The second interface is an intermediate representation in XML format (a convenient way of representing syntax trees) and do not impose any semantics on the designer. There is a DTD for this representation, which is primarily intended as documentation. In a future version, it is intended that the SALLY parser will generate an instance of such XML representation.

Eventually, the present version generates C code through standard polyhedra scanning techniques [1, 14, 3]. But the hierarchical control structure of SANDRA will require more work to generate low-level code (code compaction, code partitioning for different controlers, explicit generation of communication patterns, memory management).

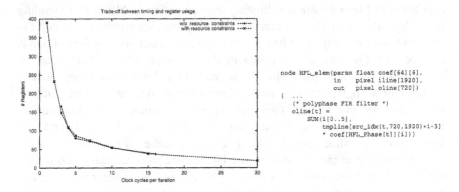

Figure 6.   Software pipelining of one HFL step.

In its standard version, YAKA does not address operation latency, real-time constraints, allocation of computations to the SANDRA controllers and low-level operators, and memory/register allocation. Theoretically, these additional constraints and tasks do fit into the YAKA model thanks to linear encodings, see e.g. [16]. Latencies and real-time deadlines are captured through additional affine constraints, and YAKA automatically converts resource constraints into artificial dependences (based on a cyclic allocation of resources to competing operations).

Most of the practical work around YAKA has to do with a better integration in the SANDRA compilation chain, both at the input and output sides. On the theoretical side, the problem of taking into account resource constraints (e.g., a limited number of adder-multipliers) has only partial solutions. In the present version, this is mainly obtained by adjusting the size of circular buffer (since the degree of parallelism cannot be higher that the amount of writable memory). This is unsatisfactory, as it needs manual arbitration between phases of the application. Another point is that the present scheduler is not modular, and the subroutines in the source have to be inlined.

4.2   Software Pipelining and Hierarchy

We are working on two possible solutions to improve the scheduling quality. Both of them are based on PILO-LORA, an existing software pipelining tool developed at INRIA that performs loop instruction scheduling (PILO) as well as loop cyclic register allocation (LORA). Unlike usual modulo scheduling algorithms, PILO implements the non iterative-DESP [18] software pipelining algorithm: it handles fine-grain resource constraints, including register types, non-uniform instruction formats and arbitrary reservation tables. PILO provides heuristics for the control of register pressure. For instance PILO-LORA can software-pipeline an elementary step of HFL (Horizontal Filter) and give the Pareto curve drawn in figure 6 for trading-off register pressure against timing in a dimensioning phase of the lowest level of the SANDRA architecture. In this example, the analysis assumed a design with 4 multipliers, 4 adders, and 3 memory ports (2 for loads and 1 for store instructions), with latencies of 20 cycles for loads and stores, and 10 cycles for multipliers and adders.

We are also considering other approaches for exploiting PILO-LORA. The first approach consists in regarding PILO-LORA as an alternative to YAKA's affine scheduling. Based on array dependences and reservation tables for every subtask involved in a process, the DESP algorithm can be applied to the innermost loops of the process. Then, application to the whole program requires a recursive application of PILO-LORA along the process hierarchy, much like hierarchical software pipelining techniques [12]. In practice, it requires an additional effort by the programmer since PILO is not able to automatically assign processes to SANDRA levels; hence a prior coarse-grain mapping has to be provided along with the SALLY source code.

Another approach — currently in progress — consists in combining the YAKA scheduler with a software pipeline "microscheduling" phase, integrating resource allocation and fine-grain rescheduling. E.g., YAKA is appropriate for detecting (possibly unlimited) parallelism in loop nests and PILO-LORA is much better at allocating resources and scheduling the innermost loops in the code generated by YAKA. Artificial scheduling constraints may be added to YAKA in order to make the innermost loop code scheduling more efficient.

4.3   Simulation of Sally Programs using SystemC

In parallel with the development of the compilation chain for native target programs, we developed a tool generating SYSTEMC simulation models directly from SALLY programs. The generation of SYSTEMC models leverages on the direct correspondence between core elements of SystemC and the SALLY process model, including time-related features (clocks, delays) and concurrency. Latency requirements information from the application source code is used to define run-time consistency checks related to deadline respect etc.

The SystemC models are generated by combining two sources of information: a SALLY program (used as a specification of application requirements, both functional and temporal), and a target system description containing the functional and temporal capabilities of elementary processes used in that SALLY program.

The granularity of the generated SYSTEMC model directly corresponds to the granularity of the processes described in the SALLY program. When the SALLY program is refined to use actual operations of the target system, the resulting SYSTEMC model will be equivalent to a compiled instruction set simulator of the target for the application specified by the SALLY program.

This approach to simulation model generation provides several advantages:

- direct representation of time (clocks, latencies) and concurrency represented in SALLY programs;

- elimination of verbosity in SystemC programming, particularly in class declarations; the SALLY process interface declarations are on average ten times smaller than the corresponding SystemC declarations;

- automatic generation of behavior cross-checking and reporting/profiling code, e.g., detection of missed deadlines, monitoring of process activity; the latter complements the analysis capabilities of HARRY, providing a dynamic feedback on the utilization rate of processes, on process network latencies, and on traffic shapes of inter-process communications.

- automatic generation of activity traces using SystemC trace generation facilities; a side-by-side analysis of the evolution of process states simplifies the analysis of synchronization errors and greatly simplifies application debugging.

5.   SANDRA System Architecture

To achieve the required degree of flexibility, all elements of the system should be configurable: functional units, their interconnection, control mechanisms, and memory subsystem. The frequency of reconfiguration of the different system elements depends on the nature of the tasks being performed, and can vary from several Hertz (mode changes, transitions between video frames) to several tens of MegaHertz (sub-pixel filtering). Centralizing the reconfiguration decisions would lead to a severe control bottleneck in the system. Instead, we propose to distribute the control and organize the system using a hierarchical approach, driven by and adapted to the characteristics of the target application domain.

The presence of explicit timing/frequency requirements in targeted applications led to another fundamental decision: instead of executing the tasks as fast as possible (driven by the intrinsic speed of hardware modules), the tasks are triggered right-on-time, synchronized with specific events. This mechanism addresses a major shortcoming of conventional interrupt-triggered architectures, which maximize average performance, and tolerate latency on "low-probability" events, expected to arrive fully asynchronously with the operation of the processor.

In our approach, the coordination of tasks operating at the same rate is performed by a single controller, which delegates the control of individual tasks to the next level of the hierarchy. This mechanism is again used to control sub-tasks inside each of the top-level tasks, and can be recursively repeated for as many levels as required. Conversely, tasks with independent clock domains may be executed by different controllers without unnecessary synchronizations. Finer-grain (thus, higher-frequency) tasks have stricter latency and bandwidth requirements: they require fast access to data and a high storage bandwidth. Conversely, coarse-grain tasks can tolerate longer latencies than the rest of the system. This fact is reflected in the memory and communication structure of SANDRA. The lowest levels of the system hierarchy use small, fast memories fully interconnected with relatively simple operators (FIR filters, etc.). Higher levels of the hierarchy offer a lower number of larger memories, and communicate through a higher-latency network. In this way, both the locality of data references and the natural synchronization of tasks at each level can be fully exploited within a unified system organization.

5.1 Functional Structures

The SANDRA hardware (see figure 3) consists of four distinct, superposed architecture layers corresponding to the different functions of the programmable system:

1. A hierarchical *control* layer managing resource activity and enforcing data dependences and real-time constraints on the three other functional structures of the SANDRA hardware.

2. A clustered *execution* layer gathering the functional units that operate on the contents of the data streams. A the lower level, each functional unit has a structure of a dedicated VLIW (Very Long Instruction Word), allowing the various operative units such as ALU, multipliers, etc to work simultaneously.

3. A heterogeneous *communication* layer tuned to the activity of each level: low latency, high bandwidth and high connectivity for the lower levels (pixel processing), higher latency and throughput achieved through larger data blocks for the higher levels (line, image processing). Because of the multiplicity of compute cores, the increased need for communication bandwidth, and the ever increasing wire cost inside systems-on-a-chip (SoC), the higher level internal communications will be implemented by networks instead of classical busses.

4. A *parameter* layer to customize the dedicated functional units of the execution layer, providing parameters for the operations on the main (pixel) computation flow. It is composed of very small RISC processors.

The execution and parameter layers are tightly coupled, but they perform radically different operations and process different kinds of data. A typical example of parameter unit is address generation: in most stream-processing algorithms, irregularities can be moved towards generating addresses, the remainder of the computation (e.g., pixel processing) following a regular flow. In the application domain considered, most of (pixel) compute kernels are similar in various algorithms; it is the way they are organized and how their parameters are computed that differ and gives the differenciating factor. Therefore, the parameter layer is the most flexible.

The architecture also distinguishes what is related to stream-processing computations from what is needed to run the SANDRA system: the execution code is split in two parts. The *application code* describes the computation on the data flow, and the *control code* schedules the application code over the hierarchical architecture. The application code is independent of the architecture instance and targets the communication, execution and parameter structures. The control code adapts the execution to a given SANDRA instance and to the dynamic part of applications.

5.2 Control Structure

The control structure of SANDRA is also composed of hierarchical layers. It illustrates the current trend in system design: systems are built by integrating components (often called IPs) (software or hardware) that are linked together by a common interface for communication and control. From the software point of view, this represents an evolution towards component-based software engineering. The hierarchical control system allows to distribute the control units near the functional units, hence to have a scalable and modular design. For example, if the higher-level controller implements a two-dimensional polyphase filter, it may decompose this task into separate horizontal and vertical filters, and delegate these subtasks to lower-level controllers. The top level does not need to know how the lower-level controllers perform the tasks, as long as they satisfy the specified time constraints. The lower level controllers can also decompose the mono-dimensional filters into blocks of vector operations, then into scalar products and additions, and so on. Each level of the computation is assigned to a controller, but several logical controllers can be folded into one physical controller. This scheme allows to have a "logical" single control flow, but in fact not centralized and supporting some asynchrony (sub-level controllers can be independent, and they can feedback to the controller just above when they have finished their task).

The controller structure is the same for each level of the hierarchy, and is a stack-based virtual machine. Since multiple reentrant control codes should be executed on one physical controller, no explicit register allocation is done.

Variables (used only for the control part of the application) are not explicitly allocated but remain on the stack. If a new task starts, it could start on top of the previous task stack as long as it eventually restores the right stack position. Using a stack-based virtual machine also eases portability across implementations of the SANDRA architecture and favors code compactness (factorization). As opposed to traditional stack based languages (Java, Forth, Postscript, OPL, etc.), we propose a *threaded* code structure where each instruction explicitly targets the next instruction to be executed; together with stacks, this improves factorization and eases reentrance and late binding. At each level (except for the lowest level), an instruction is composed of two main fields:

- the first one is dedicated to the control flow itself and its threading mechanism: instead of a "program counter", the next instruction is indicated explicitly within the current instruction, in a similar manner as the linked-task structure of a real-time OS;

- the second field manages the lower level controllers; it is composed of several slots, one for each sub-controller; thus, there is no real distinction between a slot that triggers a (lower level) controller action (in this case, it is equivalent to a subroutine call) and a slot that controls a functional unit.

This structure allows to map a code onto various instances of SANDRA, with no recompilation and a minimum load during the instantiation of the code (binding). It gives some code expansion, but it is believed to be compensated by the code factorization present in our domain-specific applications.

5.3   Dynamic Reconfiguration and Application Switching

Let us now show how our static modeling, compilation and optimization framework can cope with the dynamic features of Ambient Intelligence media applications. Stream-oriented applications can be modified on request of the user (for example, adding or moving a Picture In Picture), or due to the environment (new people entering a room, appearing in the vision field of a camera, etc.). However, these changes are slow compared to processing speed: video applications require changes at the frame rate, i.e., several milliseconds, while user interactions are at the split second level, enabling dynamic changes in system configuration.

The following paragraphs describe how dynamic reconfiguration and application switching can be mapped to the SANDRA system, while keeping the most important features of statically compiled code, such as guaranteed performances, predictability and high silicon efficiency.

In the considered application domain, there are always boundaries and limits that are given, either explicitly or implicitly: for example, given a time limit for the execution time (e.g. a frame interval) and given the hardware resources, an implicit limit on the complexity of the application can be derived. We have developed the MPPN to help determining these constraints. If an application cannot fit within the hardware or time constraints, then it has to be modified (simplified). This led to the idea of "piecewise static control": an application is split into sub-applications, each of them specifies a sub-case for a certain range of parameters. Each sub-application can therefore be statically compiled with good efficiency. Dynamic behavior within a sub-application is handled by classical methods, such as worst case dimensioning and predicated instructions. Each sub-application has known characteristics, performances, and has a better efficiency because it covers only a sub-set of the variability of the original application. The activation of the relevant sub-application is done by one controller after the analysis of the input parameters.

Task or sub-task allocation in the system is done preferably in space rather than in time (using parallelism rather than a faster clock). Although the controller can implement time-sharing of functional units, the context saving needs to be explicitly expressed in the application, and it might be costly due to the large amount of data stored in the various pixel pipelines. This is why in SANDRA we prefer to use a "space-sharing" mode: tasks are activated or deactivated at the level of the sub-controller directly in charge of the resources allocated to these tasks. In the SANDRA controller system, a task (or a sub-task) is represented by the sub-tree with the root node being the controller that directly or indirectly (by the controllers that depend on it) "covers" all the resources available for the task. Switching from one task to another one means simply deactivating one complete sub-branch of the tree and activating a new one. This is done in a very simple manner by changing the link field in the corresponding instruction of the root node.

The application domain enforces quality-of-service constraints including time requirements, and we are aiming at efficient and guaranteed usage of the architecture, rather than at "fastest possible" processing (it is useless to go faster than required!) Thus, the concept of hierarchical hardware and control provides a simple mechanism for tasks activation: if two sets of controllers, organized in a sub-tree, can control the same kind of compute elements, storage units and communication means, sub-tasks can execute indifferently on any controller. Hence, a simple compile-time scheduling and allocation is possible. The nearest controller that supervises the two tasks (i.e., the closest parent in control hierarchy) can adjust its own configuration to link the input/outputs of the new task to the rest of the system (this is made possible because all controller have an unique ID, therefore the code can know exactly where it runs). For any controller above the direct supervisor of the two tasks, the activation of one or the other task has no effect, as long as the communication schemes of the tasks are identical.

6. Conclusion

This work addresses the development of embedded systems dedicated to pervasive video applications in the Ambient Intelligence universe. The computation and bandwidth constraints of these applications exceed today's *general-purpose* processors by orders of magnitude, yet the cost of *application-specific* hardwired components becomes disproportionate with product lifetimes. To address these challenges, we stressed the need for a fast and efficient development process for *domain-specific* system solutions.

We surveyed the SANDRA approach to the architecture, compilation and language issues addressed by real-time streaming applications. The project led to promising results in four different aspects:

- The development of Multi-Periodic Process Networks — providing time and hierarchy to a restricted class of Kahn Process Networks — helps in design-space exploration, validation of resource/time properties, and in mapping onto distributed components.

- The design of SALLY, a domain-oriented language combining streams and implicit parallel constructs with non-functional properties such as time requirements and resource allocation.

- A compiler chain, using state-of-the-art algorithms for extracting parallelism, affine scheduling, software pipelining and code generation.

- A hierarchical architecture, easily tuned to the application requirements and allowing to run highly demanding algorithms at consumer price.

The proposed approach can also be applied when the resources are networked, mainly inside a SoC, and it allows to support dynamic behavior to some extend in an environment where hard time constraints are important, therefore demanding real-time streaming applications for Ambient Intelligence can be defined and efficiently mapped with our approach on embedded systems.

Further work is required before demonstrating a running prototype, and larger examples should be studied to explore the system's scalability. Nevertheless, we believe our model has matured enough to clearly state the most important directions towards a domain-specific approach to architecture and compilation development.

Acknowledgements

This project is supported by a Pierre et Marie Curie fellowship and a European Community project MEDEA+ A502 "MESA".

References

[1] C. Ancourt and F. Irigoin. "Scanning polyhedra with DO loops." In *Proc. third SIGPLAN Symp. on Principles and Practice of Parallel Programming*, pages 39–50. ACM Press, April 1991.

[2] F. Baccelli, G. Cohen, G.J. Olsder, and J.-P. Quadrat. *Synchronization and Linearity.* Wiley, 1992.

[3] C. Bastoul. "Generating loops for scanning polyhedra." Technical Report 23, PRiSM, University of Versailles, 2002.

[4] J.T. Buck, S. Ha, E.A. Lee, and D.G. Messerschmitt. "Ptolemy: A framework for simulating and prototyping heterogeneous systems". *J. Comp. Simulation*, 4, 1992.

[5] A. Cohen, D. Genius, A. Kortebi, Z. Chamski, M. Duranton, and P. Feautrier. "Multi-periodic process networks: prototyping and verifying stream-processing systems." In *Proceedings of Euro-Par 2002*, volume 2400 of LNCS, pages 299–308, Paderborn, Germany, August 2002, Springer 2002.

[6] J.B. Crop and D.K. Wilde. "Scheduling structured systems." In *EuroPar'99*, LNCS, pages 409–412, Toulouse, France, September 1999. Springer, 1999.

[7] P. Feautrier. "Parametric integer programming." *RAIRO Recherche Opérationnelle*, 22:243–268, September 1988.

[8] P. Feautrier. "Dataflow analysis of scalar and array references." *Intl. J. of Parallel Programming*, 20(1):23–53, February 1991.

[9] P. Feautrier. "Some efficient solutions to the affine scheduling problem, part II: Multidimensional time." *Intl. J. of Parallel Programming*, 21(6):389–420, 1992.

[10] G. Kahn. "The semantics of a simple language for parallel programming." In J.L. Rosenfeld, editor, *Information Processing 74: Proceedings of the IFIP Congress 74*, pages 471–475. IFIP, North-Holland Publishing Co., 1974.

[11] B. Kienhuis, E. Rijpkema, and E. Deprettere. "Compaan: Deriving process networks from matlab for embedded signal processing architectures." In *Proc. 8th workshop CODES*, pages 13–17, NY, May 3–5 2000. ACM, 2000.

[12] M.S. Lam. "Software pipelining: An effective scheduling technique for VLIW machines." In *Proc. ACM Conf. Programming Language Design and Implementation*, pages 318–328, 1988.

[13] H. Leverge, C. Mauras, and P. Quinton. "The ALPHA language and its use for the design of systolic arrays." *J. of VLSI Signal Processing*, 3:173–182, 1991.

[14] F. Quilleré, S. Rajopadhye, and D. Wilde. "Generation of efficient nested loops from polyhedra." *Intl. J. of Parallel Programming*, 28(5):469–498, October 2000.

[15] J. Rabaey, C. Chu, P. Hoang, and M. Potkonjak. "Fast prototyping of datapath intensive architectures." *IEEE Design and Test of Computers*, 8(2):40–51, 1991.

[16] L. Thiele. "Resource constrained scheduling of uniform algorithms." *J. of VLSI Signal Processing*, 10:295–310, 1995.

[17] F. Vivien. "On the optimality of Feautrier's scheduling algorithm." In *Proceedings of Euro-Par 2002*, volume 2400 of LNCS, pages 299–308, Paderborn, Germany, August 2002. Springer, 2002.

[18] J. Wang, C. Eisenbeis, M. Jourdan, and B. Su. "Decomposed Software Pipelining: A New Perspective and a New Approach." *Intl. J. on Parallel Processing*, 22(3):357–379, 1994. Special Issue on Compilers and Architectures for Instruction Level Parallel Processing.

Collaborative Algorithms for Communication in Wireless Sensor Networks

Tim Nieberg, Stefan Dulman, Paul Havinga, Lodewijk van Hoesel and Jian Wu

Universiteit Twente, Enschede, The Netherlands

{ t.nieberg, s.o.dulman, p.j.m.havinga, l.f.w.vanhoesel, j.wu } @utwente.nl

Abstract   In this paper, we present the design of the communication in a wireless sensor network. The resource limitations of a wireless sensor network, especially in terms of energy, require an integrated, and collaborative approach for the different layers of communication. In particular, energy-efficient solutions for medium access control, clusterbased routing, and multipath creation and exploitation are discussed. The proposed MAC protocol is autonomous, decentralized and designed to minimize power consumption. Scheduling of operations, e.g. for the MAC protocol, is naturally supported by a clustered structure of the network. The multipath on-demand routing algorithm improves the reliability of data routing. The approaches taken and presented are designed to work together and support each other.

Keywords   wireless sensor networks, medium access control (MAC), routing, clustering

1.   Introduction

Wireless sensor networks (WSN) are an emerging field of research which combines many challenges of modern computer science, wireless communication and mobile computing. WSNs are one of the prime examples of Ambient Intelligence, also known as ubiquitous computing. Ambient systems are networked embedded systems intimately integrated with the everyday environment and are supporting people in their activities. These systems are quite different from those of current computer systems, and will have to be based on radically new architectures and use novel protocols.

Recent advances in sensor technology, low power analog and digital electronics and low-power radio frequency design have enabled the development of cheap, small, low-power sensor nodes, integrating sensing, processing and wireless communication capabilities. Embedding millions of sensors into an environment creates a digital skin or wireless network of sensors. These massively distributed sensor networks, communicate with one another and summarize the immense amount of low-level information to produce data represen-

271

T. Basten et al. (eds.), Ambient Intelligence: Impact on Embedded System Design, 271-294.
© 2003 *Kluwer Academic Publishers. Printed in the Netherlands.*

tative of the overall environment. From collaboration between (large) groups of sensor nodes, intelligent behaviour can emerge that surpasses the limited capabilities of individual sensor nodes.

Sensor nodes collaborate to be able to cope with the environment: sensor nodes operate completely wireless, and are able to spontaneously create an impromptu network, assemble the network themselves, dynamically adapt to device failure and degradation, manage movement of sensor nodes, and react to changes in task and network requirements. Despite these dynamic changes in configuration of the sensor network, critical real-time information must still be disseminated dynamically from mobile sensor data sources through the self-organising network infrastructure to the applications and services.

Sensor network systems will enhance usability of appliances, and provide condition-based maintenance in the home. These devices will enable fundamental changes in applications spanning the home, office, clinic, factory, vehicle, metropolitan area, and the global environment. Sensor node technology enables data collection and processing in a variety of situations, for applications, which include environmental monitoring, context-aware personal assistants (tracking of location, activity, and environment of the user), home security, machine failure diagnosis, medical monitoring, and surveillance and monitoring for security.

This paper deals with networking protocols involved in a WSN. We address all traditional layers like MAC, transport, and routing, but unlike those well-known variants, we use a more integrated view. Such an integrated approach is also propagated by Min et al. [11].

1.1 Outline

In the following section, we introduce the envisioned architecture of a WSN and identify the typical communication patterns. We address a dynamic WSN in which some of the nodes are mobile, while others are fixed. These dynamics are not only due to mobility, but the characteristics of the wireless channel also include breaking connections and the creation of new links regularly. Throughout this paper, the sensor node prototype as being developed in the European research project EYES is used as a demonstrative tool. This prototype is introduced in Section 2.3. The lifetime of a WSN is directly linked to the energy consumption of each node. The sensor nodes are made aware of their energy consumption by a simple mathematical model presented in Section 2.4.

The following sections will then introduce the major parts of the communication from a bottom-up point of view, following the overview in Figure 1.

Section 3 presents EMACs, the EYES Medium Access Protocol, which uses various novel mechanisms to reduce energy consumption from the major sources of inefficiency that we have identified in existing schemes.

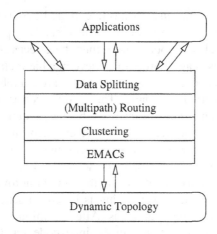

Figure 1.   Overview of the communication in the EYES network.

A clusterbased ad-hoc routing algorithm is discussed in Section 4. The clustering scheme, following a greedy approach to the independent dominating set, is used to obtain an efficient route discovery process. The trade-offs involved and some simulation results are stated to show the advantages of the proposed approach.

The nature of the wireless transmissions and the assumed redundancy in the WSN allows for multiple, different routes through the network topology. An approach to more reliable data delivery, using multipath routing and data-splitting is proposed in Section 5.

This paper ends with a short discussion on the collaboration and a look ahead at the future work.

1.2   Related Work

The current MAC designs for wireless sensor networks tackle some of the problems addressed above. Current MAC protocols can be broadly divided into contention based and TDMA protocols. TDMA protocols have the advantage of energy conservation, because the duty cycle of the radio is reduced and there is less contention-introduced overhead and collisions. However, scalability is normally not as good as that of a contention-based protocol, for example since it is not easy to dynamically change its frame length and time slot assignments.

The first step in the reservation and scheduling approaches is to define a communication infrastructure. The assignment of the channels, TDMA slots, frequency bands, spread spectrum codes to the different nodes in a way that avoids collisions is not an easy problem. One way of dealing with this complexity is to form a hierarchical structure with clusters and delegate the synchronization control to the clusterheads like in the LEACH protocol [6]. Here, issues like cluster membership, rotating clusterheads to prevent early energy depletion and intra-cluster coordination must be effectively addressed. Supporting mobile nodes is also harder to achieve in a hierarchical structure.

At first glance, the contention-based schemes are completely unsuitable for the wireless sensor network scenario. The need for constant monitoring of the channel obviously contradicts the energy efficiency requirement. On the other hand, these schemes do not require any special synchronization and avoid the overhead of exchanging reservation and scheduling information.

An example of a hybrid scheme is the so called S-MAC protocol [20] that combines scheduling and contention with the aim of improving collision avoidance and scalability. The power saving is based on scheduling sleep/listen cycles between the neighbouring nodes. After the initial scheduling, synchronization packets are used to maintain the inter-node synchronization. When a node wants to use the channel, it has to contend for the medium. The scheme used is very similar to 802.11 with physical and virtual carrier sense and RTS/CTS exchange to combat the hidden node problem. The overhearing control is achieved by putting to sleep all immediate neighbours of the sender and the receiver after receiving an RTS or CTS packet.

A WSN does not form a fully connected network, requiring multi-hop routing strategies for data to reach its destination [2]. Several different routing algorithms for WSNs have been studied until now, e.g. the Temporally Ordered Routing Algorithm [16], Dynamic Source Routing (DSR), [8], and Directed Diffusion [7]. However, these algorithms are sensitive to communication failures. To diminish the effects of node failures, multipath routing schemes have been developed on top of these algorithms, e.g. [5, 12], or as stand-alone algorithms as [9], but the resource demands are quite high.

2. Wireless Sensor Networks

In this Section, we briefly introduce wireless sensor networks and some characteristics of these. In particular, the envisioned logical architecture, typical communication patterns, and a prototype are presented. Additionally, as energy plays a key-role in WSNs, we give a mathematical model used to estimate the energy consumption of a node.

2.1 Architecture

In our approach, we define two distinct key system layers of abstraction: (1) the sensor and networking layer, and (2) the distributed services layer. Each layer provides services that may be spontaneously specified and reconfigured:

- the *sensor and networking layer* contains the sensor nodes (the physical sensor and wireless transmission modules) and the network protocols.

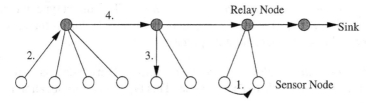

Figure 2.   WSN communication types.

Ad-hoc routing protocols allow messages to be forwarded through multiple sensor nodes taking into account the dynamic changes of the topology due to, e.g., mobility of nodes and node failures. Communication protocols must be energy-efficient since sensor nodes have very limited energy supply.

- the *distributed services layer* contains services for supporting mobile sensor applications. Distributed services coordinate with each other to perform decentralized computations and data modifications. There are two major services. The lookup service supports mobility, instantiation, and reconfiguration. The information service deals with aspects of collecting data. This service allows vast quantities of data easily and reliably accessed, manipulated, disseminated, and used in a customized fashion by applications.

This paper puts the main focus on the networking layer, the distributed services layer has to rely on the communication provided from it. On top of this architecture, applications can be built using the sensor network and distributed services.

2.2   Communication in a WSN

The purpose of a wireless sensor network is physical environment monitoring, and providing this information in an appropriate fashion to the applications in need of this data. Each node will be equipped with one or more sensors, whose readings are transported via other network nodes to a data sink.

In general, two types of nodes are recognized logically: nodes that mainly transmit their own sensor readings (*sensor nodes*), and nodes that mainly relay messages from other nodes (*relay nodes*). Sensor readings are routed from the source nodes to the sink via the relay nodes, thus creating a multi-hop topology. This logical organization implies four types of communications as shown in Figure 2, that have to be accounted for especially on the lower communication levels such as the MAC protocol.

1. *Sensor node to sensor node communication* - This direct type of communication is used for local operations, for example during the clustering process, or the route creation process

2. *Sensor node to relay node communication* - Sensor data is transmitted from a sensor node to a relay node. This type of communication is often unicast.

3. *Relay node to sensor node communication* - Requests for data and signaling messages, often multicasts, to reach a subset of the surrounding nodes at once, are spread by the relay nodes.

4. *Relay node to relay node communication* - The relay nodes form the backbone of the network. Communication between these nodes will mostly be unicast. Note that every node is equipped with a wireless transceiver and thus is able to perform the duties of a relay node.

For a sensor node, there are two operational modes to fit the possible applications arising from the WSN: 1) active polling; and 2) passive detection and notification. For a reading of a sensor, the node acting as sink can actively ask for the information (*active polling*), or request to be notified when an event is detected by one of the nodes, e.g. if a pre-determined threshold on a sensor reading is passed (*passive notification*).

2.3   EYES Sensor Node Prototype

The EYES project (IST-2001-34734, [4]) is a three year European research project on self-organizing and collaborative energy-efficient sensor networks. The goal of the project is to develop the architecture and the technology, which enables the creation of a new generation of sensors that can effectively network together so as to provide a flexible platform for the support of a large variety of mobile sensor network applications.

In the EYES project, prototype sensor nodes have been developed to demonstrate the effectiveness of our protocols. The processor used in the EYES sensor node is a MSP-430F149 [15], produced by Texas Instruments. It is a 16-bit processor and it has 60 Kbytes of program memory and 2 Kbytes of data memory. When running at full speed (5 MHz), the processor consumes approximately 1.5 mW, but it also has several power saving modes.

The communication function between nodes is realized by a RFM TR1001 hybrid radio transceiver [17] that is very well suited for this kind of application: it has low power consumption and has small size. The TR1001 supports transmission rates up to 115.2 Kbps. The power consumption during receive is approximately 14.4 mW, during transmit 16.0 mW, and in sleep mode 15.0 μW. The transmitter output power is maximal 0.75 mW.

From the above can be concluded that the processor can do thousands of operations before the energy consumptions equals that of the transceiver transmitting one single byte. Hence, throughout this document, we assume that the energy costs of transceiver are dominant.

2.4 Energy Consumption of a Node

Sensor networks are expected to be left unattended for a long period of time. Each sensor running on batteries, this requires an approach that explicitly takes energy into consideration. For this, each node is made aware of its energy requirements and usage by a model of the energy consumption.

The aim of the model is to predict the current energy state of the battery of a sensor node based on historical data on the use of the node. The model also allows for predictions on future energy consumption based on the expected task to be run in a certain upcoming time interval. The model considers the three main components of a sensor node that reduce the energy stored in the batteries: the radio, the processor, and the actual sensing device. We do not consider a reactivation of the battery by time or external cirsumstances, e.g. by battery replacement or harvesting of solar energy.

The base of the model for the energy consumption of a component is the definition of a set S of possible states $s_1, \ldots, s_k$ for the component. These states are defined such that the energy consumtion is given by the sum of the energy consumption within the states plus the energy needed to switch between different states. We assume that the energy consumption within a state s_j can be measured using a simple index t_j, e.g. execution time or number of instructions. The energy needed to switch between different states can be calculated based on a state transition matrix st, where st_{ij} dentotes the number of times the component switched from state s_i to s_j. Let P_j denote the power needed in state s_j for one time unit, and E_{ij} denote the energy consumption when switching from state s_i to state s_j. The total energy consumption of the component is given by

$$E_{\text{consumed}} = \sum_{j=1}^{k} t_j P_j + \sum_{i,j=1 \, i \neq j}^{k} st_{ij} E_{ij}.$$

In the following, we describe the state set S and the indices to measure the energy consumption within the states of the radio, processor and sensor:

- *Radio* - for the energy consumption, four different states need to be distinguished: *off, sleep, receiving,* and *transmitting.* For these four states, the energy consumption depends on the time the radio has been in this state. Thus, for the radio, we need to store information the times it has been in each state and the 4×4 state transition matrix representing the number of times the radio has switched between the four states. A hardware energy model that may be used to obtain the specific values for P_j and E_{ij} for the radio are presented in Section 2.2 of [11].

- *Processor* - in general, four main processor states can be identified: *off, sleep, idle,* and *active.* In sleep mode, the CPU and most internal peripherals are turned off, and can be woken by an external event (interrupt) only. In idle mode, the CPU is still inactive, but now some peripherals are active, for example the internal clock or timer. Whithin the active states, the CPU and all peripherals are active. In the active state, multiple substates may be defined based on clock speeds and voltages.

- *Sensor* - for a simple sensor we assume that only the states *on* and *off* are given , and that the energy consumption within both states can be measured by time. However, more powerful sensor work in different states, comparable to the processor, and need to be modeled by more states.

The energy model for the complete sensor node now consists of the consumption model for the three components, plus two additional indicators for the battery:

- for the battery, the energy state E_{old} at a time t_{old} in the past is given,

- for each component, the indices I_j characterizing the energy consumption in the state s_j since time t_{old} and the state transition matrix st indicating the transitions since time t_{old} are specified.

Based on this information, an estimate of the current energy state of the battery can be caluclated by substracting from E_{old} the sum of the estimates for each component since time t_{old}.

3.   EMACS (EYES-Medium Access Protocol)

This section proposes an energy efficient MAC protocol for wireless sensor networks (EMACS). WSNs are typically deployed in an ad hoc fashion, with individual sensor nodes to be in a dormant state for long periods, and then becoming suddenly active when something is detected (passive notification, see 2.2). Such applications will thus have long idle periods and can tolerate some latency. For example, one can imagine a surveillance or monitoring application, which will be vigilant for long periods of time, but largely inactive until something is detected. For such applications, the lifetime of the sensor nodes is critical. Once a node becomes active, data will be gathered and processed by the node, and needs to be transferred to the destination with far less latency and needs more bandwidth than in the dormant state.

Another typical use of a WSN is to have a kind of streaming data, in which little amounts of data (typically just a few bytes) are transmitted periodically

Figure 3.   Frame format of the TDMA-based MAC protocol.

(for example temperature measurements). The large number of nodes will allow taking advantage of short-range, multi-hop communication to conserve energy, especially when data aggregation is applied. Since the nodes will be deployed casually, and maybe even mobile, nodes must be able to self-configure.

The characteristics of a wireless sensor network motivate the use of a different family of MAC protocols than currently employed for wireless (ad hoc) networks (such as IEEE 802.11 [14]), in which throughput, latency, and per node fairness, are more important. Moreover, the network nodes have to operate in a self-organizing ad hoc fashion, since none of the nodes is likely to be capable of delivering the resources to act as central manager.

For the WSN, we explore a TDMA-based MAC scheme, since *code division multiple access* (CDMA) or *carrier sense multiple access* (CSMA) based protocols imply constant or very frequent listening to the radio channel. This listening to the channel consumes a large amount of energy which is certainly not available in the network nodes. The TDMA-based EMACs protocol also eases the (local) synchronization between nodes. In the next section the TDMA frame format is discussed.

3.1 Frame Format

Time is divided into so called frames and each frame is divided into timeslots (see Figure 3). Each timeslot in a frame can be owned by only one network node. This network node decides what communication should take place in its timeslot and denies or accepts requests from other nodes.

Each node autonomously selects a timeslot it wants to own. A timeslot is selected based on the already occupied timeslots as submitted by neighboring nodes. This information includes the known timeslots of the surrounding nodes of a neighbor, so that information about the second order neighborhood is respected. The radio signal has already attenuated quite severely at the third order neighbors, so that timeslots can be reused.

Nodes can ask for data or notify the availability of data for the owner of the timeslot in the *communication request* (CR) section. The owner of the slot transmits its schedule for its data section and broadcasts the above discussed table in the *traffic control* (TC) section, which tells to which other TC sections the node is listening. After the TC section, the transmission of the actual data packet follows either uplink or downlink. Both CR and TC sections consist of only a few bytes.

The designed MAC protocol does not confirm received data packets. In this way the required error control can be decided on in higher protocol layers. This is quite different from other MAC protocols, which try to reach a certain error rate guarantee per link. In a wireless sensor network a data packet is relayed via multiple hops to its destination and on its way data from other nodes will be added and processed (data aggregation and fusion). The resulting data packet will become more and more important and more resources –like energy– are spent. Hence the error rate guarantees should be adapted to the importance of the data. Possible ways to deal with unreliable links are presented in the succeeding sections, especially when presenting the data-splitting along multiple paths.

Collisions can occur in the communication request section. Although we do not expect a high occurrence of collisions, we incorporate a collision handling mechanism in the EMACS protocol. When the time slot owner detects a collision, it notifies its neighbor nodes that a collision has occurred. The collided nodes retransmit their request in the data section after a random, but limited backoff time. Carrier sense is applied to prevent the distortion of ongoing requests.

Suppose that the nodes transmit on average 50 bytes/s and receive 50 bytes/s, the expected lifetime of an EYES sensor node will be 570 days using two AA batteries, capable of delivering 2000 mAh. Although this is already quite good, the EMACS protocol supports also two low power modes. These are discussed next.

3.2   Sleeping Modes

Since transmitting and receiving are both very power consuming operations, the nodes should turn off their transceivers as often as possible. The EMACS protocol therefore supports two sleep modes of the sensor nodes:

- *Standby mode*: This sleep mode is used when at a certain time no transmissions are expected. The node releases its slot and starts periodically listening to a TC section of a frame to keep up with the network. When the node has to transmit some data (event driven sensor node), it can just fill up a CR section of another network node and agree on the data transmission, complete it and go back to sleep. It can actively be woken up by other nodes to participate in communication. Depending on the communication needs, it will start owning a timeslot.

  When we assume that a node transmits on average 50 bytes/s and receives 50 bytes/s, the expected lifetime of the node in standby mode will be 665 days. But when the node is inactive for long periods of time the lifetime will increase more rapidly than in standard operation mode.

- *Dormant mode*: This sleep mode is agreed on at higher layers. The sensor node goes to low power mode for an agreed amount of time. Then it wakes, synchronizes (rediscovers the network) and performs the communication. While in this sleep mode the synchronization with the network will be lost and all communication with the node will be impossible. This sleep mode is especially useful to exploit the redundancy in the network. In a clustered structure, the controlling instance, i.e. the clusterhead, will usually decide on the sleeping pattern of redundant nodes.

3.3   Ownership of Timeslots

Not every node in the network has to own a timeslot. It is clear that a node does not own a timeslot when it is in one of the sleep modes since being in a sleep mode is inherent to not transmitting a TC section every frame. However, event driven nodes might also not redeem their right to own a timeslot. A drawback of not owning a timeslot is that the node only being able to receive multicast messages and not messages directly addressed to it. Transmitting data to nodes that own a timeslot is not a problem. Other protocol layers in the network may invoke listening to, or transmitting in a prior agreed (and free or not owned) data section.

Before a node decides, that it does not want to own a timeslot, it should check that sufficient TC sections are transmitted by neighbors to keep the network connected and to maintain synchronization. The fact that nodes do not necessarily need to own a timeslot, eases the scalability of the network and reduces the power consumption of the nodes.

4.   Clustering and Clusterbased Routing

The TDMA-based EMACs protocol relies on several controlling mechanisms, e.g. for the assignment of time slots or for the delegation of sleeping patterns. In our approach, the controlling mechanisms are achieved using a clustered structure of the network.

Generally speaking, when dealing with large scale, ad-hoc sensor networks, clustering offers some benefits. Grouping nodes into clusters controlled by a designated node, the clusterhead, offers a good framework for the development of important features like routing and, as seen previously, channel access. In addition, the hierarchical view of the network achieved by clustering helps to decrease the complexity of the underlying network, and WSNs are expected to consist of large amounts of individual nodes.

The clustering scheme creates the clusters based on proximity, i.e. nodes that are in range of one another are grouped together. For instance, in an in-home scenario, nodes that are in the same room are likely to form cliques, separated by walls that absorb some of the radio signals and thus representing natural barriers.

4.1 Clustering Protocol

The clustering approach extends a simple protocol to set up and maintain a clustered structure of a wireless ad-hoc network by additional procedures to control the cluster size, taking into account some characteristics of such a network like link failures due to changes in the topology.

In the clustered structure of the network, the set of clusterheads to control the nodes within their neighborhood, is created to form an independent set. The main advantage of this structure comes from the fact that no two clusterheads can be direct neighbors, thus giving the clusterheads more control. Furthermore, the approach taken assures that the clusterheads also form a dominating set, so that each node is within transmission range of at least one clusterhead. The dominating set property is especially important for the MAC protocol as it is only concerned with nodes in direct transmission range of one another.

In order to get a connected structure, non-clusterhead nodes are designated as gateways to enable communication between two adjacent clusters. A node having more than one clusterhead in its direct neighborhood is called a *direct* gateway. It associates, however, with one of these as its controlling instance. A node that cannot communicate directly with a clusterhead of another cluster, but can do so with a node that is member of a different cluster, is referred to as a *distributed* gateway. For a distributed gateway, two non-clusterhead nodes are always needed to ensure connectivity between their respective clusterheads.

An example of the clustered structure showing why distributed gateways are not redundant is given in Figure 4. The network becomes disconnected when solely relying on the created clustered structure without distributed gateways.

The algorithm executed in each node to decide on its role comes from a greedy approach to the maximum weight independent set (see cf. [18]). Each node is given a weight, reflecting its ability to perform the additional duties of being the controling instance of a cluster, and its residual energy. The weights are based on the energy model presented in Section 2.4 that allows for estimating the energy that is available in the batteries of a node and also can be used to estimate the energy requirements of upcoming operations.

Initially, when a node has not determined its role, it is considered undecided. For making decision of becoming clusterhead, only information about the local neighborhood is needed. The same holds true for the decision of joining a clusterhead as part of the cluster controled by it. Therefore, the algorithm can be performed locally in a distributed fashion, e.g. as presented in [1] where also additional procedures for maintaining the clustered structure in face of

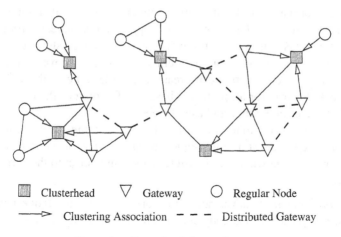

Figure 4.   Example of clustered structure.

topology changes are given. Additional procedures to control the sizes of the clusters created by this approach are presented in [13]. Obviously, when the clusterheads control more evenly sized clusters, the overall energy consumption is distributed more evenly, as well.

4.2   Clusterbased Route Discovery

The protocol to create and maintain the above described clustered structure of the network comes at the cost of additional control messages, consuming additional energy. However, the structure allows for limiting the number of transmissions for other services at the networking and other layers. For example, obtaining information for the routing process of messages in the multi-hop topology can be achieved more energy-efficient. The resulting benefits in terms of saved transmissions, and thus saved energy, show the advantage of the clustering scheme. In particular, the well known dynamic source routing (DSR, [8]) is adapted to fit the clustered structure of the network.

In principle, DSR works as follows. Suppose that a network node, the source, has to send some data packets to another node, the destination. This usually requires a multi-hop route through the network. For this purpose, each node stores information about routes previously created in its cache. When there is no stored information about a route to the destination available at the source, it initiates a route creation process. The network is flooded with route request messages. Each node along the process adds itself to a route list contained in the message, and then rebroadcast the request message. When such a request reaches the destination node, a route reply message is created, relaying the routing information back to the source, that in return can then send the data packets. DSR also offers basic routines to cope with disconnected routes.

As every node is within direct transmission range of a clusterhead, and each clusterhead has knowledge about its members, a route discovery process that reaches all clusterheads suffices to create a feasible route in the ad hoc network. This holds true for all protocols based on flooding of the network, thus especially for the route request phase of DSR.

Each node, according to its current role in the clustering scheme, can locally decide not to rebroadcast a route request message. The decision is taken according to what is known about the route request in respect to the surrounding clusterheads:

- An *ordinary node* never needs to rebroadcast a route request message since its presence is already accounted for by its clusterhead.

- A *clusterhead* rebroadcasts such a message according to the same rules as given by the original DSR. If the destination of the route creation process is a node in the vicinity of the clusterhead, it initiates the route reply process and not rebroadcast the query.

- A *gateway* only relays a route request message if it was received from a neighboring clusterhead. Additionally, if the gateway is a *distributed* gateway, and the route request was received from a gateway that is not part of the node's own cluster, it notifies its clusterhead by rebroadcasting.

- An *undecided node*, e.g. when it has recently been added to the network and has not decided on its status when receiving a route request message, participates in the route discovery process according to the flooding rules given by the original DSR.

Newly added nodes to the network, once they are synchronized with their neighbors, are immediately operational within the network routing process and do not need to settle on their role within the clustered structure before being able to communicate with the rest of the network.

During the route creation process, *detours* may occur as all routes created pass through the clusterheads of the appropriate nodes. So, in the route reply phase, when the created routes are reported back to the source, each node does not relay the route list back to the preceeding entry in the list, but checks wether it can shorten the list by sending it to a neighboring node that is also in the list, but is closer to the source of the route. A small example of this reduction process is given in Figure 5. Suppose that node l has just received the route list R_{original} from node m. Checking all entries in R_{original} preceding l together with its neighbors, node l can prune the entries j and k, and thus only sends the new route list R_{new} onwards to node i.

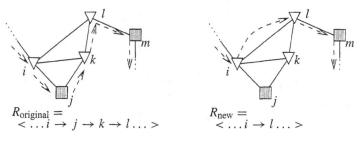

Figure 5.   Route reduction at node l during the route reply process.

The routes arriving at the source that have been created by the above process are comparable in length, i.e. hop-distance, to those created by the classic DSR. This is also confirmed by the simulations we performed and that are presented in the following part.

The ability to incorporate undecided nodes into the scheme shows that the the approach does not represent backbone-based routing as the algorithm does not rely on a strictly maintained backbone for the routing process.

4.3   Performance

We compared the additional overhead involved with creating and maintaining the clustered structure of the network and the saved messages during the routing process to the classic DSR on the flat, non-clustered network.

In the simulation setup, the nodes move in the bounded area according to the random waypoint model. After a certain time interval, a new point in the area is chosen and the node migrates there with a certain speed. Then, it waits again and starts moving towards a new chosen point. This model offers a mixture of static and mobile nodes that can be controlled by time a node remains static at each of the waypoints.

In Figure 6, these results are presented for a partially mobile network of 50 nodes placed in an area of 3 by 5 times the maximal transmission range. The waiting time of each node that reached its destination is $5s$, and the speed of the mobile nodes is fixed at the value presented by the *speed coefficient* at the horizontal axis in both graphs. The value given at the horizontal axis reflects the fixed speed of the mobile nodes with respect to their transmission range, i.e. $\frac{speed[m/s]}{range[m]}$. For example, consider the EYES wireless sensor prototype: The transmission ranges of the sensor nodes are at most 25 m in an indoor environment. Suppose the mobile nodes move with walking speed, say 1 m/s, the speed coefficient is then 0.04. The left graph presents the average number of control messages needed to construct a feasible route for a data message from a randomly chosen node to another random node, i.e. the overall number

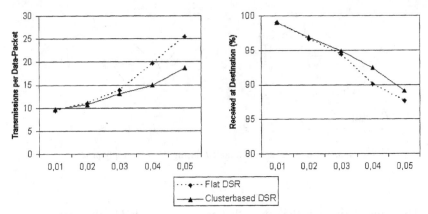

Figure 6. Control overhead and reliability of the clusterbased routing (vs. speed given by the speed coefficient).

of control messages versus the correctly received original data messages. On average, there are two data messages created every second that are to be routed through the network. The values for the clusterbased approach also include the overhead messages for set up and maintenance of the clustered structure. On the right side, the reliability is given in terms of successfully delivered messages, that is a feasible route was present, either from cache or created by the route discovery process, and a data packet was furthermore successfully routed along this route and reached the destination.

Overall, it can be stated that the clusterbased approach outperforms the classic DSR in both control message overhead and the number of successfully delivered messages reaching the destination. The effects of increasing mobility are better handled by the clusterbased scheme. Only in networks with low mobility, i.e. a rather static network, the control overhead of the clustering procedures degrades the overall performance. Note however that the performance does not drop below the performance of the classic DSR.

Even though the evaluation of the clusterbased scheme uses the route creation process of DSR as a reference point, it can easily be adapted to other algorithms that rely on flooding.

5. Multipath Routing

When using multi-hop data delivery, problems arise when intermediate nodes fail to forward the incoming messages. The resource and energy limited sensor node has many failure modes, each of which thus decreases the performance of the whole network. Usually, acknowledgements and retransmissions are implemented to recover the lost data. However, this generates additional traffic and delays in the network, and it becomes worse when the failure rates of the nodes increase. Moreover, a mobile node might be unreachable since it moved out of its region before the message arrived. Retransmissions are useless and just increase the latency and waste bandwidth and energy.

The reliability of the system can be increased by multipath routing, which allows the establishment of more than one path between source and destination and provides an easy mechanism to increase the likelihood of reliable data delivery by sending multiple copies of data along different paths. Intuitively, its drawback is the increase of traffic. However we show that this is not always true, and that using multipath routing combined with some error correction codes leads to saving energy and bandwidth while even decreasing latency.

In this part, we present a routing mechanism based on multipath routing and data splitting. The multipath routing algorithm achieves better results if combined with a clustering algorithm. The multipath routing algorithm runs on top of a clustering scheme, using the cluster-heads as the regular nodes in the basic algorithm. This way a combination between the reliability of the multipath scheme and the energy efficiency of the clustering can be achieved. Furthermore, designing the multipath routing to work together with a data-splitting algorithm reduces the overall traffic even more.

5.1   Multipath On-Demand Routing

The main goal of the Multipath On-Demand Routing algorithm (MDR) that we propose is finding multiple paths between the source and the destination while minimizing the amount of traffic in the network. The algorithm is based on the basic ideas behind the DSR algorithm, and it consists of two phases presented in Figure 7: the *route request* and *route reply* phase.

- *Route Request phase* - The data source starts the route creation process if it has a data packet to route to a destination and does not have cached route information to the the destination, the cached routes are not valid anymore or the cache contains not enough different paths to it. The source floods the network with a short and fixed length Route Request message containing the information given in Figure 8.

  After receiving a route request, a node checks in its local data structure whether it has received the route request before using the first three fields in the message. If not, it creates a new data entry in the local database to store this information and records the node ID from which the request was received. For the subsequent messages received, this node only has to store the ID of the neighbors. It can easily check and mark if the source of the message is a first order neighbor by the *lasthop* and *ack* fields. The node only forwards the first route request received (additional identical route request messages received are discarded), in which it substitutes the *ack* field with the *lasthop* value and the *lasthop* with

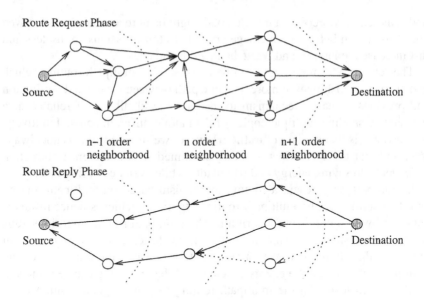

Figure 7. Overview of MDR Phases.

its own ID. After forwarding the route request message and listening to the passive acknowledgements, each node knows which neighbors are closer to the source (further referred as the *n-1 neighbor list*) and which ones are not (*n+1 neighbor list*). If the node identifies itself as being the destination of the message, it initiates the second phase of the algorithm.

- *Route Reply phase* - In this phase, several paths between the destination and the source are reported to the source. Compared to the route request message, the route reply message has three additional fields given in Figure 9.

In the previous phase each intermediate node stored information about its neighbors that forwarded the route request message, thus the complete path between the source and the destination need not be stored inside a reply message. When the source receives the first route reply, it stores the ID of the node that forwarded the message and the path length. It also sets up a timer to measure the interval that it waits for other reply messages to come. When this timer expires the data is sent using the route information obtained thus far.

On receiving a route reply addressed to it, each node will modify the *nexthop*, *ack*, *hops* and *detours* fields of the message according to the new parameters before forwarding it to the first neighbor in the *n-1 neighbor list*.

Field	Description
snodeID	source node ID
dnodeID	destination node ID
floodID	route request message ID
lasthop	ID of the ndoe forwarding this message
ack	ID of the last hop

Figure 8.   Fields of the Route Request and Route Reply message.

Field	Description
nexthop	ID of the node the message is forwarded to
hops	number of hops already traveled
detours	number of detours a message can take

Figure 9.   Additional fields of the Route Reply message.

If this list is empty and the *detours* field is not empty, it chooses the first neighbor in the *n neighbor list* (respectively *n+1 neighbor list*) and decreases the allowed *detours*. On receiving a route reply not addressed to it, a node searches its own data structure to find the entry corresponding to the first three fields. If such an entry is found, the forwarding node is removed from both the *n-1* and *n+1 neighbor lists*.

A node that forwarded a reply has to take care of two more things: first it sets a flag ensuring that it does not forward any other message for the same route and second, it waits for a passive acknowledgement. If this does not arrive, it assumes that the node to which it sent the message is no longer available, or has forwarded a message previously. The respective neighbor is deleted from the node's lists. It then tries resending the message to the next neighbor in the lists, until the lists become empty or the *detours* field becomes 0. This step of removing nodes from the list is needed to ensure that the source will receive only disjoint paths.

A route maintenance is not necessary because the energy needed to maintain the multiple paths is more than what is required to discover new routes.

MDR reduces the size of the messages considerably when compared to the original DSR as it uses fixed sizes for them. In fact we are moving the information stored inside the messages to the sensor nodes themselves. The sensor nodes are responsible to cache where the route request messages came from. The second group of modifications involves the multiple paths management. In the original DSR, if the same route request message is received several times by a node, only the first one is considered and the rest are discarded. MDR considers all these messages and uses this information. Through these changes we obtain a controlled flooding in the first phase of the algorithm by using small

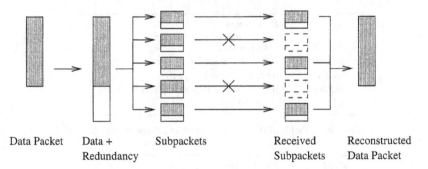

Data Packet	Data + Redundancy	Subpackets	Received Subpackets	Reconstructed Data Packet

Figure 10.   Data splitting across multiple paths.

messages with fixed length, additionally reduced by exploiting the clustered structure. The second phase also uses small fixed length messages that involve only a fraction of nodes existent between the source and the destination.

5.2   Data Splitting Across Multiple Paths

Once the multipath route discovery has finished, the data can be sent using this information. Sending the same data packet across multiple disjoint paths significantly increases the reliability of WSN. However this mechanism requires large quantities of additional network resources (such as bandwidth, and most importantly energy).

The method of reducing the amount of traffic by data splitting has been analysed in detail in [10] and [19]. In [3] we address the trade-off between the energy and the reliability of this mechanism. We also study the possibility of integrating this method with a multipath-routing algorithm.

Let us assume the route construction results in k different paths. Some of these fail in transmitting the data all the way from the source to the destination. Our approach is to split the data packet into $l \leq k$ parts (hereafter referred to as subpackets) and to send these subpackets instead of the whole data packet (see Figure 10). Based on the failure probability of each node in the network, we can estimate the number of subpackets that will reach the destination.

There exist several fast and simple forward error correction codes that allow for reconstruction of the original message that has been split up using only a fraction of the messages at the destination.

In order to obtain a predetermined reliability for the data transmission, the total number of subpackets as well as the redundancy used is a function dependent on the multipath degree and the estimated number of failing paths. As the multipath degree can change according to the positions of the source and the destination in the network, each source has to be able to decide on these

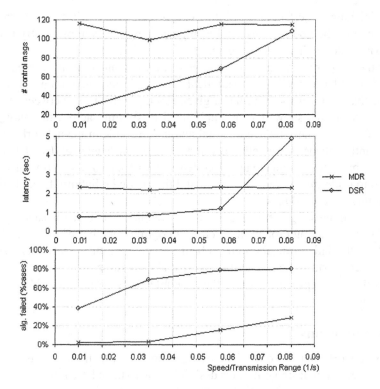

Figure 11.   Comparison MDR - DSR.

parameters before the transmission of the subpackets. The actual value for the failing probability of a node has to be obtained empirically or by using a path rating algorithm.

5.3   Performance of MDR and Data-Splitting

In Figure 11, we show the comparisons between the classic DSR and MDR using the same simulation setup as already presented in Section 4.3.

MDR performs better than DSR in terms of reliability independent of the speed coefficient. For high speed coefficient values, when DSR becomes unusable due to the error rate, MDR still works acceptable. The price for the increased reliability is paid with an increased number of control messages and with a higher latency. A closer look not just at the number of control messages, but at the traffic size, shows that the total amount of MDR traffic compared to the DSR traffic varies between 4.04:1 for low values of the speed coefficient to 1.02:1 for scenarios with high values of the speed coefficient.

It is interesting to find out that the speed of the nodes has a reduced influence on all the parameters of the algorithm than in the classic DSR. A means of diminishing the effects of mobility is usually by increasing the transmission range of the nodes. This implies higher energy consumption. By using multipath routing, this is not necessary.

For the resulting data splitting approach, several simulations have been performed in order to verify the theoretical results. In the simulations we considered that each sensor node has a given failing probability (between 0 and 0.25). The results show that by applying data splitting across the multiple paths, we substantially reduce the total amount of traffic in the network while significantly enhancing the reliability and reducing the latency of data delivery. The possibility to determine values on-demand for the error correction offers a method to trade-off between the traffic and the desired reliability by adjusting the parameters accordingly.

Another interesting result that we have obtained was that for each upper bound of node failing probability there is an optimal number of paths needed, for which the failed transmission probability gets to a minimum. Increasing the number of paths, the probability of error also increases. Therefore it is not always the best approch to use all the available paths given by the MDR, but only a subset.

6. Conclusions

Sensor networks may be one of the best examples in which the pervasiveness of energy efficient design criteria is desirable, due to the inherent resource limitation, which makes energy the most valuable resource. Sensor nodes should be able to establish self-assembling networks that are incrementally extensible and dynamically adaptable to mobility of sensor nodes, changes in task and network requirements, device failure and degradation of sensor nodes. Therefore, each sensor node must be autonomous and capable of organising itself in the overall community of sensors to perform co-ordinated activities with global objectives.

All these required capabilities for sensor nodes are not trivial mainly because the sensor nodes are resource poor: they must be deployed spontaneously to form efficient ad-hoc networks using tiny sensor nodes with limited energy, computational, storage, and wireless communication capabilities. Because the nodes are resource poor, and operate in a time-varying environment it is important that all these issues are addressed together. We presented energy efficient solutions to an integrated approach for wireless communication in a sensor network. We addressed in particular the even more challenging area of a dynamic topology, and propose solutions which are robust against these dynamics.

The proposed TDMA-based MAC protocol is designed to support the different communication types often used in wireless sensor networks. The protocol minimizes the utilization of the transceivers of the nodes in order to save energy. The data requests are made efficiently to reduce the bandwidth used in the network. Latency in the network is reduced by allowing transmissions in not owned or released data sections. The traffic control section can be deployed to make wake-up calls to sleeping nodes.

As the envisioned WSN topology consists of a large number of nodes, clustering is used to reduce the complexity and ease the maintenance of the network by assigning clusterheads to control a certain number of surrounding nodes. We showed that the benefits of a clustered structure, in terms of energy consumption, can be harvested in the routing process alone.

Multipath routing and data splitting offer ways to exploit the given redundancy in a WSN to increase the reliability of data delivery. One such approach was presented.

We have shown that the development of energy efficient communication protocols requires a systemwide view of the whole network protocol stack. Future work will focus on the integration of these protocols for medium access control, clustering and multipath routing in a physical testbed of EYES-nodes.

References

[1] S. Basagni. "Finding a maximal weighted independent set in wireless networks." *Telecommunication Systems, Special Issue on Mobile Computing and Wireless Networks*, 18(1/3):155–168, September 2001.

[2] A. Chandrakasan, R. Min, M. Bhardwaj, S.-H. Cho, and A. Wang. "Power aware wireless microsensor systems." In *Keynote Paper ESSCIRC*, Florence, Italy, September 2002.

[3] S. Dulman, T. Nieberg, J. Wu, and P. Havinga. "Trade-off between traffic overhead and reliability in multipath routing for wireless sensor networks." In *Proceedings of the Wireless Communications and Networking Conference*, 2003.

[4] Energy-Efficient Sensor Networks (EYES). http://eyes.eu.org.

[5] D. Ganesan, R. Govindan, S. Shenker, and D. Estrin. "Highly-resilient, energy-efficient multipath routing in wireless sensor networks." *ACM SIGMOBILE Mobile Computing and Communications Review*, 5(4):11–25, 2001.

[6] W.R. Heinzelman, A. Chandrakasan, and H. Balakrishnan. "Energy-efficient communication protocol for wireless microsensor networks." In *HICSS*, 2000.

[7] C. Intanagonwiwat, R. Govindan, and D. Estrin. "Directed diffusion: A scalable and robust communication paradigm for sensor networks." In *Proc. Sixth Annual International Conference on Mobile Computing and Networks*, 2000.

[8] D.B. Johnson and D.A. Maltz. "Dynamic source routing in ad hoc wireless networks." In Imielinski and Korth, editors, *Mobile Computing*, volume 353. Kluwer Academic Publishers, 1996.

[9] S. Lee, W. Su, and M. Gerla. "On-demand multicast routing protocol in multihop wireless mobile networks." *Mobile Networks and Applications*, 7(6): 441–453, December 2002.

[10] S.J. Lee and M. Gerla. "Split multipath routing with maximally disjoint paths in ad hoc networks." In *Proc. Intl. Conf. on Comm. (ICC)*, 2001.

[11] R. Min and A. Chandrakasan, "Energy-Efficient Communication for High Density Networks", this volume, 2003.

[12] A. Nasipuri and S. Das. "On-Demand Multipath Routing for Mobile Ad Hoc Networks". In *8th Intl. Conference on Computer Communications and Networks (IC3N 99)*, 1999.

[13] T. Nieberg, P. Havinga, and J. Hurink. "On the advantages of clusterbased routing in wireless sensor networks", 2003. preprint.

[14] Standards Committee of the IEEE Computer Society. IEEE standard 802.11-1999, 1999. Wireless LAN Medium Access Control (MAC) and Physical Layer (PHY) specifications.

[15] Texas Instruments. MSP430x1xx family user's guide.

[16] V.D. Park and M.S. Corson. "A highly adaptive distributed routing algorithm for mobile wireless networks". In *Proceedings of IEEE INFOCOM'97 Conf.*, April 1997.

[17] RFM. TR1001 868.35 MHz hybrid transceiver.

[18] S. Sakai, M. Togasaki, and K. Yamazaki. "A note on greedy algorithms for maximum weighted independent set problem." *Discrete Applied Mathematics*, 126(2-3): 313-322, 2003.

[19] A. Tsirigos and Z.J. Hass. "Multipath routing in the presence of frequent topological changes." *IEEE Comm. Magazine*, pages 132–138, 2001.

[20] W. Ye, J. Heidemann, and D. Estrin. "An energy-efficient mac protocol for wireless sensor networks." In *Proceedings of IEEE INFOCOM 2002 Conf.*, June 2002.

Energy-Efficient Communication for High Density Networks

Rex Min and Anantha Chandrakasan

Massachusetts Institute of Technology, Boston, MA, USA

{ rmin, anantha } @mtl.mit.edu

Abstract   New generations of AmI devices will need to deliver unprecedented system lifetime from increasingly smaller and constrained energy sources. Energy-efficient
system design, and energy-efficient wireless communication in particular, will
therefore be key enablers of next-generation AmI devices. Using the microsensor
network as our design driver, we present the impact of accurate system models,
dynamic energy vs. performance trade-offs, and application-specific design, each
applied to the protocol and application layers. Comprehensive models of wireless communication hardware reveal surprising discrepancies between expectations and realities of the energy consumption of multihop routing. An abstraction
layer encourages performance trade-offs at the protocol level that translate into
energy agility at the hardware level. Application-specific design, applied to protocols, enables an efficient protocol for data propagation in microsensor networks.

Keywords   energy, energy efficiency, wireless communication, energy scalability, energy
agility, middleware, energy models, multihop routing, application-specific protocol, unaddressed forwarding, microsensor networks

1.   Introduction

Emerging devices for AmI are characterized by shrinking size and increasing
spatial density. From the perspective of wireless communication, a high density of nodes implies competition over limited bandwidth, and the devices'
small physical size suggests a limited battery capacity. It is therefore imperative that communication in AmI devices utilize these resources efficiently.
This paper focuses on the management and conservation of the latter resource–
energy–as wireless devices scale into ambient intelligence.

1.1   Distributed Microsensor Networks

The distributed microsensor network is one of the most anticipated applications of ultra-high-density wireless networking. A spatially dense network of
microsensor nodes offers a rich, high-resolution, multi-dimensional picture of

T. Basten et al. (eds.), Ambient Intelligence: Impact on Embedded System Design, 295-314.
© 2003 *Kluwer Academic Publishers. Printed in the Netherlands.*

the environment that is not possible with single sensor or small group of sensors. The sheer number of nodes naturally leads to the network's robustness and fault-tolerance to the loss of individual nodes, making maintenance unnecessary. As the nodes self-organize into ad hoc networks, deployment can be as easy as sprinkling nodes about the region of interest or dropping them by air, and setting up a conveniently located base station to which the nodes will relay their observations. These advantages, and the nodes' diminutive size, make sensor networks ideal for any number of inhospitable or unreachable locations where deployment is difficult, wires impractical, and maintenance impossible. The cramped confines of an appliance, the surface or interior of the human body, facilities that produce toxic radiation or chemical vapors, and the lands of extreme desert or Arctic climates, are excellent candidates for microsensor networks [16, 13, 5].

Small, densely placed nodes raise serious energy consumption considerations. Small nodes imply limited physical space for batteries, and their dense distribution implies that periodic battery replacement will be tremendously inconvenient—and more likely, impossible. Given that wireless communication will be responsible for a substantial portion of the node's energy consumption, it is imperative that the nodes' communication subsystems be designed with maximum energy efficiency. Overheads that may have been neglected before must be addressed and mitigated.

1.2   The Top Five Myths of Energy-Efficient Wireless Communication

This paper advocates three design principles to facilitate energy-efficient communication. First, it is crucial that energy-efficient communication software be based on sound models of the hardware on which it will operate. While this may seem an obvious point, incomplete or inaccurate energy models appear often and invariably lead to surprising discrepancies between designers' expectations and system realities. Second, *energy agility* enables a graceful adaptation of energy consumption in response to dynamically varying environmental conditions. An energy-agile system consumes just enough energy to achieve only the level of performance required at any given time. Finally, application-specific design reduces the overheads of overly general solutions. Any unnecessary functionality present at any level of the system hierarchy, from processor to protocol, is an inefficiency by design.

By applying these fundamental principles across the system hierarchy, we dispel five commonly held myths about its energy consumption.

1. **"Communication energy scales as d^n."**
   While communication energy is commonly modeled as d^n, the transmission distance raised to some constant power, this transmission *path loss*

does not dominate the energy consumption of short-range wireless communication. The constant and distance-independent power consumption of the radio and digital electronics often exceeds the distance-dependent power consumed by the amplifier.

2. **"Multihop saves energy."**
   Due to the dominance of distance-independent energy terms for short-range wireless communication, reducing the distance through intermediate hops does not save energy.

3. **"Wireless performance can't scale gracefully."**
   The correct reception of bits is indeed an all-or-nothing proposition, but the performance of wireless communication may be parameterized and gracefully degraded. The idea of *energy agility*, long advocated for digital systems, is therefore applicable to communication as well.

4. **"Abstractions lead to inefficiency."**
   Abstraction is all too often viewed as a trade-off between simplicity and efficiency. Power aware abstractions, however, offer a bridge between energy-scalable hardware and performance-scalable protocols, thereby providing an abstraction layer that *encourages* the efficient use of energy.

5. **"The protocol is independent of the application and hardware."**
   The traditional view of isolated protocol layers is giving way to the realization that cross-layer collaboration is essential for efficient communication. This is the overarching theme that unites each of the topics presented in this paper. In particular, we advocate that application-specific design at the protocol layer to minimize overhead and unnecessary functionality.

We begin our discussion with a careful consideration of the energy consumption in a wireless communication system, culminating in a surprising result about the real-world energy consumption of multihop communication. Next, Section 3 proposes an abstraction that encourages energy-agile communication. Finally, Section 4 introduces an efficient, application-specific protocol for microsensor networks.

2. Identifying Communication Inefficiencies

This section summarizes the sources and nature of energy consumption across the communication system hierarchy, with an emphasis on attaining comprehensive and accurate hardware energy models. While many of the energy models are reassuringly familiar, considering them in tandem provides a surprising insight into the energy consumption of multihop routing protocols.

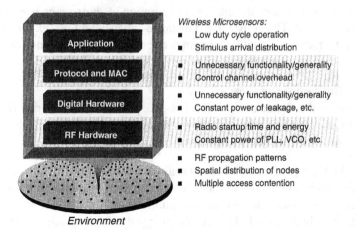

Figure 1.   Summary of communication overheads, inefficiencies, and non-idealities.

2.1   Non-Idealities of Wireless Communication

Enhancing the efficiency of any system requires particular attention to its non-ideal behavior. The discrepancies between ideal models and non-ideal, real-world behavior are the most common sources of inefficiencies.

Ideal models of wireless communication consider spherical radiation with an attenuation that follows a power law with distance, resulting in an attenuation of d^n. Unfortunately, non-ideal behaviors within a real wireless system cause the actual performance of communication to differ substantially from this simplified model. Inefficiencies arise from a number of sources at many levels of the system hierarchy; several prominent examples are illustrated in Figure 1.

The nodes' environment, hardware, and communication protocols are fundamental sources of inefficiencies. Uneven terrain, obstacles, and interferers will cause radio propagation patterns to be aspherical. Nodes will rarely be placed at regular spatial intervals; many visions of ambient intelligence assume nodes that are dropped by air or sprayed onto a surface. Radio hardware cannot shut down and restart immediately. A non-negligible startup time is required, during which energy and time are consumed. Analog subcomponents of the radio require substantial bias currents, resulting in constant power dissipation while powered [15]. Constant power is also present in digital circuits in the form of leakage currents [4].

A less obvious but equally insidious source of overhead is unnecessary functionality or generality. A general-purpose microprocessor may be several orders of magnitude less efficient than an application-specific circuit [6] due to the former's overhead of decoding instructions, maintaining caches, and so

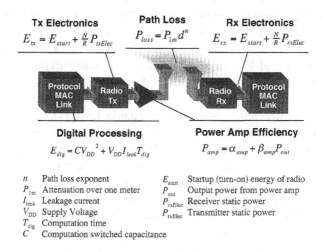

Figure 2. Real-world energy model for wireless communication.

forth. This effect is present in media access and routing protocols as well. Many applications, such as multimedia streaming, do not require explicit guarantees of reliable end-to-end communication for all packets. If a protocol that implements reliable communication through persistent retries is used in conjunction with such applications, energy and bandwidth are wasted due to the retransmissions and associated control channel (non-data) messages.

Finally, each application contributes a unique set of inefficiencies to the system. In the case of wireless microsensors, a primary concern is the infrequent and non-uniform arrival of stimuli that activate the nodes. Infrequent events would cause the nodes to be idle for long periods of time. The resulting low duty cycle would worsen the overheads of idling and waking components.

2.2 Hardware Energy Models

Accurate energy models for communication hardware are essential for analyzing the energy consumption of communication protocols. This section reviews comprehensive models for the energy consumption of wireless communication hardware.

The energy of wireless communication can be modeled as the sums of the energies of three components: digital electronics, the core radio transceiver electronics, and the output power amplifier. These blocks, and models for their energies, are illustrated in Figure 2.

Digital computation is typically used for digital signal processing of gathered data and for implementation of the protocol stack. Energy consumed by digital circuits consists of dynamic and static dissipation. The dynamic energy of digital computation is the energy required to charge and discharge the parasitic capacitances of the digital circuits. The dynamic energy per processed bit is described by $C_{bit}V_{DD}^2$, with C_{bit} representing the switched capacitance per processed bit and V_{DD} the supply voltage [1]. Static dissipation originates from the undesirable leakage of current from power to ground at all times. The leakage current is exponential with V_{DD} and depends upon the fabrication technology with which the transistors are manufactured [4].

The energy consumption of the radio consists of static current dissipated by the analog electronics (analogous to leakage in the digital case, except that these bias currents serve the useful purpose of stabilizing the radio), and the energy required by the power amplifier to radiate RF waves. The static energy consists of the startup energy E_{start} and a term that is directly proportional to the length of time that the radio is on. This model applies to both the transmitter and receiver. The transmitter model must also include a power amplifier, whose non-negligible efficiency losses are often ignored in analyses of communication energy. The simplest model, illustrated by Figure 2, is a linearization of the inefficiency. While not as accurate as a higher-order model, a linear model provides a reasonable trade-off between accuracy and complexity.

The output power required for node-to-node communication is determined by the distance between nodes. Neglecting fading and other time-variant effects, the mean *path loss attenuation* P_{loss} between a transmitter and receiver over a distance d is described as $P_{1m}d^n$, where P_{1m} is the attenuation over one meter and n is the path loss exponent, which typically falls between 2 and 4.

The above terms can be summarized as an energy consumption per bit of $\alpha + \beta d^n$, where α is a distance-independent term that accounts for the overheads of transmitter and receiver electronics and digital processing, and β models linear losses of the power amplifier and its associated antenna. This simplification accounts for all of the above terms except the radio startup energy.

2.3   Hardware and Multihop Routing

Since path loss attenuation scales with distance d as d^n, the use of several shorter transmissions via intermediate relay devices is commonly proposed to conserve energy. Figure 3 illustrates this idea of *multihop routing*, which reduces path loss attenuation from d^n to $h(d/h)^n$ for an h-hop relay [10]. At first glance, multihop is especially appealing for microsensor networks, for the high spatial density of nodes implies the availability of many neighbor nodes as potential relays. Figure 4(a) illustrates the ideal result: as the total transmission distance increases, energy is saved by adding intermediate relays to limit the distance covered by each hop.

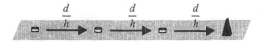

Figure 3. Multihop routing.

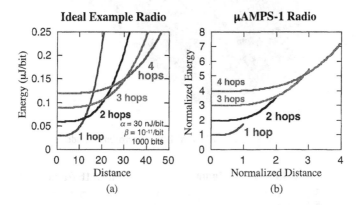

Figure 4. (a) Ideally, intermediate hops ideally save energy for longer range communication. (b) The distance-independent overhead of a real-world radio cancels the benefits of multihop.

Unfortunately, Figure 4(a) does not reflect the energy consumption of real-world radios. Figure 4(b) reflects the actual energy consumption of a 2.4 GHz radio present in the MIT μAMPS-1 microsensor node [9]. The μAMPS-1 node, pictured in Figure 5, incorporates environmental sensing, digital signal processing, and a 1 Mbps radio into an aggressively power-managed hardware prototype. Using the energy consumption models above with parameters from [9, 11], it is clear that multihop does not save energy within the range of output powers supported by this radio (1-100 mW).

The reason for this surprising result is hinted by the preceding discussion in this section. For virtually all of today's short-range radios, the distance-independent α term substantially exceeds the distance-scalable βd^n, *even at the radio's maximum output power* [12]. For instance, the Cisco Aironet 350 card [2], an 802.11b transceiver, has $\alpha = 14.8$ nJ/bit while $\beta d^n = 11$ nJ/bit at the radio's maximum output power of 100 mW. As the d^n term never dominates, multihop does not save energy. Additional media access (MAC) layer and routing protocol overheads would further reduce the efficiency of multihop. Figure 6 plots the energy consumed by h-hop communication for the Cisco Aironet radio and a simpler, on-off keyed radio that operates at 915 MHz [14].

Hence, while multihop remains useful for long-range radios that output several Watts of power, we must reconsider this technique for high-density AmI applications. To avoid surprises, energy-efficient communication protocols

Figure 5. The MIT μAMPS-1 microsensor node.

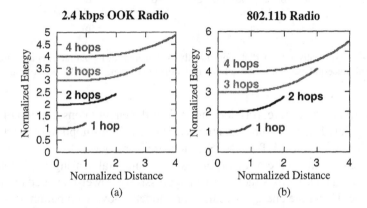

Figure 6. Multihop does not save energy for: (a) a commercial, low-rate, on-off-keyed (OOK) radio, and (b) an 802.11b radio.

must be designed around accurate energy models of the targeted hardware. Multihop saves energy only when the path attenuation dominates the static energy consumption of hardware, a case that occurs less frequently than is commonly believed.

3. Abstraction for Energy-Agile Communication

Energy agility refers to the ability of circuits, architectures, and algorithms to save energy through graceful trade-offs between performance and quality. A node designed or optimized for any one operating point will necessarily be inefficient for all the others. Perhaps the nodes are accidentally deployed more closely together than expected, stimuli are fewer and farther between, or the

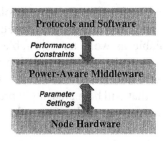

Figure 7.   Abstractions for energy-agile communication.

user chooses to tolerate a few additional milliseconds of communication delay. A node that cannot respond to relaxed performance demands with energy reductions is utilizing energy in an inefficient manner. Energy agility, then, suggests a keen awareness of the exact performance demands of the user and the environment. Such a system consumes just enough energy to achieve only that level of performance.

This section introduces an abstraction that enhances the energy agility of communication software and illustrates the energy savings possible by this abstraction on the μAMPS-1 microsensor node introduced in Section 2.3.

3.1   Power-Aware System Hierarchy

The design of an energy-agile communication system can be divided into three steps. First, the digital and analog communication hardware are designed to provide energy and performance trade-offs. Second, protocols and applications are designed to expose their explicit communication performance needs. Finally, a power-aware abstraction layer links communication quality with hardware energy consumption. Figure 7 illustrates the resulting hierarchy.

Power aware hardware reacts gracefully to constantly changing operational demands. As performance demands increase or decrease, power aware hardware scales energy consumption accordingly to adjust its performance on-the-fly. Graceful energy scalability is highly desirable for any energy-constrained wireless node since the operational demands on a real-world node constantly change, and the peak performance of the node is rarely needed. Energy agility is effected by key parameters that act as "knobs" to adjust energy and performance simultaneously.

Several such knobs are available in a power aware wireless communication subsystem. The communication subsystem consists of digital processing for error correction and protocol handling, and a radio transceiver for the actual transmission and reception. The performance of digital processing can be adjusted with dynamic voltage scaling. Reducing processor voltage slows computation, permitting a graceful exchange of energy for latency. The workload on the processor can be varied through adaptive forward error correction

(FEC) coding. "Stronger" codes that are more resilient to errors generally require more processing, and therefore more computation energy. The radio transceiver is energy-scalable as well; an adjustable power amplifier in the transmitter allows energy to scale with transmission range. Voltage scaling, convolutional code strength, and radio transmission power are three crucial knobs for power awareness that will be increasingly present in modern wireless nodes. The μAMPS-1 microsensor node offers all of these knobs for energy agility.

While power-aware hardware is a significant stride toward energy-efficient communication, application designers that utilize wireless communication typically do not wish to concern themselves with low-level parameters such as processor voltage or transmit power. Since energy conservation is so crucial to wireless nodes, however, we must provide a way for the application to take advantage of hardware energy scalability. Figure 7 illustrates our approach. We introduce a middleware abstraction layer that bridges the gap between these low-level knobs for energy scalability and performance metrics more relevant to an application. Performance metrics for communication are expressed by the application and translated by middleware into energy-efficient parameter settings for the communication hardware.

3.2   Middleware Abstraction Layer for Energy Agility

Energy agility for wireless communication can be achieved once the notion of communication "quality" is defined. We can define communication quality by four of its fundamental metrics: *distance, reliability, delay* and *throughput*. Protocols become energy-agile by specifying the required communication performance for each transmission in terms of these metrics. Explicit specifications of performance ensure that the hardware layer does not waste energy by providing performance in excess of what is required by the software layers.

To bridge the gap between these performance parameters and the actual hardware knobs for energy scalability, we suggest a power-aware middleware layer in between the hardware and the communication software. Following the discussion in Section 2.2, the middleware must be empowered with accurate hardware energy models for the digital processing circuits and radio transceiver, allowing this layer to select the minimum-energy hardware settings for the performance level commanded through the API [11].

Conventional wisdom states that abstractions provide organization and ease of use, in exchange for a loss of efficiency. In this case, however, our energy-agile API and middleware layers can actually *increase* energy efficiency. Unlike traditional abstractions that introduce overhead, our abstractions encourage the efficient utilization of energy-agile hardware by encouraging explicit requests for communication performance from the protocol layer.

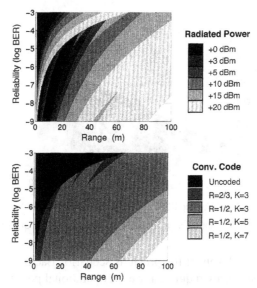

Figure 8.   Minimum-energy radiated power and convolutional code policy for the μAMPS microsensor node.

A concrete illustration of this middleware abstraction is provided by the μAMPS-1 node. Again, using the energy consumption models presented in Section 2.2 and the measured parameters from [9, 11], Figure 8 illustrates the minimum-energy operational policy for communication between two μAMPS-1 nodes. The minimum-energy selections of two hardware knobs (radiated power and convolutional coding scheme) are shown as a function of two high-level performance metrics (reliability and distance). On μAMPS-1, radio transmission power may be scaled between 0 dBm and 20 dBm in six steps, and transmission may either be uncoded or convolutionally coded at various combinations of code rate R and constraint length K.

Expending additional energy increases the range and reliability of communication in one of two ways. First, the range and reliability of communication increases as more power is radiated from the transmitter. Second, stronger (lower R, higher K) error-correcting codes may increase error resilience over longer distances at the expense of longer transmit and receive times in the case of lower R or higher decoding energy in the case of higher K. μAMPS-1 clearly favors increased transmission power over convolutional encoding. This is due to the use of a general-purpose SA-1110 processor, which causes the energy cost of Viterbi decoding to exceed that of additional radio output power.

When the hardware is driven by the minimum-energy operational policy for each quality level, the system's energy consumption increases monotonically with communication quality. Figure 9 computes the energy required for the

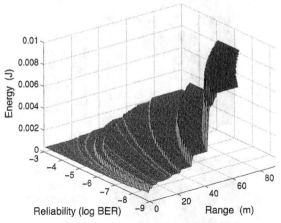

Figure 9.   Energy consumption as a function of communication quality for the μAMPS microsensor node.

range and reliability levels of the previous scenario. Each "step" in the graph corresponds to the energy required for each operational policy selected in Figure 8. Energy agility enables energy consumption to scale gracefully with quality.

Considering the relative inefficiency of Viterbi decoding on the SA-1110 processor and the communication quality gains achievable with coding, it is worthwhile to move the Viterbi decoding operation onto an application-specific integrated circuit (ASIC) dedicated to this purpose. An ASIC trades flexibility for energy savings; a typical ASIC consumes several orders of magnitude less energy than a microprocessor [7]. With this in mind, Figure 10 illustrates the minimum-energy policies for coding and output power if the digital circuitry for Viterbi decoding consumed 1000 times less power than a SA-1110 implementation. As one might reasonably expect, the operational policy now favors complex convolutional codes. Now, the use of $K=7$ convolutional codes, the most complex codes considered in this section, is preferred to raising the output power beyond +5 dBm. The energy of decoding has been lowered to the point that the only impact of coding is the increase in communication delay and the radio on-time.

4.   An Application-Aware Protocol for Microsensor Networks

4.1   The Need for Application Awareness

Many contemporary wireless communication protocols harbor inefficiencies caused by excessive generality. For example, the 802.11b media access layer [8] incorporates a channel reservation and acknowledgment mechanism that requires four transmissions for each actual data packet sent. TCP incorpo-

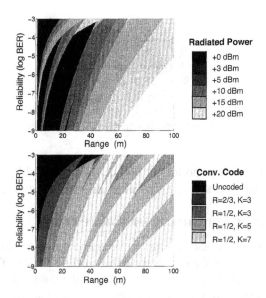

Figure 10. Minimum-energy radiated power and convolutional code policy for the μAMPS microsensor node with an ASIC-based Viterbi decoder.

rates acknowledgments and retransmissions for reliable transport and back-offs that encourage bandwidth fairness among multiple nodes. While each of these mechanisms is potentially useful in applications such as file transfer, their presence adds communication overhead to applications that do not require them. For instance, latency-critical, real-time applications such as streaming may not benefit from a reliable transport layer, and a network dominated by low-bandwidth streams may not require arbitration for fairness.

It is therefore unreasonable to assume that one media access layer and protocol can *efficiently* support the entire range of AmI applications from environmental microsensing to two-way voice and on-demand video. Instead, protocols should be tuned to the target hardware and application, just as an ASIC is tuned for application-specific processing.

One particular example is a protocol tuned for energy-efficient, unidirectional data propagation in high-density microsensor networks. We begin with the observation from Section 2.2 that a radio receiver that is on and idle consumes a substantial amount of power—often as much as transmission. Given that many nodes will likely be in the receiving range of any node's transmission, it is desirable to shut down the radio receiver in the majority of idle nodes. Unfortunately, most wireless protocol and MAC layers in use today utilize unique addresses to route packets to specific destinations, with the expectation that these destinations are actively listening for packets. With radio

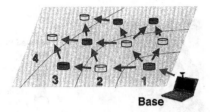

Figure 11.   Assignment of distance metrics for unaddressed forwarding.

shutdown, this assumption no longer holds, and routing tables—whether they contain source routes or next-hop information—may become unstable.

4.2   Unaddressed Forwarding for Unidirectional Communication

For the microsensor application, communication in a microsensor network is decidedly one-way, from the observer nodes to a base station. There are many data sources and relays, but few actual sinks. As the individual relays have no need for the data they are relaying, the entire notion of addressing a packet to a specific relay node is unnecessary. Our only concern is that packets move progressively closer to a base station.

We thus propose that microsensor nodes not utilize explicit addresses at the protocol or MAC level, but rather a metric of their approximate distance to the nearest base station. This metric could be propagated across the network by "increment-and-forward" flooding. The base stations initially broadcast a zero metric, and all nodes that successfully receive this transmission assign themselves a metric of one (the value of the metric heard plus one). Each of these nodes, in turn, rebroadcasts its own metric, so that metrics propagate across the network. Figure 11 reflects metric assignment using this approach.

In a high-density microsensor network, many nodes are within radio range of the receiver. Hence, under the above scheme, many adjacent nodes are likely to share the same metric. As receivers are shut down, it becomes increasingly likely that no active receiver with a lower distance metric is within range of the sender. A natural way to increase the spatial resolution of the metrics is to broadcast distance metric messages with a lower radio transmission power than data messages. This is yet another way that variable output power may benefit the protocol layer.

Once the metrics are established, a node with a packet destined for the base station simply broadcasts its packet with its current distance metric. Nodes that receive the packet compare their own distance metric to that of the packet. To minimize the number of hops, the receiving node that is closest to the base station and farthest from the originating node would relay the packet onward. Such behavior may be implemented by the following timer at each node:

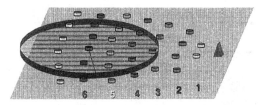

Figure 12. Unaddressed propagation of broadcast packets toward a base station.

$$T_{forward} = (\frac{T_0}{d_{sender} - d_{myself}})(1 + kU[0, 1])$$

where d_{sender} is the distance metric embedded into the data packet by the sender, d_{myself} is each receiver's own distance metric, $U[0, 1]$ is a uniform random variable between zero and one, and T_0 and k are constants. The local choice of forwarding timer value at each node ensures that nodes farther from the sender will initiate relay transmissions first. The random term allows for arbitration among nodes with the same metric. In this manner, the node with the lowest delay forwards the packet; an acknowledgment by the initial sending node follows. The forwarding timers of the other nodes are cancelled when they hear either of these messages. For instance, in the example of Figure 12, the distance-3 node would forward the packet before all others.

Unaddressed forwarding allows any active node—rather than one that is specifically addressed—to relay a packet. This technique enables decentralized, protocol-independent radio receiver shutdown. Additional noteworthy work on unaddressed forwarding is presented by [17].

4.3 Simulation Results

To demonstrate the performance of unaddressed forwarding, we compare our protocol to a generic addressed multihop protocol based on the principles of Bellman-Ford routing [3]. Both protocols are evaluated in simulation using a custom wireless network simulator that was co-developed with the protocol implementations. This interactive wireless simulation tool written in Java is an extensible, energy-conscious, and high-performance simulator for high-density wireless networks. Figure 13 illustrates the simulation of unaddressed forwarding using this tool.

The simulated environment consists of 500 nodes distributed randomly over a 300-meter by 150-meter region. The packet source nodes and the base station are at geographically opposite ends of the scene, and power limitations on the nodes' transmitters require multihop transmission from source to sink. The performance of addressed and unaddressed forwarding are compared under two impairments to successful packet transmission: an increase in the variance of the path loss, and the periodic shutdown of radio receivers in the network.

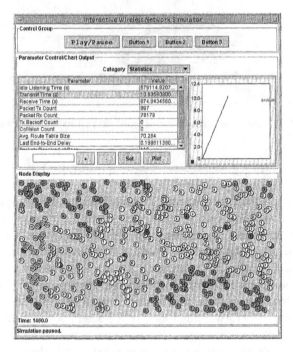

Figure 13.   Screenshot of the interactive wireless network simulator.

Path loss is simulated as $\alpha_F d^n$, where α_F is a log-Gaussian random variable. Figure 14 plots two metrics of communication performance, the end-to-end packet delay and the total network energy consumed per packet, versus the standard deviation of α_F. For an "ideal" channel with no fading, addressed forwarding achieves a lower delay than unaddressed forwarding. Unaddressed forwarding incurs additional delay due to the delay-based forwarding timer utilized to select a next-hop. As the channel becomes increasingly variable, however, the end-to-end delay of addressed forwarding rises while the delay of unaddressed forwarding actually falls.

As the standard deviation of the attenuation random variable increases, addressed forwarding requires progressively shorter hops and more packet retransmissions to achieve end-to-end communication. Hence, both the transmit and receive times increase. Unaddressed forwarding requires additional transmissions as well, but for a very different reason: channel variations cause the packets to 'fan out' across the network due to the loss of some acknowledgment and suppression messages. As more nodes hear (and forward) the packets, the transmission and receive totals increase.

For unaddressed forwarding, there is a slight *decrease* in delay apparent in Figure 14 as the variance increases. Unaddressed forwarding achieves *longer*

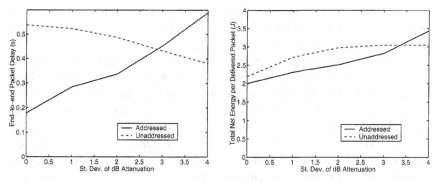

Figure 14. End-to-end delay and energy for addressed and unaddressed forwarding under variable path loss.

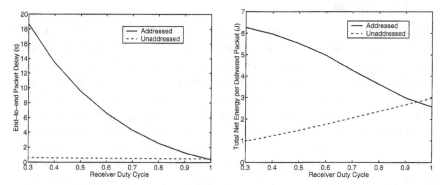

Figure 15. End-to-end delay and energy for addressed and unaddressed forwarding under variable radio duty cycle.

hops when variations in path loss are in the transmitter's favor. Addressed forwarding, which relies on routing tables, is less capable of exploiting such rapid variations in channel conditions.

To investigate the impact of receiver shutdown on protocol performance, all node radios (except for the base station and the initial packet sources) are duty cycled at a fixed on-time period of 40-80 seconds with a random phase offset to simulate worst-case, uncoordinated shutdown. The standard deviation of α_F is fixed at 2 dB.

Figure 15 illustrates the delay and energy of both forwarding techniques as radio duty cycles are reduced. In a result more dramatic than that for channel quality above, the delay and energy consumption of addressed forwarding suffer greatly with lower radio duty cycles. A lower duty cycle increases the likelihood that a packet will be addressed to a node whose radio is off, resulting in a retransmission timeout and the selection of a new route (which also may or may not be listening). The delay of unaddressed forwarding, with the lack of explicit relay selection, is immune to radio duty cycling. Moreover, the total energy per delivered packet is reduced as the radio duty cycle is reduced, reflecting a savings in idle and receive energy.

Application-specific protocols improve communication efficiency by tailoring their behavior to the expected needs of the application. Unaddressed forwarding offers a simple and elegant solution to the problem of forwarding sensor data to a base station. The unaddressed forwarding protocol was simulated and compared to traditional addressed routing under two dimensions of operational diversity: unpredictable channel characteristics and radio receivers that shut down frequently and arbitrarily to conserve energy. Unlike addressed forwarding, the unaddressed technique offers resilience and energy-efficiency that was immune to the operational diversity presented.

5.   Conclusion

The shrinking size and increasing density of next-generation wireless devices imply reduced battery capacities, meaning that emerging wireless systems must be more energy-efficient than ever before. This exploration of wireless communication for high-density wireless networks has revealed three general conclusions regarding energy-efficient communication. First, it is crucial that energy-efficient communication software layers be based upon sound models of the hardware on which they will operate. Incomplete or inaccurate energy models lead to surprising discrepancies between designers' expectations and system realities, as we have shown through the example of multihop routing. Second, energy agility enables graceful trade-offs between communication performance and energy consumption. The abstraction layer that bridges notions of communication performance and hardware energy consumption is one that—unlike traditional abstraction layers–*increases* efficiency. Finally, application-specific design, a technique long advocated by hardware designers for energy-efficiency, may be applied to protocols as well. Our unaddressed forwarding protocol offers a simple and robust solution for unidirectional data propagation in microsensor networks.

As ambient intelligence becomes reality, we foresee the emergence of new, application-specific protocols for the many modalities supported by AmI devices, and increasing cooperation between hardware and protocols to enable energy- and quality-scalable communication. These trends, guided by accurate characterizations of the energy usage of hardware and applications, will ensure that AmI devices deliver the battery life required by the most demanding users.

Acknowledgments

The authors wish to thank Nathan Ickes, Kevin Atkinson, and Fred Lee for their implementation and models of the μAMPS microsensor node; Twan Basten for numerous suggestions with the text; and Curt Schurgers for insightful discussions. This research is sponsored by the Defense Advanced Research Projects Agency (DARPA) and Air Force Research Laboratory, under agreement number F33615-02-2-4005; by the Army Research Laboratory (ARL) Collaborative Technology Alliance, through BAE Systems, Inc. subcontract RK7854; and by Hewlett-Packard under the HP/MIT Alliance. Rex Min is supported by an NDSEG Fellowship. The U.S. Government is authorized to reproduce and distribute reprints for Governmental purposes notwithstanding any copyright annotation thereon.

References

[1] A.P. Chandrakasan and R. Brodersen, "Minimizing power consumption in digital CMOS circuits," in *Proceedings of the IEEE*, pp. 498–523, 1995.

[2] Cisco Systems, *Cisco Aironet 350 Series Client Adapter Data Sheets*, 1991.

[3] T. Cormen, C. Leiserson, R. Rivest, and C. Stein, *Introduction to Algorithms*, MIT Press, Cambridge, MA, 2001.

[4] V. De and S. Borkar, "Technology and design challenges for low power and high performance," in *ISLPED'99, Proceedings*, pp. 163–168, Aug. 1999.

[5] D. Estrin, R. Govindan, J.S. Heidemann, and S. Kumar, "Next century challenges: Scalable coordination in sensor networks," in *MobiCOM'99, Proceedings*, pp. 263–270, 1999.

[6] J. Goodman and A. Chandrakasan, "An energy-efficient IEEE 1363-based reconfigurable public-key cryptography processor," in *ISSCC 2001, Proceedings*, San Francisco, CA, USA, pp. 330–331, 461–462, Feb. 2001.

[7] J. Goodman, A. Dancy, and A. Chandrakasan, "An energy/security scalable encryption processor using an embedded variable voltage dc/dc converter," *Journal of Solid State Circuits*, vol. 33, no. 11, pp. 1799–1809, Nov. 1998.

[8] G.J. Holzmann, "IEEE standard for wireless LAN-medium access control and physical layer specification, p802.11," Nov. 1997.

[9] N.J. Ickes, "Hardware and software for a power-aware wireless microsensor node," M.S. thesis, Massachusetts Institute of Technology, May 2002.

[10] L. Kleinrock, "On giant stepping in packet radio networks," Packet Radio Temporary Note #5 PRT 136, UCLA, Los Angeles, CA, Mar. 1975.

[11] R. Min and A. Chandrakasan, "A framework for energy-scalable communication in high-density wireless networks," in *ISLPED'02, Proceedings*, Monterey, CA, USA, Aug. 2002.

[12] R. Min and A. Chandrakasan, "Top five myths about the energy consumption of wireless communication," *to be published in ACM Sigmobile Mobile Communication and Communications Review (MC2R)*.

[13] J. Rabaey et al., "Picoradio supports ad hoc ultra-low power wireless networking," *Computer*, vol. 33, no. 7, pp. 42–48, 2000.

[14] RF Monolithics, Inc., *TR1000, 916.50 MHz Hybrid Transceiver*, 2003, Product datasheet available at http://www.rfm.com/products/data/tr1000.pdf, last checked on 06/2003.

[15] D. Shaeffer and T. Lee, Eds., *The Design and Implementation of Low-Power CMOS Radio Receivers*, Kluwer Academic Publishers, Norwell, MA, USA, 1999.

[16] B. Warneke, B. Atwood, and K.S.J. Pister, "Smart dust mote forerunners," in *14th IEEE Int'l Conference on MEMS, Proceedings*, pp. 357–360, Jan. 2001.

[17] F. Ye, S. Lu, and L. Zhang, "GRAdient Broadcast: a robust, long-lived sensor network," *UCLA CS IRL technical report*, Sept. 2001.

Application Re-mapping for Fault-Tolerance in Ambient Intelligent Systems

Phillip Stanley-Marbell, Nicholas H. Zamora, Diana Marculescu and Radu Marculescu

Carnegie Mellon University, Pittsburgh, PA, USA

{ pstanley, nhz, dianam, radum } @ece.cmu.edu

Abstract   As technology advances, devices become smaller and cheaper, making it possible to build systems containing large numbers (possibly hundreds, or more) of miniature processing elements. Such platforms, although superficially similar to traditional distributed systems, pose additional unique challenges. Due to the desire to minimize costs, coupled with the sheer numbers of devices, it will be difficult to perform manufacture-time testing, and runtime-failures will likewise be common. One challenge is to efficiently harness the capabilities of large numbers of low power (and relatively low performance) processing elements, in the presence of failures such as depleted battery resources, as well as those due to unpredictable sources (e.g., electrical and mechanical failures). It will however be possible to employ a fraction of the multitude of resources as *redundant* or *spare* devices, *re-mapping* applications onto them, from failing ones.

This paper investigates the use of *code migration* as a general means of performing such application re-mapping, in the presence of intermittent communication and device failures, as well as limited battery resources. A new technique, *Pre-Copying with Remote Execution (PCRE)*, an extension of code migration which enables more efficient application re-mapping in the presence of energy and communication constraints for *symmetric applications*, is presented.

It is shown that PCRE provides a 28.6% improvement in system lifetime and 9.8% improvement in energy efficiency for the applications investigated, over the baseline code migration strategy. Naturally, the re-mapping of applications involves overheads in computation and communication, and PCRE reduces these overheads to within 10% of the ideal case of doubled energy resources.

Keywords   ambient intelligent systems, low power, sensor networks, fault-tolerance, application re-mapping

1.   Introduction

Embedding large numbers of computational devices into everyday environments poses many design challenges. Rather than considering them as collections of independent processing elements, as in the case of traditional networks of embedded systems, the sheer numbers of devices which collectively form a

T. Basten et al. (eds.), Ambient Intelligence: Impact on Embedded System Design, 315-335.

given "system" must be considered as a single computational substrate, over which applications will execute. The number of available computational devices in an environment might far exceed the number of applications being executed at any given time, and the excess or *redundantly deployed* devices could be used to improve the system's reliability, employed in enhancing performance, or both.

Developing applications for execution on such a hardware substrate requires taking into account several unique static and runtime concerns. It will be necessary at compile time, to have a means of partitioning applications for execution on the multitude of hardware resources, taking into consideration possibly defective devices. During the execution of applications, it will likewise be necessary to tradeoff performance for system lifetime and reliability in the presence of failures in both devices and their interconnection networks. In contrast to the traditional application cycle—*design, implement, compile (test)*, and *run*—ambient intelligent systems will need to be adaptive, to appropriately account for their failure-prone nature and possibly harsh environments. They will thus entail a slightly different approach—*design, implement, compile (test), run, and monitor*. This work investigates means of performing the runtime remapping of applications in the event of exceptional conditions, specifically, low battery levels.

1.1 Related Work

The use of *code migration* has been successfully applied in the field of mobile agents [2, 6, 12]. Mobile agents are an evolution of the general ideas of process migration [5]. They can be thought of as autonomous entities that determine their motion through a network, moving their code as well as state as they do so. Process migration has traditionally been employed in distributed systems of servers and workstations, primarily for load distribution and fault-tolerance [5]. Unlike the traditional implementations of process migration, the application remapping techniques employed in this work are of *significantly lighter weight*, taking into consideration the special properties enforced on applications.

The use of *remote execution* to reduce the power consumption in mobile systems has previously been investigated [3, 8]. The goal in these prior efforts was to reduce power consumption by offloading tasks from an energy constrained system (e.g., a mobile computer or PDA) to a server without constraints in energy consumption. The tradeoff involved determining when it was worthwhile to transfer data to the remote sever, given the overheads involved, versus performing computation locally. The systems investigated were mobile computers and PDAs operating with a fast, reliable wireless network. Such environments contrast the *ultra low power, unreliable* network sensors with possibly *faulty communication and computation* which are of interest in this work.

1.2   Contributions of this Work

This work introduces techniques for the *robust* execution of applications in distributed, embedded, failure-prone environments, such as ambient intelligent systems. The proposed techniques, based on the idea of code migration, enable the re-mapping of failing applications in environments with redundantly deployed hardware.

It is shown how *lightweight code migration* can be used, in the general case, to improve the lifetime of applications in the presence of failures, by as much as 30% for the driver applications studied. For the special case of gracefully degrading applications, a new technique, *Pre-Copying with Remote Execution (PCRE)* is shown to provide added benefits over the simple code migration approach, increasing system lifetime by an additional 28.6%.

The experimental evaluation employs two driver applications, *beamforming* and *software radio*. Beamforming is a symmetric application, simple to partition for execution on multiple processing units. Software radio, on the other hand, is asymmetric and more challenging to partition. Both applications were partitioned for execution on a network of embedded processing elements and their executions simulated using a cycle-accurate simulation framework that faithfully models a network of computational elements, communication, power consumption and battery discharge characteristics.

The remainder of this paper is organized as follows. The proposed techniques for performing application re-mapping are presented in Section 2. A description of the beamforming and software radio driver applications is given in Section 3. Section 4 describes the simulation infrastructure and details the setup employed in subsequent evaluations. Section 5 presents experimental results and analysis and Section 6 concludes the paper with a summary of the contributions of this work.

2.   Application Re-mapping Techniques

In the presence of exceptional conditions, such as critically low levels of energy resources or increasingly rampant intermittent failures, it is desirable to re-map application execution from one failing device (or set of devices) to another. Deciding *if* and *when* to perform re-mapping involves tradeoffs. For example, with low energy resources, re-mapping should occur early enough so as to have sufficient energy to complete. Acting too early can however lead to unused energy resources going to waste, while acting too late may result in failure.

One possibility for re-mapping is the use of *code migration*, in which executing code is transferred (*re-mapped*, or *migrated*) to available nodes when the battery depletes past a predetermined threshold, or when an inordinate number of runtime failures are detected. In the case of migration due to depleted energy resources, the energy resource threshold must be set conservatively so as to ensure successful re-mapping, even in the presence of other adverse conditions, such as intermittent communication failures.

2.1　Lightweight Code Migration

In traditional computing systems, such as distributed systems of workstations and servers, code migration is a non-trivial endeavor. This is because it generally involves migration of executing *processes*. Processes executing in a multiprogramming operating system generally have a fair amount of system and machine state. For example, associated with a given process might be open file descriptors, semaphores, open network sockets, etc. Likewise, machine state associated with a process may include several machine register contents. Process migration is therefore not a trivial task.

A key observation made in this work is that, if applications are constructed in a manner in which they may be *asynchronously restarted*, then only state specific to an application, such as variables that are updated over the course of its lifetime, and its current stage of execution, need to be persistent across restarts.

The requirement of asynchronous restartability implies that applications must maintain some notion of their current state of execution, such that upon a reset, they can resume execution with minimal overhead. This requires that applications keep track of the current stage of execution as execution progresses—there is a tradeoff between how accurately this state is maintained, and the amount of computation that must be redone in the event of a restart.

For example, an application that consists of an initialization phase followed by periodic processing of incoming data samples would include as part of its state to be saved, the data structures which are set by the initialization phase, as well as an indicator of whether the initialization has been performed. During migration, it is therefore sufficient to migrate only application code together with this state, in order to enable resumption of the migrated application.

Figure 1 illustrates the memory layout of a system employing the proposed solution. Each computational device consists of a processor and memory. The memory space is partitioned between that used for the application, and memory dedicated to the device's firmware. The memory region in which the application runs is occupied by the different parts of the running application: program code (text), initialized data (data), uninitialized data (bss), stack (grows downwards from the top of the memory region) and heap (grows upwards from bss), as illustrated by the blow-up in Figure 1. By placing information that must be persistent across restarts in the data and bss segments of the application, it is possible to maintain state across migration while only transferring the text, data and bss segments. The level of detail in the foregoing dis-

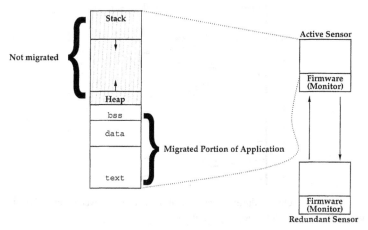

Figure 1.   Lightweight migration of application code and state. Only program code, data and uninitialized data are transferred during migration. Applications are constructed to ensure important state is in data and bss segments, as opposed to being in the stack or the heap.

cussion reflects the level of detail at which the eventual evaluation is carried out. The evaluation framework is described in detail in Section 4.

Each node has pre-loaded into it firmware referred to as the *monitor*. This firmware is responsible for receiving applications and loading them into memory. Under normal execution, the monitor program spends most of its time with the processor in a sleep mode, awakening on interrupts generated by events such as incoming network traffic.

During migration, each application will transmit its text, data and bss segments, and when these are loaded by a remote monitor into its device's memory, the migration is complete. The logical sequence of messages that are exchanged between a node attempting to migrate and a redundant one (executing the monitor code) that receives it is illustrated in Figure 2. The focus of this work is *not* on defining network protocols; the migration sequence shown in Figure 2 [1] could be made more robust by employing acknowledgments for each transmission, or by performing the message exchange over a reliable transport layer.

In later experiments, each node attempts to migrate its application code when its remaining energy resources fall below a given threshold (migration due to excessive numbers of failures, though a desirable avenue of investigation, is not considered in this work). The energy threshold must be chosen

---

[1]Note that this is the logical sequence of messages. The actual sequences of messages transpiring between the two parties must include provision for lost messages, and could thus include, e.g., further acknowledgements and timeouts for both communicating parties.

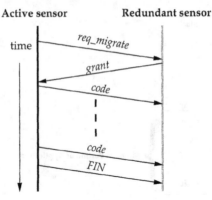

Figure 2.   Message exchange during migration of application between a node hosting a migrating application and a redundantly deployed one.

conservatively so as to ensure that each node can successfully migrate in the presence of communication errors such as link-layer collisions or carrier-sense errors, *before* they completely deplete their energy resources. The energy costs and characteristics of code migration in the modeled system are described in further detail in [10].

Remote execution is a possible alternative to code migration. In a system employing remote execution, each application that will be run on *any* device in the system must be duplicated on *every* device. Using remote execution, applications simply need to transfer their application state and subsequently transfer control of execution to the destination node at a desired point in time to complete migration. Transfer of state is inexpensive compared to transferring both state and code, as the text segment need not be transmitted. This is particularly useful when communication is prohibitively expensive, as is the case in wireless networks. The disadvantage is the requirement of ample memory on each device to store every application, which, given the need to employ physically tiny, cheap, low-power devices, is not a desirable constraint. The tradeoffs in using remote execution versus code migration therefore include the added cost for larger memories and the increased power consumption due to the additional hardware in the case of remote execution, versus the overhead of having to transmit larger amounts of data in the case of code migration. A more detailed comparison of the tradeoffs between remote execution and code migration is provided in [11].

2.2   Pre-Copying with Remote Execution (PCRE)

When employing code migration to re-map applications dynamically, it becomes critical to judge the point at which migration should be initiated. Specifically, for re-mapping initiated due to low battery resources, if migration occurs too early, then optimal use would not have been made of the local energy store before seeking out fresh resources. On the other hand, if migration is initiated too late, then there might be insufficient energy resources remaining locally for migration to complete. Further, due to the significantly larger currents drawn from the battery subsystem during migration, it is desirable to schedule the compute- and power-intensive components of re-mapping separately from the actual handoff from failing devices to redundantly deployed ones [4].

Pre-Copying with Remote Execution (PCRE) is proposed in this work as a means of scheduling the communication- and power-intensive portions of code migration, while performing the final handoff as late as possible. The primary insight is to distribute the time at which migration of application code occurs without actually transferring execution of applications (i.e., terminating them locally and activating them remotely), until energy resources fall to a critical threshold. At this point, a final message is sent to the destination where the application code was previously pre-copied, and execution is transferred. This enables the threshold for migration to be set much lower, resulting in more complete usage of energy resources. It also enables performing the bulk of the work involved with migration earlier in the life-cycle of the application, significantly reducing the chances of not having sufficient energy to migrate and enabling the migration scheduling to consider the effects of battery discharge nonlinearities over time [4].

The *scheduling* of PCRE may require global synchronization. This problem is mitigated by the fact that due to the event-driven nature of many of the applications of interest, all devices can be synchronized to the first system-wide event. Such event synchronization might however not be possible for all applications.

2.2.1 PCRE Formulation

The *lifetime* of a system is defined as the amount of time during which the system is functional. Assume that N is the minimum number of nodes required to keep the system alive, and P processing nodes are available in the architecture. For example, one system might be considered alive only if at least 3 or more of its nodes are functional [2], whereas in other applications, every node might be required to keep the system alive. If each node in the system contains an equal amount of limited battery energy, the system lifetime can be increased by a factor of at most P/N, compared to the case of not performing application re-mapping in the event of depleted energy resources. When the number

---

[2]Such systems that can tolerate failure of a fraction of their components, at the cost of a reduction in performance, are often referred to as *gracefully degrading systems*.

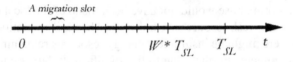

Figure 3.   Determining the migration slots for $N = 14$.

of all available processing nodes, P, is large, this potential increase in system lifetime can be considerable.

Assume the worst case (shortest) lifetime across all nodes is T_{SL} and that the worst case migration cost (in time units) is T_M. If each of the N active nodes in a system, has associated with it an identification number, i, ranging from 0 to $N-1$, then, a *migration slot* is assigned to each node, beginning at time T_{P_i}, to begin code pre-copying:

$$T_{P_i} = i \cdot W \cdot \left(\frac{T_{SL}}{N} \right)$$

W is a factor ranging from 0 to 1 that, when large, designates to each active node a lengthy migration time slot for pre-copying its code to a redundant node. A small W will force all the pre-copying for the N active nodes to occur quickly. The determination of migration slots for a case with $N = 14$ is illustrated in Figure 3. In this case, for $W = 0.7$, the 14 slots are equally divided between $t = 0$ and $W \cdot T_{SL}$.

After pre-copying, the node resumes executing the application until its battery falls below a critically low threshold, denoted B_{vlow}. At this time, the final step to complete migration is to have that active node send a final message to the redundant node to which it previously pre-copied its code, indicating to it that it must initiate execution of the re-mapped application. A compromise in employing PCRE is that once pre-copying has occurred, any further state accrued at a device will be lost when the final handoff occurs.

When migration occurs over a shared communication bus, the condition:

$$T_M \leq W \cdot \left(\frac{T_{SL}}{N} \right) \tag{1}$$

must hold for simultaneous attempts to communicate on the shared medium to be avoided.

3.   Driver Applications

The applications employed in the subsequent experimental evaluation are *beamforming* and *software radio*. These two particular applications were chosen to highlight the potential of the proposed ideas, when applied to two very different classes of applications—on one hand, gracefully degrading applications in which any of the components of the partitioned application may fail without catastrophic failure of the entire system (i.e., systems that degrade gracefully with failing components) and, on the other hand, heterogenous applications in which failure of any individual component of the partitioned application, leads to the failure of the system as a whole. Beamforming has the properties of degrading gracefully with failing nodes, as well as being trivially partitioned for execution over a collection of devices. Software radio, on the other hand, is dependent on each active node during execution and is also less trivial to partition. Both software radio and beamforming have direct applicability to ambient intelligent systems.

3.1 Beamforming

The objective in beamforming is to achieve *spatial filtering* of signals, effectively providing a sort of active antenna that focuses on specific incoming signals. It is of particular interest in ambient intelligent systems as it may form the basis of a variety of applications. Such applications include active microphone arrays which perform speaker location and spatial filtering, and active antennas, for example, to improve the performance of Ultra-Wide Band (UWB) radios.

One implementation of a beamformer consists of an array of sensors working in conjunction with a centralized processor, with all the signal processing being performed on the single central processor. Since the samples obtained from each sensor may be processed independently, it is possible to partition the beamforming application such that each of the sensors can be equipped with local processing capabilities, to perform portions of the requisite computation locally. The system as a whole must still operate in synchrony, with samples obtained by the sensors (henceforth, "slave nodes") and processed each sample period, before forwarding the partial computation results to a central aggregator (henceforth, "master node"). Different applications of beamforming will require different sampling rates. For example, for voice signals, a sampling rate of 8 KHz is sufficient, while positional estimation of slow moving objects (e.g., walking humans) is possible with sampling rates on the order of 10 Hz.

The beamforming implementation considered in this work employs repetitive *rounds* in which all computational nodes participate in accomplishing the overall goal of signal recovery (Figure 4). At the beginning of a round, the *master* node sends a broadcast message to all *slave* nodes instructing them to begin sampling. Next, each slave obtains a sample from its attached sensor and performs a 32-tap FIR filter operation on this sample. The slave node then waits for a predetermined time-slot, based on its identification number, to send the result to the master node for analysis. During this analysis step, the master

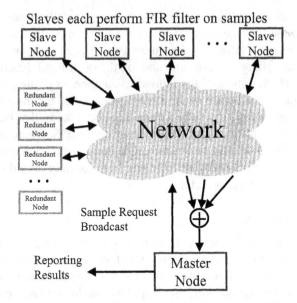

Figure 4. Organization of beamforming application. Application is partitioned across master and slave nodes. The arrows in the Figure indicate communication.

node combines the individual samples to obtain a composite sample for that sampling period. Finally, at the completion of a round, the master node waits for the beginning of the next round to send the next sample request broadcast and continue operation.

It should be noted that, despite the notion of different kinds of nodes logically, the device which is employed as the master is no different physically from the other devices in the system—it does not require larger memory resources or a faster processor [3]. For this reason, and also for reasons of scalability, application re-mapping schemes which employ the master as a central point of coordination, e.g., for keeping copies of the code to be migrated, are untenable.

3.2 Software-Defined Radio

Software-defined radio or *software radio* employs a general purpose RF front-end, controlled by software, to achieve the implementation of a variety of wireless transmission protocols with a single piece of hardware. Software radio is of interest in this investigation as it is likely to play a pivotal role in future

---

[3]Our simulations assume no energy constraints on, and hence no re-mapping of, the master, only for the purpose of removing an unnecessary added dimension in complexity—in reality, there will be no such thing as a device with unlimited resources.

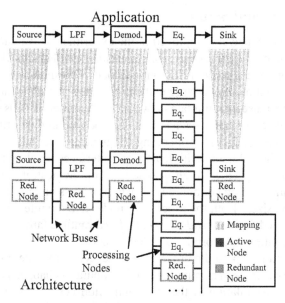

Figure 5.   Organization and mapping to hardware of software radio application. Application is partitioned into five main stages, with the fourth stage further partitioned into eight stages (LPF = Low-pass filter, Demod. = Demodulation, Eq. = Equalization).

wireless communication systems, which may employ software-defined radios to implement a variety of communication technologies, at low cost.

The implementation of software radio employed in the subsequent experimental evaluations can be divided into five primary components (as shown in Figure 5). The first step is the acquisition of the modulated signal, perhaps with the use of an antenna (possibly augmented with the previously discussed beamforming technique, to achieve better spatial filtering and hence better performance). The second step is the decimation filter, followed by demodulation of the signal, transforming the signal from the carrier frequency down to the baseband frequency. Following demodulation is equalization. The final step is the sink, which simply is the collector of the resulting samples, for example, to drive a loudspeaker.

The software radio implementation in this work maps each of these five steps onto a single computational node in the distributed architecture, except for the equalization step, which is further partitioned eight ways and mapped onto eight processing nodes as shown in Figure 5, in order to provide a better balance of the CPU occupancy across nodes, as well as to increase performance. In Figures 4–5, redundant nodes are the ones to which the applications executing on active nodes will be transferred, during re-mapping.

4. Experimental Setup

The simulation framework employed in the experimental evaluation models a network of embedded processors, each consisting of a Hitachi SH3 microcontroller, memory and a communication interface. The simulation environment permits the instantiation of a large number of nodes as well as interconnection links (it has been used to simulate networks on the order of 100s of nodes).

The modeled processor in each system may be configured to run at different voltages and, hence, different operating frequencies. Each simulated processor can be configured with multiple network interfaces. Each of these network interfaces may be attached to an instantiated communication link. The communication links can be configured for data transmission at different bit rates and link frame sizes. Both the processing nodes and the interconnection links can be configured for *independent* or *correlated* runtime failures with different failure probabilities, and correlation coefficients.

The simulation of both the processor elements and the interconnection network is cycle-accurate. The simulation of computation happens at the instruction level. The simulation of the interconnection network is at the data-link layer of the Open System Interconnection (OSI) model. The power estimation performed for each individual processor is based on an instruction level power model. The power-estimating simulation core has been verified to be within 6.5% of measurements from the actual hardware [9]. The power estimation for the communication links assigns configurable costs for transmitting and receiving. Each individual processor, or groups of processors, may be attached to batteries of configurable capacity. The battery simulation is based on a discrete-time battery model [1], and it models the battery electrochemical characteristics of a Li-Ion cell, attached to a DC-DC converter. The specific parameters for the processing elements, failure rates, communication links and the battery subsystem are detailed in the next section.

Failures are modeled in the links that interconnect the processing elements. These failures are modeled after intermittent electrical failures, with a parameterizable probability of failure (referred to as the *failure rate* in Section 5). The modeled failures follow a uniform failure distribution. These probabilities of failure are with respect to one simulation time step. For example, a stated failure rate of 1E-8 implies a probability of failure of 1E-8 per simulation step. The correlation coefficient is the likelihood that a link error will cause a node error on a node connected to that link.

4.1 Simulation Setup

Several relevant simulation variables, and their associated values, are given in Table 1. The software radio application is a much more demanding application, requiring a much higher communication link speed and was accordingly configured with higher transmit and receive power dissipations.

All the beamforming experiments employ an implementation with 1 master node and 10 slave nodes. In every experiment, half the slave nodes are inactive and used as targets for migration. The software radio application employed 12 nodes, one for each of the source, low-pass filter, demodulator and sink nodes, and 8 nodes for the equalizer stage.

Table 1.   Simulation variables and their associated values for both driver applications.

Simulation Parameter	Beamforming	Software Radio
Operating Modes (each node)	60 MHz, 3.3V	60 Mhz, 3.3V
	15 MHz, 0.85V	15 Mhz, 0.85V
Battery Size (each node)	1 mAh	1 mAh
Trans. and Rec. Power	100 mW	250 mW
Link Speed	200 Kb/s	10 Mb/s
Frame Size	1024 bits	8192 bits
Frame Headers	288 bits	296 bits
Node Failure Probability	1E-8	1E-8
Correlation Coefficient	0.1	0.1
Baseline Migration Threshold, B_{low}	0.6	0.2
PCRE Handoff Threshold, B_{vlow}	0.06	—

As can be seen from Table 1, each node is attached to a battery with a capacity of 1.0 mAh. For the design alternative of employing a larger battery per node, each node is attached to a battery with twice the capacity, or 2.0 mAh. These battery capacities are on the order of 10 times smaller than a common wristwatch battery [7]. *This size of battery was chosen to limit the simulated system's lifetime and make simulation possible in a reasonable amount of real time.* The battery threshold for the final PCRE handoff, B_{vlow}, in Table 1 is 10 times lower than that for baseline migration in the beamforming experiments, because only one message needs to be transmitted to complete handoff after pre-copying.

Thirteen experiments were conducted to compare the proposed PCRE scheme, with two other design solutions: using the baseline lightweight code migration scheme as defined in Section 2.1, and using a larger battery (twice as large) per processing node instead of performing any application re-mapping. PCRE is well suited for re-mapping the beamforming application because beamforming degrades gracefully with the removal of nodes—the absence of one or more of the slave nodes leads to reduced system effectiveness, but does not render the system completely non-functional. Therefore, using non-overlapping migration time-slots as in PCRE permits graceful degradation of performance during migration, and during each such PCRE time slot, the performance of the system will be slightly reduced due to the absence of one of the slave nodes.

Software radio, in contrast, is highly dependent on the functioning of each of its components. Since the absence of any of the nodes in the system will lead to system failure[4], employing the PCRE scheme with non-overlapping time slots would imply that during the entire duration when pre-copying was taking place, the system would be non-functional. Migration for each active node occurring simultaneously therefore provides the best application lifetime, making the baseline lightweight migration technique better suited for software radio and similar non-gracefully-degrading applications.

Table 2.   Details showing how each experiment was setup (L.E. = Link Errors, N.E. = Node Errors, Ind. = Independent, Cor. = Correlated, BF = beamforming application).

	Topology	Errors?	CPU MHz
1	Dual-Bus (BF)	None	60 MHz
2	Dual-Bus (BF)	L.E.	60 MHz
3	Dual-Bus (BF)	N.E.	60 MHz
4	Dual-Bus (BF)	Ind. L.E. + N.E.	60 MHz
5	Dual-Bus (BF)	Cor. L.E. + N.E.	60 MHz
6	Dedicated Migration Link (BF)	None	60 MHz
7	Dedicated Migration Link (BF)	L.E.	60 MHz
8	Dedicated Migration Link (BF)	N.E.	60 MHz
9	Dedicated Migration Link (BF)	Ind. L.E. + N.E.	60 MHz
10	Dedicated Migration Link (BF)	Cor. L.E. + N.E.	60 MHz
11	Dual-Bus (BF)	None	20 MHz
12	Software Radio	None	60 MHz
13	Software Radio	Cor. L.E. + N.E.	60 MHz

The details of each configured experiment are shown in Table 2. Two topologies for interconnecting the master, slave and redundant nodes were considered for the beamforming application, and are depicted in Figure 6. In one topology, referred to as *Dual-Bus*, shared communication buses are employed, one for exchanging samples between the master and slave nodes and the other for migration of slaves to redundant nodes. In the second topology, the single shared bus for the exchange of samples is maintained, but *Dedicated Migration Links* are employed for migrating the slaves to the redundant nodes. In Figure 6, the redundant nodes are labeled R and the slave nodes labeled S.

To investigate the robustness of PCRE in the presence of faults, simulations were performed with intermittent failures in the slave and redundant nodes, failures in the communication links, as well as independent and correlated fail-

---

[4] Any of the nodes in the equalizer stage may indeed be re-mapped without complete system failure, but this is not exploited in this work.

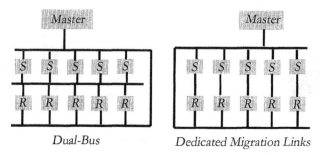

Figure 6. Two topologies considered in experiments: Dual Bus and Dedicated Migration Link topologies (S = slave node, R = redundant node).

ures in both the nodes and the links[5]. Table 1 shows the node and link failure probabilities per simulation cycle.

5. Experimental Evaluation

5.1 Re-mapping in Software Radio Application

The variation in times at which the individual stages in the software radio application would have to be re-mapped is shown in Figure 7. For each of the source, low pass filter, demodulator and equalizer stages, plots are provided for the variation of key battery system indicators with time, as obtained from the simulation environment, which models both the battery electrochemical properties and DC-DC converter characteristics in detail. The instantaneous battery current, *Ibatt* is the current as seen at the terminals of the battery cell: in order to maintain a constant voltage at the terminals of the battery subsystem, the DC-DC converter will draw increasing amounts of current from the battery cell, as the battery state of charge (indicated by the Vc plot in Figure 7) and hence the battery terminal voltage decreases. Due to the time constant associated with the electrochemical processes in the battery cell, the changes in the instantaneous battery current are in essence filtered, and the effective discharge rate is indicated by the plot *Vrate* in Figure 7.

The time at which the energy stores of the nodes executing the various stages is depleted, is a function of the computational intensity and current draw profile of each stage. On the average, the fraction of time spent with the CPU busy versus idle for each of the 5 stages was 0.22, 0.67 0.46, 0.76 and 0.54 for the source, low-pass filter, equalizer, demodulator and sink node respectively. These occupancy ratios, coupled with the battery discharge current induced by

---

[5]During communication link failures, devices attempting to transmit data on the failed link incur a *carrier-sense error*, and must retry their transmissions at a later time. During a node failure, execution at the device is suspended for the duration of the failure.

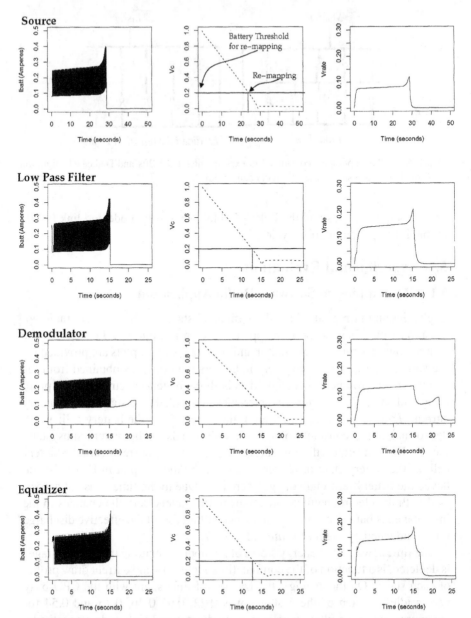

Figure 7. Variation in instantaneous battery current (*Ibatt*), normalized discharge rate (*Vrate*) and normalized state of charge (*Vc*) of batteries for the signal source, low pass filter, demodulator and equalizer stages of the software radio application. For each stage, the point at which re-mapping would be initiated for a re-mapping battery threshold of 20% is shown. For clarity, these battery profiles do not include the effects of migration, but rather only show the point at which migration should be initiated.

the application behavior lead to different times at which the individual batteries are depleted, as illustrated in Figure 7.

Since the software radio pipeline is broken when any of its stages are inactive, the system halts completely during re-mapping. If PCRE were employed with this application therefore, during the entire period when pre-copying was taking place for each of the stages, the system would not be functional. Using the baseline lightweight migration scheme, in which each stage of the application was re-mapped when the low battery threshold (B_{low}) was reached, re-mapping enables the system lifetime to be extended by 30%.

The software radio application was also executed in a system in which nodes were attached to batteries with twice the capacity in the previous discussion. The system employing re-mapping (lightweight code migration) to extend lifetime witnessed a system lifetime that was 65.1% of that which would have been achieved had batteries of twice the capacity been employed. The inability of re-mapping to achieve the lifetime attained when employing doubly-sized batteries is due in part to the overheads involved—transmitting on the migrating nodes end, and receiving on the redundant node all consume power. Further, even when the redundant node is idle, prior to re-mapping, it consumes a small non-zero amount of power.

The evaluation of the benefits of the code migration scheme were repeated for a modeled hardware platform with correlated node and communication link failures, with failure rates of 1E-8 and a correlation coefficient of 0.1. For this hardware configuration, the lifetime of the system employing code migration, relative to that of a system employing doubly-sized batteries was reduced further, to 52.6%. This further decrease is due to the additional communication overhead in the presence of failures, which leads to an increased occurrence of link layer collisions, thus nodes spend more time (and energy) in communication attempts.

5.2 Re-mapping in the Beamforming Application

For the baseline migration implementation, as is apparent in the solid line of Figure 8, nodes begin migration at approximately 15 seconds into the experiment and completely resume normal operation by 20 seconds. The remaining usable battery energy as plotted (dashed line in Figure 8), is the difference between the total amount of energy available to the application (in both the active and redundant nodes) during the experiment and the energy consumed—when migration occurs before the energy resources in the active node are completely depleted, this is reflected in the measure as a drop in the energy remaining, since it in essence leads to usable energy being discarded. The dashed lines show that, for the baseline migration scheme, the amount of usable battery energy decreases considerably in the region of 15 to 20 seconds because, in this

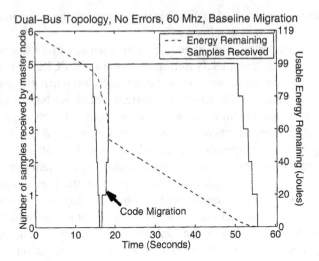

Figure 8.   Performance of baseline migration scheme in Dual Bus Topology, no errors, 60 MHz processors.

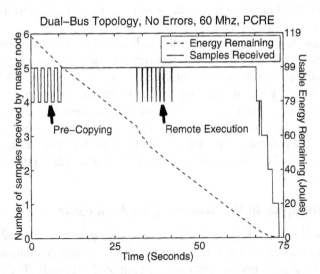

Figure 9.   Performance of PCRE in Dual Bus Topology, no errors, 60 MHz processors.

region, each slave node is abandoning what remaining energy it may have left by performing the migration.

Figure 9 shows the number of samples received by the master node in the beamforming application as well as the remaining usable battery energy for the beamforming application as it evolves, for a system implementation employing PCRE. The solid line in Figure 9 represents the number of samples received by the master node in the beamforming application during each sampling period.

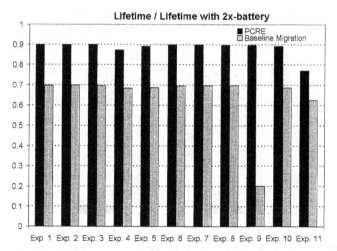

Figure 10.   System lifetime for two migration schemes, normalized to the system lifetime resulting by simply using batteries with twice the capacity.

The "saw-tooth" shape in the first 10 seconds is a manifestation of each of the scheduled pre-copying steps, which result in one of the active nodes ceasing to send samples to the master, and performing pre-copying of its code to a redundant node. At approximately 35 seconds, all nodes send their final *FIN* message (see also Figure 2) to their corresponding redundant nodes to perform the final handoff. The amount of usable energy remaining using PCRE also drops during the final migration stage, but *not* as drastically as in the baseline migration case—a smaller amount of useful energy is discarded, hence the smaller drop.

In the above experiments, the system lifetimes are consistently very small. This is the result of deliberately choosing a small battery capacity for simulation tractability.

5.3   Variation of Lifetime and Collision Rates

Summarizing results for all the beamforming experiments are included in Figures 10–11. Figure 10 shows these lifetimes, normalized to the system lifetime of the "ideal" configuration where each node has twice the battery capacity and no migration occurs. As it can be seen, PCRE achieves, in most cases, a system lifetime within 10% of the ideal case while baseline migration is usually 30% worse than the ideal system lifetime. For baseline migration, re-mapping failed in Exp. #9 resulting in the small observed lifetime relative to that of a doubly-sized battery. Removing this experiment from consideration results in an average system lifetime improvement of 28.6% for PCRE compared to baseline migration. The energy efficiency is also improved by 9.8% using

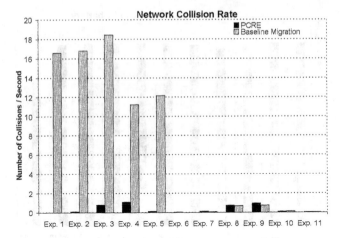

Figure 11.   Average number of collisions per second for each experimental setup.

PCRE in comparison to the baseline migration technique in the beamforming application.

The final metric of interest is the network collision rate during execution of the beamformer. The average number of network collisions per operating second is given in Figure 11. In general, the baseline migration scheme witnesses a significantly larger number of link layer collisions, since re-mapping is often initiated for many nodes at approximately the same time. The PCRE scheme on the other hand explicitly schedules the migration slots to be in non-overlapping time ranges, resulting in significantly smaller number of collisions. On the other hand, the use of dedicated migration links in the beamforming application (Exp. #6–10) results in a significant reduction in the number of collisions for the baseline migration scheme as to be expected. In Exp. #8 and Exp. #9, PCRE witnesses a slightly larger number of link layer collisions than the baseline migration scheme. This is a result of the simulated random failures. On the average, over a large number of experiments, there should be no significant difference between the number of collisions witnessed by PCRE versus baseline migration. This is because dedicated links are employed for migration in both cases, and collisions only occur during the sample exchange with the master, which is independent of the re-mapping scheme, be it PCRE or baseline migration.

6.   Summary and Conclusions

This paper presented the use of *lightweight code migration* as a means of performing application re-mapping to counteract the effects of runtime failures, such as intermittent communication and device failures, or limited battery resources. By constructing applications in such a manner that they may be *asynchronously restarted*, and by appropriately managing application state, the amount of state that must be transferred during migration is minimized. A new technique, *Pre-Copying with Remote Execution (PCRE)*, that decouples the compute- and communication-intensive portions of code migration from the final handoff, enabling flexible scheduling of the phases of migration was introduced. PCRE enables more efficient application re-mapping in the presence of energy and communication constraints for *symmetric applications*. It was shown that PCRE provides a 28.6% improvement in system lifetime and 9.8% improvement in energy efficiency over the baseline *lightweight code migration* strategy for the applications investigated.

References

[1] L. Benini, G. Castelli, A. Macii, E. Macii, M. Poncino, and R. Scarsi, "A discrete-time battery model for high-level power estimation," in *Proceedings of the conference on Design, Automation and Test in Europe (DATE '00)*, January 2000, pp. 35–39.

[2] D. Johansen, R. van Renesse, and F. Schneider, "Operating system support for mobile agents," in *5th IEEE Workshop on Hot Topics in Operating Systems*, 1995.

[3] U. Kremer, J. Hicks, and J. Rehg, "Compiler-Directed Remote Task Execution for Power Management," in *Workshop on Compilers and Operating Systems for Low Power (COLP'00)*, October 2000.

[4] D. Marculescu, N.H. Zamora, P. Stanley-Marbell, and R. Marculescu, "Fault-Tolerant Techniques for Ambient Intelligent Distributed Systems," Tech. Rep. 03-06, Center for Silicon System Implementation, Dept. ECE, Carnegie Mellon, May 2003.

[5] D. Milojičić, F. Douglis, Y. Paindaveine, R. Wheeler, and S. Zhou, "Process Migration," *ACM Computing Surveys*, vol. 32, no. 3, pp. 241–299, September 2000.

[6] D. Milojičić, W. LaForge, and D. Chauhan, "Mobile objects and agents," in *USENIX Conference on Object-oriented Technologies and Systems*, 1998, pp. 1–14.

[7] "Panasonic Coin Type Li Ion Battery (Part no. BR1216)," Digi-Key Catalog, http://www.digikey.com.

[8] A. Rudenko, P. Reiher, G.J. Popek, and G.H. Kuenning, "The Remote Processing Framework for Portable Computer Power Saving," in *ACM Symposium on Applied Computing*, February 1999, pp. 365–372.

[9] P. Stanley-Marbell and M. Hsiao, "Fast, flexible, cycle-accurate energy estimation," in *Proceedings of the International Symposium on Low Power Electronics and Design*, August 2001, pp. 141–146.

[10] P. Stanley-Marbell and D. Marculescu, "Exploiting Redundancy through Code Migration in Networked Embedded Systems," Tech. Rep. 02-14, Center for Silicon System Implementation, Dept. ECE, Carnegie Mellon, February 2002.

[11] P. Stanley-Marbell, D. Marculescu, R. Marculescu, and P.K. Khosla, "Modeling, Analysis and Self-Management of Electronic Textiles," *To appear, IEEE Transactions on Computers*, vol. 52, no. 8, August 2003.

[12] J. White, "Mobile Agents," in J.M. Bradshaw (Ed.), *Software Agents*, pp. 437-472, MIT Press, 1997.

Contributing Authors

Emile Aarts is Vice President and Scientific Program Director of the Philips Research Laboratories Eindhoven, The Netherlands. He holds an MSc. and PhD. degree in physics. For almost twenty years he has been active as a research scientist in computing science. Since 1991 he holds a teaching position at the Eindhoven University of Technology as a part-time professor of computing science. He also serves on numerous scientific and governmental advisory boards. He holds a part-time position of senior consultant with the Center for Quantitative Methods in Eindhoven, The Netherlands. Emile Aarts is the author of five books and more than one hundred and forty scientific papers on a diversity of subjects including nuclear physics, VLSI design, combinatorial optimization and neural networks. In 1998 he launched the concept of Ambient Intelligence and in 2001 he founded Philips' HomeLab. His current research interests include embedded systems and interaction technology.

Alejandro Alonso graduated in 1985 and received a Ph.D. in 1994 in Computer Science in the Technical University of Madrid (Universidad Politécnica de Madrid). He is currently an Associate Professor in the Department of Telematic Systems Engineering at the School of Telecommunication Engineering of the Technical University of Madrid (DIT-UPM). His current research interests are in real-time systems design and implementation, including QoS, design methods, software architectures, and real-time operating systems and languages. He has participated in a number of ESPRIT funded projects such as ARES, and COMITY, and also in several national government and industry funded research projects. He teaches courses on Operating Systems and Real-Time Systems. He is a member of IEEE, ACM, CEA-IFAC, Ada-Spain and Ada-Europe.

Twan Basten is an assistant professor in the Department of Electrical Engineering at the Eindhoven University of Technology. In 1993, he received a Master's degree (with honors) in Computing Science from the Eindhoven University of Technology; in 1998, he received his PhD degree in Computing Science from the same university, with a thesis titled "In Terms of Nets: System Design with Petri Nets and Process Algebra." His current research interest is the design of complex, resource-constrained embedded systems, based on a solid mathematical foundation, with a special focus on multiprocessor systems. Twan Basten has publications in the areas of workflow management, formal methods, embedded systems, and system design. He is the topic chair for ambient intelligence and ubiquitous computing in the program committee of DATE 2004. He is a member of the IEEE. You can reach him via http://www.ics.ele.tue.nl/~tbasten/.

337

page
31 **Luca Benini** received the Dr. Eng. degree in electrical engineering from Università di Bologna, Italy, in 1991, and the M.S. and Ph.D. degrees in Electrical Engineering from Stanford University in 1994 and 1997, respectively. From 1997 to 2000 he has been first a research associate, then Assistant Professor at Università di Bologna and a post-doctoral fellow at Stanford University. He is currently an Associate Professor at Università di Bologna. He also holds a position as visiting scientist at the Hewlett-Packard Laboratories, Palo Alto, CA. His research interests are in all aspects of computer-aided design of digital circuits, with special emphasis on low-power applications.

page
183 **Jan Bormans**, Ph.D., has been a researcher at the Information Retrieval and Interpretation Sciences laboratory of the Vrije Universiteit Brussel (VUB), Belgium, in 1992 and 1993. In 1994, he joined the VLSI Systems and Design Methodologies (VSDM) division of the IMEC research center in Leuven, Belgium. Since 1996, he is heading IMEC's Multimedia Systems group. This group focuses on the efficient design and implementation of embedded systems for advanced multimedia applications, with an emphasis on MPEG-4, MPEG-21 and wavelet-based systems. Jan Bormans is the Belgian head of delegation for the ISO/IEC MPEG and SC29 standardization committees. He is also MPEG-21 requirements editor and chairman of the MPEG liaison group.

page
159 **Reinder J. Bril** received a B.Sc. and an M.Sc. (both with honors) from the Department of Electrical Engineering of the University of Twente, The Netherlands. Since 1985, he has been with Philips. He has worked in both Philips Research and Philips' Business Units, on various topics, including fault-tolerance, formal specifications, and software architecture analysis, and in different application domains. The last four years, he worked at Philips Research Laboratories Eindhoven (PRLE), The Netherlands, in the area of Quality of Service (QoS) for consumer devices, with a focus on dynamic resource management in receivers in broadcast environments (such as digital TV-sets and set-top boxes). He is currently detached for a period of a year by Philips to the Technische Universiteit Eindhoven (TU/e), Department of Mathematics and Computer Science, Group System Architecture and Networking (SAN).

page
251 **Zbigniew Chamski** is a senior consultant at the Electronic Design and Tools group in Philips Research. His research interests include high-performance parallel systems, parallelizing programming paradigms, domain-specific languages, and system modeling. Before joining Philips, he held research and software engineering positions at the University of Rennes 1, Inria, and the University of Manchester, England. He graduated from the University of Rennes 1 in 1989, and received his PhD, also from University of Rennes 1, in 1993.

page
295 **Anantha P. Chandrakasan** received the Ph.D. degree in EECS from the University of California, Berkeley, in 1994 and is currently a Professor of Electrical Engineering and Computer Science at the Massachusetts Institute of Technology. His research interests include the ultra low power implementation of custom and programmable digital signal processors, distributed wireless sensors, ultra wideband radios, and emerging technologies. He is a co-author of "Low Power Digital CMOS Design" by Kluwer Academic Publishers and "Digital Integrated Circuits" (second edition) by Prentice-Hall. He is also a co-editor of "Low Power CMOS Design" and "Design of High-Performance Microprocessor Circuits" from IEEE Press. He has served as the technical program co-chair for the 1997 International Symposium on Low-power Electronics and Design (ISLPED), the technical program chair for ISSCC 2003, and an associate editor for the IEEE Journal of Solid-State Circuits from 1998 to 2001.

page
203 **Adrian Chirila-Rus** received his degree of Microelectronics Engineer in 1998 from the "Politehnica" University of Bucharest, Romania. Next year, 1999, he received the MSc degree in Microelectronic System Design and CAD, from the "Politehnica" University of Bucharest, Romania. In 1999, he also joined the Interuniversity Micro Electronics Center (IMEC) in Leuven, Belgium, the group of the Multimedia Image Compression Systems (MICS), which is a part of the Design Technology for Integrated Information and Communication Systems (DESICS) division of IMEC. Currently, Adrian Chirila-Rus is working on the FlexWave-II architecture, a wavelet-based still-image compression engine with MPEG-4 functionality. He is also working on the Unequal Error Protection of scalable streams for transmission over unreliable channels.

page
251 **Albert Cohen** is a research scientist in the A3 group at INRIA. His research interests include program optimization for high-performance architectures and embedded systems, static analysis and automatic parallelization. He graduated from École Normale Supérieure de Lyon, and received his PhD from the University of Versailles in 1999.

page
91 **Jon Crowcroft** is the Marconi Professor of Networked Systems in the Computer Laboratory of the University of Cambridge. Prior to that he was professor of networked systems at UCL in the Computer Science Department. He is a Fellow of the ACM, a Fellow of the British Computer Society, a Fellow of the IEE and a Fellow of the Royal Academy of Engineering, as well as a senior member of the IEEE. He was a member of the IAB; was general chair for the ACM SIGCOMM 95-99. He is on the editorial team for COMNET, and on the program committee for ACM SIGCOMM and IEEE Infocomm. He has published 5 books—the latest is *TCP/IP and Linux Protocol Implementation*, published by Wiley in 2001.

page
229 **Robertas Damaševičius** received the MSc degree in informatics from Kaunas University of Technology, Lithuania in 2001. Currently he is an assistant at the Software Engineering Department and PhD candidate at the Informatics Faculty, Kaunas University of Technology. His research interests include metaprogramming, software reuse, software generation and program transformation, as well as hardware design with VHDL and SystemC.

page
183   **Geert Deconinck** is Senior Lecturer (hoofddocent) at the Katholieke Universiteit Leuven (Belgium) since 2003. He is a staff member of the research group ELECTA (Electrical Energy and Computing Architectures) of the Department of Electrical Engineering (ESAT), where he guides a group of about 8 PhD students. His research interests include the design and assessment of software-based solutions to meet dependability real-time, and cost constraints for embedded applications on distributed systems. In this field, he has authored and co-authored more than 100 publications in international journals and conference proceedings. He received his M.Sc. in Electrical Engineering and his Ph.D. in Applied Sciences from the K.U.Leuven, Belgium in 1991 and 1996 respectively. He was a visiting professor (bijzonder gastdocent) at the K.U.Leuven since 1999 and a postdoctoral fellow of the Fund for Scientific Research - Flanders (Belgium) in the period 1997-2003. In 1995-1997, he received a grant from the Flemish Institute for the Promotion of Scientific-Technological Research in Industry (IWT). He is a Certified Reliability Engineer (ASQ), a member of the Royal Flemish Engineering Society, a senior member of the IEEE and of the IEEE Reliability, Computer and Power Engineering Societies.

page
vii   **Hugo De Man** is professor in electrical engineering at the Katholieke Universiteit Leuven, Belgium since 1976. He was visiting associate professor at U.C.Berkeley in 1975 teaching semiconductor physics and VLSI design. His early research was devoted to the development of mixed-signal, switched capacitor and DSP simulation tools as well as new topologies for high speed CMOS circuits which led to the invention of NORA CMOS. In 1984 he was one of the cofounders of IMEC (Interuniversity Microelectronics Center), which, today, is the largest independent semiconductor research institute in Europe with over 1200 employees. From 1984-1995 he was Vice-President of IMEC, responsible for research in design technology for DSP and telecom applications. In 1995 he became a senior research fellow of IMEC, working on strategies for education and research on design of future post-PC systems. His research at IMEC has led to many novel tools and methods in the area of high level synthesis, hardware-software codesign and C++ based design. Many of these tools are now commercialized by spin-off companies. His work and teaching also resulted in a cluster of DSP oriented companies in Leuven now known as DSP Valley where more than 1500 DSP engineers work on design tools and on telecom, networking and multimedia integrated system products. In 1999 he received the Technical Achievement Award of the IEEE Signal Processing Society, The Phil Kaufman Award of the EDA Consortium and the Golden Jubilee Medal of the IEEE Circuits and Systems Society. Hugo De Man is an IEEE Fellow and a member of the Royal Academy of Sciences in Belgium.

page
271   **Stefan Dulman** received the BS and MS degrees from the Department of Electronics and Telecommunications, Gh.Asachi-University, in Iasi, Romania, in 2001 and 2002 respectively. He is currently a Ph.D. student at the Department of Computer Science, University of Twente, The Netherlands. His research interests include fault tolerance aspects in distributed wireless sensor networks.

page
251
Marc Duranton is a principal scientist in the Embedded System Architectures on Silicon group in Philips Research. His research interests include parallel and high performance architectures for video and image processing, system modeling and validation, and software optimization. He worked within Philips Semiconductors on several video coprocessors for the TriMedia and Nexperia platforms. He graduated from Ecole Nationale Supérieure d'Electronique et de Radioélectricité de Grenoble, from Ecole Nationale Supérieure d'Informatique et de Mathématiques Appliquées de Grenoble and received his PhD from Institut National Polytechnique de Grenoble in 1995.

page
251
Christine Eisenbeis received a PhD degree on loop optimization for array-processors at University of Paris VI in 1986. She is currently a senior research associate at INRIA Rocquencourt and heads the A3 INRIA research team whose main topic is program analysis for code optimization. Her research interests include compiler technologies for high performance general purpose and embedded microprocessors, instruction scheduling, software pipelining, register allocation, data dependence analysis and data locality optimization. C. Eisenbeis has coauthored more than 40 publications in journals and international conferences and in 1990, she was awarded the first prize of "CONCOURS CRAY FRANCE" for her collaborative work on instruction scheduling.

page
251
Paul Feautrier is a graduate of Ecole Normale Supérieure and Paris University, where he got his Doctorat d'Etat on a subject in Computational Astrophysics in 1968. In 1969, he was appointed professor of Computer Science at University Pierre et Marie Curie in Paris. He moved to University of Versailles in 1992, and joined Ecole Normale Supérieure de Lyon in 2002. His research interests include computer architecture, operating systems, parallel programming, automatic parallelization and program optimization.

page
159
Marisol García Valls, born in Castellón (Spain) in 1973, studied Computer Science Engineering at the University Jaume I de Castellón in Spain, where she graduated in 1996. She received her PhD degree from the Technical University of Madrid (Spain) in 2001. Since 2001, she is at the University Carlos III of Madrid (Spain), where she is currently an Associate Professor. She has been involved in several national and European research projects. Also, she has been a member of several program committees in international conferences on the fields of real-time systems and middleware. Her main research interests are on the fields of QoS management in real-time embedded multimedia systems and real-time support in Java middleware for distributed applications.

page
1
Marc Geilen graduated in 1996 (with honors) from the Information Technology program at the Faculty of Electrical Engineering, Eindhoven University of Technology. At the same department, at the Information and Communication Systems Group, he received his Ph.D. in 2002. He is currently a researcher in that group working in the European IST Ozone project. His research interests include validation and (formal) verification, real-time systems, system-level modelling, simulation and programming paradigms for concurrent systems.

Daniela Genius is an assistant professor in the LIP6 department, ASIM Laboratory at the University Paris-VI. Her research interests include system-on-chip, hardware/software co-design and compilers. She graduated from Technical University of Aachen, received her PhD from the University of Karlsruhe in 2000 and worked from 2000 to 2002 as a postdoctoral researcher at Philips France and INRIA.

Harmke de Groot holds an M.Sc. degree in Electrical Engineering from the Eindhoven University of Technology, The Netherlands. She has worked from 1997 through 2001 at Philips Semiconductors as a system engineer and project leader working on several communication systems like ADSL and Bluetooth. In this position she contributed to the system specification of the second-generation Bluetooth solution from Philips Semiconductors. She was also a contributer to the Bluetooth standard. At the moment she works as a senior architect and project coordinator in the Information Processing Architecture department of the sector Information and Software Technology at Philips Research. She is currently involved in the IST Ozone project. Main research interests include the impact of ambient intelligence on embedded systems, quality of service in embedded distributed systems and heterogeneous system integration.

Sunil Hattangady is the Program Manager for OMAP(tm) Wireless Security in TI's Wireless Terminals Business Unit. TI's OMAP applications processing platform delivers voice and multimedia-enhanced applications for 2.5G and 3G mobile phones, PDAs and advanced mobile Internet appliances. Sunil leads a team chartered with developing software and hardware security solutions for the OMAP platform to enable secure wireless services on OMAP-enabled devices. He has a Ph.D. in engineering from North Carolina State University, and an M.B.A. from the University of Texas at Austin.

Paul Havinga received his BS degree in computer engineering in 1985. Since 1985 he has been working at the Computer Science Department of the University of Twente, the Netherlands. His research has been on parallel processing, ATM switching, and mobile computing and wireless communication. He has been a visiting researcher at the University of Pisa in 1998, and the Communications Research Laboratory in Yokosuka, Japan in 2000. He received his Ph.D. on mobile multimedia systems in 2000, and was awarded with the DOW Dissertation Energy Award for this work.

Lodewijk van Hoesel received the M.Sc. degree in electrical engineering from the University of Twente, Enschede, The Netherlands, in 2002. He is currently working towards a Ph.D. degree at the University of Twente in the area of energy efficient wireless communication mechanisms. Research topics include medium access protocols, wake-up radio and signal processing.

page
3, 203
Gauthier Lafruit was a research scientist with the Belgian National Foundation for Scientific Research from 1989 to 1994, being mainly active in the area of wavelet image compression implementations. Subsequently, he was a research assistant with the VUB (Free University of Brussels, Belgium). In 1996, he became the recipient of the Scientific Barco award and joined IMEC (Interuniversity Micro Electronics Center, Leuven, Belgium), where he was involved as Senior Scientist with the design of low-power VLSI for combined JPEG/wavelet compression engines. He is currently the Principal Scientist in the Multimedia Image Compression Systems Group with IMEC. His main interests include progressive transmission in still image, video and 3D object coding, as well as scalability and resource monitoring for advanced, scalable video and 3D coding applications.

page
159
Sjir van Loo (1949) graduated in physics at Eindhoven University of Technology in 1974. He joined Philips in 1985, and Philips Research Laboratories in 1992. He has been working as a systems architect for over 25 years, and was involved in the design and realization of many industrial and research systems. His current research interests are systems architectures for Ambient Intelligence and resource management for high-volume electronics.

page
315
Diana Marculescu is an Assistant Professor of Electrical and Computer Engineering at Carnegie Mellon University. She received her Dipl. Eng. in Computer Science from "Politehnica" University of Bucharest, Romania in 1991 and her Ph.D. in Computer Engineering from University of Southern California in 1998. Diana Marculescu is a recipient of a National Science Foundation Faculty Career Development Award (2000) and of an ACM-SIGDA Technical Leadership Award (2003). She serves as the Technical Program Chair of International Workshop on Logic and Synthesis (2004) and is an executive board member of the ACM Special Interest Group on Design Automation (SIGDA). Her research interests include energy aware computing, CAD for low power systems and emerging technologies (such as electronic textiles or ambient intelligent systems).

page
315
Radu Marculescu received his Ph.D. from the University of Southern California (Los Angeles) in 1998. In 2000 he joined the Electrical and Computer Engineering faculty of Carnegie Mellon University, where he is now an Assistant Professor. His research interests include SOC design methodologies, NOCs and fault-tolerant communication, and ambient intelligent systems. Radu was awarded the National Science Foundation's Career Award and has recently received three best paper awards in the area of systems design methodologies. He was also awarded the 2002 Ladd Research Award from Carnegie Institute of Technology.

Bart Masschelein received his degree of Industrial Engineer in 1998 from the Katholieke Hogeschool Brugge-Oostende, in Belgium. That year he started a part-time MSc-course in Electronic System Design, in collaboration with Leeds Metropolitan University. In 1998, he also joined the Interuniversity Micro Electronics Center (IMEC) in Leuven, Belgium, more specifically the group of the Multimedia Image Compression Systems (MICS), which is a part of the Design Technology for Integrated Information and Communication Systems (DESICS) division of IMEC. Currently, Ing. Masschelein is working on the FlexWave-II architecture, a wavelet-based still-image compression engine with MPEG-4 functionality. He is implementing the Local Wavelet Transform as decorrelator of the system, which is an instruction-based custom processor for block-based wavelet transform with the same net result as the global wavelet transform.

Ivo Mentuccia received his M.Sc. in Computer Engineering from "La Sapienza" University of Rome in 2002. He moved to the Autonomous Laboratory for COgnitive Robotics (AL-COR) in 2001, working on the reactive control of the mobile manipulators ArmHandOne, Mr.ArmHandOne, and ArmHandX. He has published some papers on the behavior control of autonomous mobile robots in unknown environments. BICLazio has awarded him in 2003 for the best business idea derived from a research project. He has also been awarded, together with the other members of the Laboratory, at the Edmonton AAAI-02 conference for the innovative robot architecture presented. His current research interests include cognitive mobile robots and in particular motion control for autonomous vehicles using fuzzy logic.

Rex Min received the B.S. and M.Eng. degrees in Electrical Engineering and Computer Science from the Massachusetts Institute of Technology (MIT) in 1998 and 1999 respectively, and the Ph.D. degree at MIT in June 2003. His research interests center on the power aware design of hardware and software for large-scale embedded systems, with an eye towards emerging wireless applications. He is a member of Eta Kappa Nu, Sigma Xi, and Tau Beta Pi, and was a National Defense Science and Engineering Graduate (NDSEG) Fellow from 2000-2003. In September 2003, he will continue his research career with the Laboratory for Telecommunications Sciences in College Park, Maryland.

Tim Nieberg studied mathematics at the University of British Columbia, Vancouver, Canada, and at the University of Osnabrück, Germany, where he received his Diplom in 2002. He is currently working towards a Ph.D. in applied mathematics at the University of Twente, Enschede, The Netherlands, where he is also with the European project EYES on energy efficient sensor networks.

Clara M. Otero Pérez, born in Pondevedra (Spain) in 1975, studied Physics at the University of Santiago de Compostela in Spain. She came to the Netherlands in 1997 with an interchange Erasmus program of 6 months, after which she followed a post-graduate master class in Computer Science at the Eindhoven University of Technology. Since 1999, she works as a Research Scientist at Philips Research Laboratories Eindhoven on Quality of Service Resource Management for consumer terminals. Her main research interest is on the fields of resource scheduling, streaming architectures and real-time. She has participated in several major European projects and real-time conferences.

page 183 **Nam Pham Ngoc** received the MEng degree in electrical engineering from Hanoi University of Technology, Vietnam, in 1997 and the MEng degree in artificial intelligence from Katholieke Universiteit Leuven, Belgium, in 1999. He is currently pursuing a PhD degree in electrical engineering at Katholieke Universiteit Leuven. His research interests focus on terminal QoS and resource management for interactive multimedia applications. He is a student member of the IEEE.

page 131 **Fiora Pirri** PhD (très bien avec félicitation du jury), Université Pierre et Marie Curie, Paris 6, is associate Professor at the University of Rome "La Sapienza", and she has obtained tenure for full professorship in March 2002. Her current research efforts focus on various areas of Artificial Intelligence (AI) and in particular on Formalisms and Languages for Automated Reasoning, Knowledge Representation, Theories of Actions and Perception, and Cognitive Robotics. She has published the results of her research in the most relevant AI Journals and Conference Proceedings. She has been member of several program committees and she is currently Co-chair of Tableaux-2003, and member of the Steering Committee of Cognitive Robotics. She is leading the project PEGASO funded by the Italian Space Agency. In 1998 she founded the Autonomous Laboratory for COgnitive Robotics (ALCOR).

page 31 **Massimo Poncino** received the Dr.Eng. degree in electrical engineering in 1989 and the Ph.D. degree in computer engineering in 1993, both from Politecnico di Torino. From 1993 through 1994 he was a Visiting Faculty at the University of Colorado at Boulder. From 1995 to 2001 he was an assistant professor at Politecnico di Torino. He is currently an Associate Professor at Università di Verona. His research interests include various aspects of the design automation of digital systems, with emphasis on power modeling, estimation and optimization.

page 51 **Anu Purhonen** is a senior research scientist at VTT Technical Research Centre of Finland. She received her MSc degree in computer engineering in 1991 and LicTech degree in embedded systems in 2002, both from the University of Oulu, Finland. In 2003 she is a visiting researcher at the University of Groningen, the Netherlands. Her research interests include design and assessment of embedded software architectures.

page 103 **Jean-Jacques Quisquater** is a professor of cryptography and multimedia security in the Department of Electrical Engineering at UCL (Louvain-la-Neuve, Belgium), where he is the director of the Microelectronics Laboratory. His research interests include smart cards (protocols, implementations, and side channels); secure protocols for communications, digital signatures, pay TV, and copyright protection; and security tools for electronic commerce. J.-J. Quisquater has a degree in engineering (applied mathematics) from UCL and received a PhD in Computer Science from LRI (University of Orsay, France) in 1987. He was the main designer for the Philips Corsair and Fame coprocessors, developed for fast computations of public-key cryptosystems for smart cards. He is a director of the International Association for Cryptologic Research (IACR). He is a member of IEEE and ACM. He received the chait Franqui in 2000, the Montefiore prize in 2001 and a doctorate honoris causa from the university of Limoges (France) in 2003. He is listed in the "Who's who in the world" (Marquis). He published about 170 papers in the field of cryptography, security and graph theory. He holds 17 patents. URL: http://uclcrypto.org.

page
103
Anand Raghunathan received the B. Tech. degree in Electrical and Electronics Engineering from the Indian Institute of Technology, Madras, India, in 1992, and the M.A. and Ph.D. degrees in Electrical Engineering from Princeton University, Princeton, NJ in 1994 and 1997, respectively. Dr. Raghunathan is currently a Senior Research Staff Member at the NEC Laboratories America in Princeton, NJ, where he leads several projects related to the research and development of System-on-Chip architectures, design methodologies, and design tools, with emphasis on high-performance, low power, and testable designs. His is responsible for the overall security architecture of NEC's application processors for 3G mobile terminals. Dr. Raghunathan has co-authored a book (High-level Power Analysis and Optimization), and around 100 publications in leading conferences and journals. He holds or has filed for 14 U.S. patents in the areas of advanced system-on-chip architectures, design methodologies and VLSI CAD. He has received three best paper awards at the IEEE International Conference on VLSI Design (one in 1998 and two in 2003), two awards at the ACM/IEEE Design Automation Conference (1999 and 2000), and three best paper award nominations at the ACM/IEEE Design Automation Conference (1996, 1997, and 2003). He received the patent of the year award (an award recognizing the invention that has achieved the highest impact) from NEC in 2001. Dr. Raghunathan is a Senior Member of the IEEE. He has served on the committees of several leading IEEE and ACM conferences, and on the Editorial Board of IEEE Transactions on VLSI, and IEEE Design & Test. He was a recepient of the Meritorious Service Award from the IEEE Computer Society, and was elected a Golden Core Member in 2001, in recognition of his contributions.

page
103
Srivaths Ravi received the B. Tech. degree in Electrical and Electronics engineering from the Indian Institute of Technology, Madras, India, in 1996, and the M.A. and Ph.D. degrees in Electrical Engineering from Princeton University, Princeton, NJ, in 1998 and 2001, respectively. He is, at present, a Research Staff Member with NEC Labs America, Inc., Princeton. Dr. Ravi's current activities include the research and development of advanced embedded architectures (such as NEC's MOSES security processor), as well as enabling methodologies and tools for system-level test and low power design. Dr. Ravi has several publications in leading ACM/IEEE conferences and journals on VLSI/CAD, including invited contributions and talks on wireless security at ISSS 2002, VLSI Design 2003 and DATE 2003. He holds or has filed for several patents in the above areas. Dr. Ravi's papers have received awards at the International Conference on VLSI design in 1998, 2000 and 2003. He received the Siemens Medal from the Indian Institute of Technology, Madras, India in 1996. He is a member of the program committees of the VLSI Test Symposium (VTS) and the Design Automation and Test in Europe conference, and has served as the Compendium Chair in the 20th Anniversary Committee of the VTS.

page
11
Raf Roovers was born in Borgerhout, Belgium, in 1967. He received the master's degree in electrical and mechanical engineering and the Ph.D. degree in electrical engineering from the Katholieke Universiteit Leuven, Belgium, in 1990 and 1996 respectively. From 1990 to 1995 he was a research assistant in the ESAT-MICAS Laboratory of the Katholieke Universiteit Leuven. In 1995 he joined Philips Research Laboratories, Eindhoven, The Netherlands where he is a principal engineer in the Mixed-Signal Circuits and Systems group. His research interests are in analog-to-digital and digital-to-analog conversion circuits, high frequency integrated circuit design and circuits and architectures for new communication systems.

page
159
José F. Ruíz has a Telecommunication Engineering Degree from the Technical University of Madrid (UPM). He is finishing his Ph.D. within the Software Engineering and Real-Time Systems group at UPM, and the topic of the thesis is the design and implementation of a hierarchical and decentralized QoS and resource management architecture for embedded systems. José F. Ruíz joined ACT Europe in October 2002, where he is working on embedded system technologies.

page
91
Frank Stajano (Dr. Ing., PhD) is a faculty member at the Laboratory for Communication Engineering of the University of Cambridge. Before this post he worked for several years as a research scientist for Olivetti, Oracle, AT&T and Toshiba, and was made a Toshiba Fellow in 2000. For the past few years his main research themes have been computer security and ubiquitous computing, covered in his latest book *Security for Ubiquitous Computing* (Wiley, 2002). He maintains strong links with industry and academia across the world and is often in demand as a security consultant or guest lecturer.

page
315
Phillip Stanley-Marbell is a PhD student in the department of electrical and computer engineering at Carnegie Mellon.

page
159
Liesbeth Steffens graduated in physics from Utrecht University in 1972, took a post-graduate computer-science course at Grenoble University, and joined Philips Research in 1973, where she spent most of her career. Her main focus is on real-time, quality-of-service (QoS), and resource-management issues for streaming and control in consumer-electronics devices and systems. She contributed to the design of a distributed real-time operating system, a video-on-demand system, and a QoS-based resource-management system for digital video.

page
9, 159
Peter van der Stok (1949) holds an M.Sc. (1973) and a Ph.D. (1982) degree in physics from the University of Amsterdam. He has worked from 1973 till 1989 at CERN where he became responsible for the operating system software of the distributed control system of the LEP accelerator. He worked from 1989 at the Eindhoven University of Technology where he is associate professor. His main interests are distributed fault-tolerant real-time systems. Since September 2000 he works for 90% of his time at Philips Research on the subject of In-Home Digital Networks.

page
131
Sandro Storri is a research associate of ALCOR group at D.I.S., University of Rome "La Sapienza", and he has been developing the network communication layer for the cognitive robot ArmHandX. He is also project manager at EIDOS Software Company, and has been working on network architectures for automated and robotic systems for several major software companies, participating in and leading different network oriented software solutions. His research is currently focused on remote and networked robot control and speech recognition.

page 229 **Vytautas Štuikys** is currently a professor at the Software Engineering Department of Kaunas University of Technology. He is a teacher and researcher as well as a leader of the Research Group "Design Process Automation". His research interests include software and hardware design methodologies. He has published more than 100 papers (more than 20 in recent years) in the area. He is an author of several books and monograph. In 2002, he has published Research Report "Metaprogramming Techniques for Program Generation and Soft IP Design" presented for habilitation in Technological Sciences (Informatics Engineering), thus gaining the title of doctor habilitatis.

page 51 **Esa Tuulari** is a research scientist at VTT Technical Research Centre of Finland. He received his MSc degree in biophysics 1987 and LicTech degree in embedded systems 2000, both from the University of Oulu. Tuulari was guest researcher at Philips Research Laboratories Eindhoven, The Netherlands, in the Media Interaction group from 2002 to 2003. His research interests include ubiquitous computing and ambient intelligence in general and sensor-based context awareness and user interfaces in particular.

page 271 **Jian Wu** received the B.S. degree in Electrical Engineering from Beijing University of Posts and Telecommunications, China, in 1999, the M.S. degree in Computer Science from University of Twente, the Netherlands, in 2002. Currently he is a Ph.D. student in Computer Science department at the University of Twente and his main research interest is the communication and networking of Wireless Sensor Networks.

page 315 **Nicholas H. Zamora** earned the Bachelor of Science Degree in Electrical Engineering and Computer Science with Honors from the University of California at Berkeley in 2001, and the Master of Science Degree in Electrical and Computer Engineering from Carnegie Mellon University in 2003. He held a UCB Chancellor's Scholarship and received the Eta Kappa Nu Lifetime Achievement Award in 2001. He is currently working as an SRC graduate fellow, focusing on formal methods for system-level design.